Constraint Decision-Making Systems in Engineering

Santosh Kumar Das
Sarala Birla University, Ranchi, India

Nilanjan Dey
Techno International New Town, Kolkata, India

A volume in the Advances in Computational Intelligence and Robotics (ACIR) Book Series

IGI Global
PUBLISHER of TIMELY KNOWLEDGE

Published in the United States of America by
IGI Global
Engineering Science Reference (an imprint of IGI Global)
701 E. Chocolate Avenue
Hershey PA, USA 17033
Tel: 717-533-8845
Fax: 717-533-8661
E-mail: cust@igi-global.com
Web site: http://www.igi-global.com

Library of Congress Cataloging-in-Publication Data

Names: Das, Santosh Kumar (Professor of computer science), editor. | Dey, Nilanjan, 1984- editor.
Title: Constraint decision-making systems in engineering / edited by Santosh Kumar Das, Nilanjan Dey.
Description: Hershey, PA : Engineering Science Reference, 2023. | Includes bibliographical references and index. | Summary: "It helps to derive new and innovative conclusions based on problems. It helps to make the system robust and efficient. It helps to deal with linear and non-linear applications. It gives new insight into the area of research"-- Provided by publisher.
Identifiers: LCCN 2022045613 (print) | LCCN 2022045614 (ebook) | ISBN 9781668473436 (h/c) | ISBN 9781668473443 (s/c) | ISBN 9781668473450 (eISBN)
Subjects: LCSH: Decision support systems. | Constraints (Artificial intelligence)
Classification: LCC T58.62 .C66 2023 (print) | LCC T58.62 (ebook) | DDC 620.00285/63--dc23/eng/20221128
LC record available at https://lccn.loc.gov/2022045613
LC ebook record available at https://lccn.loc.gov/2022045614

This book is published in the IGI Global book series Advances in Computational Intelligence and Robotics (ACIR) (ISSN: 2327-0411; eISSN: 2327-042X)

British Cataloguing in Publication Data
A Cataloguing in Publication record for this book is available from the British Library.

All work contributed to this book is new, previously-unpublished material.
The views expressed in this book are those of the authors, but not necessarily of the publisher.

For electronic access to this publication, please contact: eresources@igi-global.com.

Advances in Computational Intelligence and Robotics (ACIR) Book Series

ISSN:2327-0411
EISSN:2327-042X

Editor-in-Chief: Ivan Giannoccaro University of Salento, Italy

MISSION

While intelligence is traditionally a term applied to humans and human cognition, technology has progressed in such a way to allow for the development of intelligent systems able to simulate many human traits. With this new era of simulated and artificial intelligence, much research is needed in order to continue to advance the field and also to evaluate the ethical and societal concerns of the existence of artificial life and machine learning.

The **Advances in Computational Intelligence and Robotics (ACIR) Book Series** encourages scholarly discourse on all topics pertaining to evolutionary computing, artificial life, computational intelligence, machine learning, and robotics. ACIR presents the latest research being conducted on diverse topics in intelligence technologies with the goal of advancing knowledge and applications in this rapidly evolving field.

COVERAGE

- Evolutionary Computing
- Natural Language Processing
- Pattern Recognition
- Computational Logic
- Cyborgs
- Fuzzy Systems
- Adaptive and Complex Systems
- Cognitive Informatics
- Robotics
- Artificial Life

IGI Global is currently accepting manuscripts for publication within this series. To submit a proposal for a volume in this series, please contact our Acquisition Editors at Acquisitions@igi-global.com or visit: http://www.igi-global.com/publish/.

The Advances in Computational Intelligence and Robotics (ACIR) Book Series (ISSN 2327-0411) is published by IGI Global, 701 E. Chocolate Avenue, Hershey, PA 17033-1240, USA, www.igi-global.com. This series is composed of titles available for purchase individually; each title is edited to be contextually exclusive from any other title within the series. For pricing and ordering information please visit http://www.igi-global.com/book-series/advances-computational-intelligence-robotics/73674. Postmaster: Send all address changes to above address.

Titles in this Series

For a list of additional titles in this series, please visit:
www.igi-global.com/book-series/advances-computational-intelligence-robotics/73674

Principles and Applications of Socio-Cognitive and Affective Computing
S. Geetha (VIT University, Chennai, India) Karthika Renuka (PSG College of Technology, India) Asnath Victy Phamila (VIT University, Chennai, India) and Karthikeyan N. (Syed Ammal Engineering College, India)
Engineering Science Reference • © 2023 • 330pp • H/C (ISBN: 9781668438435) • US $270.00

Revolutionizing Industrial Automation Through the Convergence of Artificial Intelligence and the Internet of Things
Divya Upadhyay Mishra (ABES Engineering College, Ghaziabad, India) and Shanu Sharma (ABES Engineering College, Ghaziabad, India)
Engineering Science Reference • © 2023 • 300pp • H/C (ISBN: 9781668449912) • US $270.00

Convergence of Big Data Technologies and Computational Intelligent Techniques
Govind P. Gupta (National Institute of Technology, Raipur, India)
Engineering Science Reference • © 2023 • 335pp • H/C (ISBN: 9781668452646) • US $270.00

Design and Control Advances in Robotics
Mohamed Arezk Mellal (M'Hamed Bougara University, Algeria)
Engineering Science Reference • © 2023 • 320pp • H/C (ISBN: 9781668453810) • US $305.00

Handbook of Research on Applied Artificial Intelligence and Robotics for Government Processes
David Valle-Cruz (Universidad Autónoma del Estado de México, Mexico) Nely Plata-Cesar (Universidad Autónoma del Estado de México, Mexico) and Jacobo Leonardo González-Ruíz (Universidad Autónoma del Estado de México, Mexico)
Information Science Reference • © 2023 • 434pp • H/C (ISBN: 9781668456248) • US $315.00

IGI Global
PUBLISHER of TIMELY KNOWLEDGE

701 East Chocolate Avenue, Hershey, PA 17033, USA
Tel: 717-533-8845 x100 • Fax: 717-533-8661
E-Mail: cust@igi-global.com • www.igi-global.com

List of Reviewers

Samiran Bera, *IIT(ISM), Dhanbad, India*
Arun Prasad Burnwal, *GGSESTC, Bokaro, India*
Santosh Kumar Das, *Sarala Birla University, Ranchi, India*
Preeti Gupta, *Inderprastha Engineering Technology, Ghaziabad, India*
Praphula Kumar Jain, *IIT(ISM), Dhanbad, India*
Ambedkar Kanapala, *IIT(ISM), Dhanbad, India*
Abhishek Kumar, *National Institute of Technology, Jamshedpur, India*
Ranjit Kumar, *Galgotias College of Engineering and Technology, India*
Shalini Mahato, *Indian Institute of Information Technology, Ranchi, India*
Ajay Kumar Mallick, *IIT(ISM), Dhanbad, India*
Manoj Kumar Mandal, *Jharkhand Rai University, Ranchi, India*
Nabajyoti Mazumdar, *Indian Institute of Technology, Allahabad, India*
Divya Mishra, *Swami Vivekanand Subharti University, Meerut, India*
Vishwas Mishra, *Swami Vivekanand Subharti University, Meerut, India*
Anand Mohan, *MIT Art, Design, and Technology University, Pune, India*
Amitesh Kumar Pandit, *Dr. Rammanohar Lohia Avadh University, India*
Jitesh Pradhan, *IIT(ISM), Dhanbad, India*
Amalendu Rana, *Jadavpur University, India*
Asish Kumar Roy, *NIST, Berhampur, India*
Sagar Samal, *NIST, Berhampur, India*
Sourav Samanta, *University Institute of Technology, The University of Burdwan, India*
Chandan Kumar Shiva, *SR University, Warangal, India*
Amit Kumar Singh, *SRM University, India*
Rakesh Ranjan Swain, *Indian Institute of Technology, Kanpur, India*

Table of Contents

Detailed Table of Contents

An intrusion detection system (IDS) is a system that monitors network traffic for suspicious activities and issues alerts when such activities are discovered. If an intruder is able to exploit any kind of vulnerability then he/she is able to steal, modify, delete personal data of a person. The primary function of IDS is anomaly detection and reporting, but few intrusion detection systems are capable of taking actions when malicious activity or anomalous traffic is detected, including blocking traffic sent from suspicious IP addresses. The continued ability to detect malicious network intrusions has become an exercise in scalability, in which data mining techniques are playing an increasingly important role. The objective of this chapter is to simulate IDS using data mining technique on some public and private datasets such as virtualized, synthesized, and realistic.

the accuracy in order to save energy and make better use of the available bandwidth for data transmission. Three performances are assessed by the decision blending function: detection accuracy (DA), sensitivity, and rate of error. Different methods, such as k-nearest neighbor (KNN), enhanced extreme learning machine (EELM), enhanced support vector machine (ESVM), and enhanced recurrent extreme learning machine, are used in the belief function approach (ERELM). Here, the authors applied decision blending approaches in WSNs to emulate these techniques for improving belief function.

Wireless sensor networks (WSN) can be referred to as a group of spatially dispersed and dedicated sensors that are self-capable of collecting environmental information, processing it as per requirement, and forwarding it to a base station for different application. In this chapter, the authors focus on secure data transformation in wireless sensor node. The intrusion detection system using sensor node is complicated as sensors lifetime depends upon power consumption and for which the network's lifetime also get affected. Security as well as network lifetime can be enhanced by relocating the computational load of the operation from the sensor nodes to the base station. At base station by encrypting and compressing the data using cryptography and artificial neural network (ANN), the authors increase the security and increase the throughput of the network. On the server side it is decompressed and then decrypted.

Underwater wireless sensor networks (UWSN) comprise of numerous relatively low-cost, self-organizing, and static as well as mobile sensor nodes. The wireless sensor node is used to sense natural phenomenon in terms of humidity, temperature, etc. BS is used to collect required information with proper location information from the sensor nodes and process it for future references. The problem of localization in underwater sensor nodes has led to proposal of many techniques over the past few decades. In this chapter, an intelligent scan model method has been proposed for localization system under UWSN. The localization system is used to identify the actual location of sensor node for identifying sense data properly trough Data MULE (mobile ubiquitous LAN extensions), which is an unmanned floating vehicle that has a sufficient amount of storage area and rechargeable energy to store the data of the sensors in the field. Finally, the proposed method is compared with existing method and outperforms the results efficiently.

The authors use the manifold technique to demonstrate that the issue combining a nonlinear integro-differential operator and a term with critical exponent has three nontrivial weak solutions. The operator considered in the problem covers the case of the fractional p-Laplacian operator and is possibly degenerate. The outcomes include a sign-changing solution and a positive and a negative solution. Also, presence of critical exponent term increases the complexity of the problem as compactness property fails in this case.

Women entrepreneurship development is indeed an essential part of human resource development. In India the growth of women entrepreneurship is very low mostly in rural areas. Entrepreneurship is one of the most important issues under business paradigm. Women entrepreneurship, under this phenomenon is the recent concern. Women are very much aware about their skill, knowledge, capacity, rights, and work situations. The progress is more visible in upper class families than middle class society. This paper focuses on understanding of capacity of Indian women and their role effectiveness in the business development area under industrial revolution 4.0. It is the current trend of automation, it provides data exchange in the field of manufacturing technologies. With the application of Artificial Intelligence the women entrepreurship has taken a place of role playing to expand country's economy.

The internet of things (IoT) is one of the new technologies based on new requirements and innovations. It is a set of equipment that consists of several sensor nodes and devices to make efficient communication. It is an embedded system that contains several sensor nodes, software, and exchanging equipment. There are several applications of IoT such as smart homes, health monitoring systems, smart factories, wearable health monitoring systems, logistic tracking, etc. Although, it has several applications and advantages, but it has also some limitations regarding data privacy, data security, data volume and data complexity, etc. Hence, it requires innovating and modeling of security of IoT based on different techniques. Therefore, in this chapter, a details review is discussed based on the fusion of edge computing and blockchain.

Single layer graphene is called monolayer graphene which exhibits some unique characteristics with a special arrangement of atoms in its crystal. In contrary, graphene can be sometimes modified for numerous application specific uses. This book chapter presents brief details of graphene structure, along with formation of sp2 hybridization, critical parameters of the unit cell, formation of σ and π bonds, different properties, synthesis processes and their advantages and disadvantages over one another, potential applications of graphene in different fields. Out of different synthesis processes, mechnical exfoliation, chemical exfoliation, reduced graphene oxide, chemical vapour deposition, and epitaxial growth are the most well-known processes, and are detailed and described here. Graphene application in integrated circuit, optical electronics generally in OLED, energy application generally for the storage purpose is elaborately discussed here in this chapter. Throughout the chapter the authors described everything with suitable diagrams as per the requirements.

The notions of different types of compactness in a spacetime manifold has reviewed in this article. Also, their relations with Cauchy hypersurfaces, which play very important role in globally hyperbolic spacetimes has been discussed. For example, A be a closed subset of a spacetime M having a compact intersection with all the Cauchy hypersurfaces of it, then $A \subset J(C)$ for some compact set $C \subset M$ and conversely.

Preface

In the last decade, there are several techniques are innovated along with invention. The main cause of this initiation is the conflicting demand of the customer as well as the user. It affects the requirement of society based on real-life application and its motivation. This type of decision making is also used in the area of entrepreneurship management for industry revolution 4.0. It also helps in industry 4.0 management. Entrepreneurship is also used as women's entrepreneurship to solve some issues in society. There are several new techniques along with areas are designed such as cloud computing, machine learning, data science, internet of things, supply chain management, information security, etc. Each of the technique directly or indirectly involved with artificial intelligence. It requires constraint decision making system for the purpose of modeling different information. It also helps to model several techniques based on objectives along with some data and parameters. Hence, the main motive of this book is to enhance the application of constraint decision making systems in the area of real-life applications.

THE OBJECTIVE OF THE BOOK

The objective of this book is to increase the usages of constraint decision making system in society. It also helps to increase the application of this field in different areas where smart computing is used efficiently. It helps to solve intelligent and constraint problems based on some features and facilities. It helps different users such as students, research scholars, academicians, industry people, etc.

ORGANIZATION OF THE BOOK

The book contains 15 chapters that are organized into three sections as follows. Section 1 indicates different aggregation and troubleshooting based on different areas and their applications. Section 2 indicates the modeling and optimization of several decisions based on some objective to resolve different issues. Section 3 indicates different performances and analyses of some operations based on some evaluation within the context of some applications.

Section 1: Aggregation and Troubleshooting (Chapters 1-5)

This section is based on some problem-solving system based on troubleshooting and aggregation methods.

Chapter 1

This chapter deals with intrusion detection system based on some modeling of network analysis. It is based on data mining technique that helps to model several applications with the context of intrusion analysis.

Chapter 2

This chapter helps to model several information based on knowledge based system. It helps to model several applications based on wireless sensor networks. This chapter presents the review literature on knowledge management and demonstrates the creating, managing, sharing, and utilizing in organizations.

Chapter 3

This chapter is based on water monitoring system for the agriculture management system. The work is based on wireless device system analysis and modeling that enhances the methodology of agriculture. It helps farmers based on a device which helps farmers do so in some easy way. Visual surveillance and machine learning help to model this application.

Chapter 4

This chapter helps to detect magnetic resonance in the NV centre of the diamond. The work is based on theoretical analysis for the purpose of the microwave antenna. The proposed work helps to present a simulated model of a microwave antenna and proposes further certain modifications to increase the intensity of the microwave field.

Chapter 5

This chapter is based on a theoretical investigation of EMHD flow through a microchannel with a porous medium. The impact of the Hall Effect on the nanofluid flow transport is analyzed by considering the Brinkman momentum equation to govern the flow performance through the porous microchannel.

Section 2: Optimization and Modeling (Chapters 6-10)

This section helps to analyze some information based on optimization and modeling. The work of this section helps to deal with several constraints that help to meet the objective based on the requirement.

Chapter 6

This chapter deals with decision blending system for modeling and detection of different types of faults. The work is the fusion of three parameters such as detection accuracy, sensitivity and rate of error. Some techniques are used for modeling this method such as KNN, EELM, ESVM, and ERELM.

Chapter 7

This chapter is based on secure data transformation in the wireless sensor node. The intrusion detection system using sensor nodes is complicated as sensors lifetime depends upon power consumption and for which the network's lifetime is also gets affected. The security, as well as network lifetime, can be enhanced by relocating the computational load of the operation from the sensor nodes to the base station.

Chapter 8

This chapter used an intelligent scan model for the localization system under UWSN. The localization system is used to identify the actual location of the sensor node for identifying sensitive data properly through Data MULE which is an unmanned floating vehicle. It has a sufficient amount of storage area and rechargeable energy to store the data of the sensors in the field.

Chapter 9

This chapter is based on multiple nontrivial solutions for nonlocal critical exponent system. In this work, the author deals to analyze the existence of three nontrivial weak solutions by using parameters involving a nonlinear integro-differential operator and a term with a critical exponent.

Chapter 10

This chapter is based on women's entrepreneurship that helps to model industry revolution 4.0. The proposed chapter made a discussion of the bravery and challenges of women entrepreneurs. It helps to break traditions with innovative ideas in the business environment and the way to become role models for others

Section 3: Performance Analysis and Evaluation (Chapters 11-15)

This section helps to model the application based on several parameters. The purpose of this modeling is to analyze several performances and evaluate them. It helps to enhance the model and compatibility of the system.

Chapter 11

This chapter presents security analysis of IoT based on fusion of edge computing and blockchain. The work is based on review paper that helps to illustrates several applications and operations for modeling security of IoT. It helps to design several illustrations that help to model some architecture and estimate the issues.

Chapter 12

This chapter presents brief details of graphene structure, along with the formation of sp2 hybridization, critical parameters of the unit cell, formation of σ and π bonds, different properties, synthesis processes and their advantages and disadvantages over one another, and potential applications of graphene in different fields.

Chapter 13

This chapter has reviewed the notions of different types of compactness in a spacetime manifold. It also manages several relations with cauchy hypersurfaces that play a very important role in globally hyperbolic spacetimes that have been discussed.

Chapter 14

This chapter discusses b-open sets in the cofinite topology. It also studied the information that is related with this topology. The author also characterized the b-open sets in the cofinite topology over . Finally, several results are discussed based on performance evaluation system.

Chapter 15

This chapter is based on some investigations based on turbulent flow inside a corrugated microtube. It simulated numerically at a low Reynolds number. The effect of irregular roughness of the wall of the microtube has been investigated and the turbulent flow model has been simulated to measure the state of turbulence.

Santosh Kumar Das
Sarala Birla University, Ranchi, India

Nilanjan Dey
Techno International New Town, Kolkata, India

Chapter 1
An Efficient Data Mining Technique for an Intrusion Detection System in Network

Santosh Kumar Das
https://orcid.org/0000-0001-6852-1361
Sarala Birla University, Ranchi, India

Sagar Samal
National Institute of Science and Technology, Berhampur, India

Priya Ranjan
National Institute of Science and Technology, Berhampur, India

Shom Prasad Das
Birla Global University, Bhubaneswar, India

ABSTRACT

An intrusion detection system (IDS) is a system that monitors network traffic for suspicious activities and issues alerts when such activities are discovered. If an intruder is able to exploit any kind of vulnerability then he/she is able to steal, modify, delete personal data of a person. The primary function of IDS is anomaly detection and reporting, but few intrusion detection systems are capable of taking actions when malicious activity or anomalous traffic is detected, including blocking traffic sent from suspicious IP addresses. The continued ability to detect malicious network intrusions has become an exercise in scalability, in which data mining techniques are playing an increasingly important role. The objective of this chapter is to simulate IDS using data mining technique on some public and private datasets such as virtualized, synthesized, and realistic.

DOI: 10.4018/978-1-6684-7343-6.ch001

INTRODUCTION

In last few decades, the applications of wireless network increase rapidly due to its various autonomous and efficient facilities (Bosch et al., 2020). A wireless network is the data communication system amongst multiple nodes. It is a method by which a house, organization or any system can reduce the costliest wired connection throughout their buildings. As its wireless so there is no medium in between rather than air. In the physical layer of the Open System Interconnection (OSI) model it has been implemented. Examples of wireless networks include cell phone networks, Wireless Local Area Networks (WLANs) (Raschellà et al., 2020), Wireless Sensor Networks (WSNs) (Das et al., 2013; Das et al., 2014; De et al., 2020), Wireless Ad-hoc Network (WANET) (Das & Tripathi, 2019; Das & Tripathi, 2018, Das & Tripathi, 2018; Das & Tripathi, 2017; Das et al., 2020) satellite communication networks, terrestrial microwave networks, etc. There are several components of wireless network that help to achieve the goal such as user devices, radio Network Interface Cards (NICs), Access Points (AP), routers, repeaters, antennae, etc. There are several applications of wireless network in terms of vehicles, emergency, business, replacement of wired network, location dependent service, entertainment, education, etc. There are several limitations of wireless network compared to wire network such as file-sharing transfer speeds are normally slower with wireless networks. Setting up a wireless network often difficult for non-experienced people. Speed also less compare to wired communication. There is some vulnerability by exploiting which an intruder can cause harm or that simply known as intrusion like: weak password, denial of service, insecure host, weak firewall, no gateway protection, neglected security software maintenance.

An Intrusion Detection System (IDS) is a system that monitors network traffic for suspicious activity and issues alerts when such activity is discovered (Mahdavi et al., 2020; Krzysztoń & Marks, 2020). It is a software application that scans a network or a system for harmful activity or policy breaching. Any malicious venture or violation is normally reported either to an administrator or collected centrally using a Security Information and Event Management (SIEM) system. A SIEM system integrates outputs from multiple sources and uses alarm filtering techniques to differentiate malicious activity from false alarms.

Although, intrusion detection systems monitor networks for potentially malicious activity, they are also disposed to false alarms. Hence, organizations need to fine-tune their IDS products when they first install them. It means properly setting up the intrusion detection systems to recognize what normal traffic on the network looks like as compared to malicious activity. Intrusion prevention systems also monitor network packets inbound the system to check the malicious activities involved in it and at once sends the warning notifications.

The rest of the paper is divided as follows. Section 2 highlights some existing work related to some routing techniques. Section 3 shows the basic preliminaries. Section 4 shows details of main proposed method. Section 5 indicates simulation analysis part. Finally, in Section 6 conclude the paper.

RELATED WORKS

In last few decades, several works have been proposed based on intrusion detection system in WSN. Some of them describe as: Qu et al. (2018) proposed a method for intrusion detection system in the WSN using unsupervised method. In this paper, the basic key element is ANN system that used unsupervised learning technique by the fusion of Mean Shift Clustering (MSC) system for handling threat of the network. At last, by using iterated learning method, the network lifetime is improved and network channel is also enhanced. Devi et al. (2017) prepared an Intrusion Detection System for 5G wireless communication network using adaptive neuro-fuzzy interface. In this paper main focus is on implementing 5G network with ease. Due to the high demand of faster communication in this era the communication should be safe and secure thus using neuro-fuzzy algorithm, this IDS is developed which will keep track of attack on relay. Thaseen et al. (2017) implemented an Intrusion Detection System for the rapidly growing huge amount of data. In this paper the focus is to build a multiclass support vector machine with chi-square feature selection, which can be used for reducing false alarm by improved dimensionality reduction. Mazini et al. (2019) prepared a model to reduce the false alarm in the Intrusion Detection System. In this paper the main model used is artificial bee colony (ABC) with AdaBoost algorithms to improve the result by gaining a high detection rate and low false positive rate. Mukkamala et al. (2005) prepared a model to use with Intrusion Detection System to deal with increasing cyber-attacks. In this paper the main approach is the ensemble learning which can be availed with the help of Artificial Neural Network (ANN) (Chatterjee et al., 2017; Dey et al., 2015), Support Vector Machine (SVM) and Multivariate Adaptive Regression Splines (MARS). Khan et al. (2020) proposed a concept to overcome from the situations like security violation, breaking the firewall etc in the Ad hoc network. In this paper the main focus is on IDS System which will act as the second level security in the network by exploring more details on the Ad hoc network. Manan et al. (2019) implemented a security system which will work for IP Multimedia Subsystem (IMS). In this paper main focus is on the protocol that works in this IMS System. Session Initiation Protocol (SIP) which is used to establish the connection in case of the IMS. So to find the vital vulnerabilities focus is imposed in the DDoS (Distributed Denial of Service) over the IMS with registered SIP Protocol. Ozyer et al. (2007) proposed an intelligent intrusion detection system using classification and association rule mining. In this paper the main focus is how to apply the above-mentioned machine learning techniques in the IDS so that it acts smartly. For this to happen there are two tasks to be done. One is to generate candidate rules using association rule then in second phase a boosted genetic algorithm imposed in each class of the candidate rule. Elbasiony et al. (2013) proposed a model to overcome from the difficulties of the rule-based system. In this paper the problem is divided into two parts Misuse based and Anomaly based. For Misuse based random forest algorithm is used whereas for Anomaly based k-mean clustering is used. For improving the performance in case of anomaly based weighted k-mean algorithm is used this is nothing but a modified version of k-mean clustering algorithm. Khalili et al. (2015) proposed a model which can be used to secure the Cyber Physical System (CPS). In this paper focus is about anomaly detection which is classified in to 3 parts such as: normal states, normal transitions between the normal states, and normal time-intervals for transitions. To do this task a special IDS is used known as Industrial IDS which will work efficiently to detect the above anomalies. Idhammad et al. (2018) proposed a machine learning model to be implemented in the cloud computing. It this paper it has been observed that due to rapid grow of data in today's world causes problem or challenge to the security system, so a system is implemented using random forest-based classifier to use it in IDS to improve the performance. Mohammad et al. (2011) proposed an intelligent machine leaning model to be used with existing IDS. In this paper he used a Weka tool which is a very dynamic tool developed using java and having predefined data pre-processing techniques, model design and implement and visualization. Nadiammai and Hemalatha (2014) proposed a system that can detect the network vulnerability present over a network. In this paper, four major points are taken into the consideration like: classification of data, high level of human interaction, lack of labelled data, and effectiveness of distributed denial of service attack, which are being solved using the proposed algorithms like EDADT algorithm, Hybrid IDS model, Semi-Supervised Approach and Varying HOPERAA Algorithm respectively. Duque and Omar (2015) proposed a model which can be used for reducing the no of false positive values in the network. In this paper after some observation, they found that using k-means data mining algorithm followed by signature-based approach is effective in reducing the false negative rate. Borkar et al. (2019) proposed a support vector machine classifier in the IDS for the wireless sensor network. In this paper an adaptive SVM classifier and RRF algorithm are effectively utilized to classify and to reduce the extra features within the sensor nodes. It will help to impose a high security IDS during transmission. Hidayanto et al. (2017) proposed an IDS sys-tem with associates with association rule mining in the machine leering. In this paper the model is an unsupervised where algorithms used are Frequent Item Set Mining Algorithm FP-Max and Apriori Algorithm. It will to find out malware in the network. Bekri and Diouri (2019) proposed a model which can be used to accurately identify the network behaviour. In this paper the main focus is imposed on how IDS will be capable of extracting the accurate network behaviour while being intelligent enough to detect new intrusions. Data mining used for learning and swarm optimization used for search and detect. Koucham et al. (2018) proposed a model which can be used with IDS to help industrial sector. In this paper the main focus is relying on automatically mined process specifications to detect attacks on sequential control systems. The specifications are synthesized as monitors that read the execution traces and report violations to the operator. Fang et al. (2020) proposed a model to ensure the network security in IDS. In this paper he studied the VC dimension theory and structural risk minimization principle of statistical learning theory and apply it using Lagrange multiplier method and satisfy the Karush-kuhn-Tucher condition. Mohammad et al. (2011) proposed an intelligent machine leaning model to be used with existing IDS. In this paper he used a Weka tool which is a very dynamic tool developed using java and having predefined da-ta pre-processing techniques, model design and implement and visualization. Some of the security-based works also proposed for the purpose of intrusion detection and encryption purpose. It helps to enhance the performance of existing model by using cryptography system and its related methodologies (Roy et al., 2016; Dey et al., 2017; Das & Giacalone, 2022).

Preliminaries

In this section, some basic preliminaries are used that help to understand the work of this paper efficiently. Short description of this given as.

Intrusion Detection System

An IDS is a system that monitors network traffic for suspicious activity and issues alerts when such activity is discovered. It is a software application that scans a network or a system for harmful activity or policy breaching. Any malicious venture or violation is normally reported either to an administrator or collected centrally using SIEM system. A SIEM system integrates outputs from multiple sources and uses alarm filtering techniques to differentiate malicious activity from false alarms. There are two types of IDS given in Figure 1.

NIDS stands for network intrusion detection system. It implemented at a pre-planned point within the network to evaluate the traffic from all devices on that network. HIDS stands for host intrusion detection system. It run on independent hosts or devices on the network. There is no such pre-planned points to implement.

Figure 1. Types of IDS

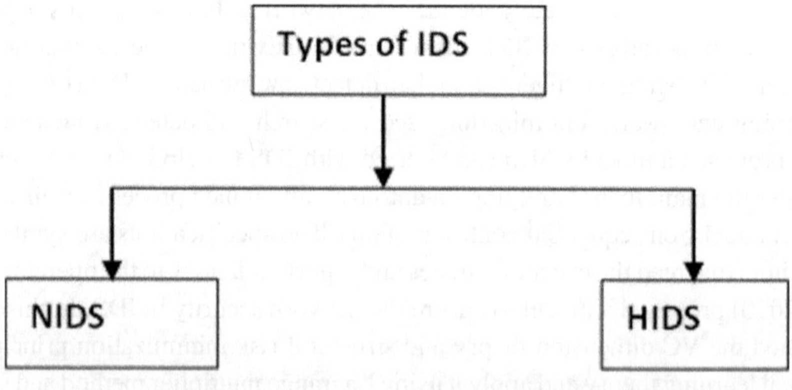

An ID is detected based on two methods such as signature based method and anomaly based method. First method is used to detect the attacks on the basis of the specific patterns such as number of bytes in the network traffic. It also detects on the basis of the already known malicious instruction sequence that is used by the malware. Second method is used in machine learning for creating a trustful activity model and anything coming is compared with that model and it is declared suspicious if it is not found in model. Phases of IDS as follows.

(a). **Data Collection:** This module passes the data as input to IDS. Data is recorded into file then analyzed.
(b). **Feature Selection:** To select the particular feature large data is available in the network and they are usually evaluated for intrusion.
(c). **Analysis:** The data is analyzed to check the correctness. Incoming traffic is checked against predefined signature in case of signature-based IDS and behavior is studied in case of anomaly-based IDS.
(d). **Action:** It defines about the reaction and attack of the system. It can inform system administrator through an alert message

Support Vector Machine

In data science, Support Vector Machines (SVM) is defined as a supervised machine learning model, associated with different learning algorithms that analyze the data used for different techniques of supervised learning like: regression, classification etc. The working principle of SVM is based on the hyperplane concept. The main use of this hyper plane concept is for the outlier detection purpose

Confusion Matrix

It is generally a statistical table used in the machine learning field to predict the result of the model by comparing actual test dataset with the predicted dataset. It is also known as error matrix. There are four basic terms used in the confusion Matrix such as: true positive, true negative, false positive, false negative. The following is an example of the all above terms mentioned.

(a). **True positive:** Prediction- Yes Actual- Yes
(b). **True negative:** Prediction- No Actual- No
(c). **False positive:** Prediction- Yes Actual- No (Type I error)
(d). **False negative:** Prediction- No Actual- Yes (Type II error)

PROPOSED METHOD

This is main section of the proposed work. In this section proposed method is described in chronological order. Details of the method shown in Algorithm 1. In this paper, there are two datasets are used for anomaly detection and misuse operation named as "Data_Anomaly.csv" and "Data_Misuse.csv" to achieve the task of intrusion detection system. In first dataset, the class or prediction variable is either in "Normal" or "Attack" case where first category represents secure case (i.e., no attack) an second category represents an abnormal case (i.e. attacked). Detail of anomaly dataset shown in Figure 2. Figure 2 consists of 42 attributes that represent different properties of network. The class variable is AttackType which have two categories: Normal and Attack. For our observation we take 1000 records to predict our result using a model. Here the values of the attributes are numer-ic and non-text format. Contrary to the anomaly detection data set, the misuse detection data set has a class variable Normal or Name of the attack which represents a specific type of attack such as Smurf, Nmap, Rootkit, etc. We carry out data cleaning on Data_Anomaly.csv and Data_Misuse.csv using Weka to obtain Selected_Data_Anomaly.csv and Selected_Data_Misuse.csv which has less attributes that help speed our Model. Details of misuse dataset shown in Figure 3. In case of Misuse Dataset, we have also 42 columns and our class attribute are AttackType. It has 11 different categories: Normal and 10 types of attacks like Back, BufferOverflow, FTPWrite, GuessPassword, Neptune, Nmap, PortSweep, Rootkit, Satan, Smurf.

Algorithm 1: The proposed intrusion detection system.

Step 1: Data collection from Kaggle (KDD dataset).

Step 2: Data exploration and understanding the dataset.

Step 3: Preparation of the dataset to a standard format (CSV).

Step 4: Attribute selection using WEKA tool with the help of RemoveUseless () filter.

Step 5: R Studio is used to create a model.

Step 6: Reading cleaned dataset from WEKA.

Step 7: Random Sampling of data into training set and testing set.

Step 8: SVM classifier used for our training dataset.

Step 9: Prediction will be done using classifier with test dataset.

Step 10: Evaluation of model accuracy using Confusion Matrix.

Figure 2. Anomaly dataset

	A	B	C	D	E	F	G	H	I	J	K	L	M	N	O	P	Q	R	S	T	U
1	duration	protocol_	service	flag	src_bytes	dst_bytes	land	wrong_fr.	urgent	hot	num_faile	logged_ir	num_corr	root_shel	su_attem	num_root	num_file	num_she	num_acce	num_outl	is_ho
2	0	0	0.12	0.1	0	0	0	0	0	0	0	0	0	0	0	0	0	0	0	0	
3	0	0	0.12	0.07	0	0	0	0	0	0	0	0	0	0	0	0	0	0	0	0	
4	0	0	0	0	0.00209	0.15642	0	0	0	0	0	0.1	0	0	0	0	0	0	0	0	
5	0	0	0	0	0.00245	0.01651	0	0	0	0	0	0.1	0	0	0	0	0	0	0	0	
6	0	0	0	0	0.00253	0.01906	0	0	0	0	0	0.1	0	0	0	0	0	0	0	0	
7	0	0	0	0	0.00242	0.02164	0	0	0	0	0	0.1	0	0	0	0	0	0	0	0	
8	0	0	0	0	0.00234	0.13187	0	0	0	0	0	0.1	0	0	0	0	0	0	0	0	
9	0	0	0.03	0	9.00E-05	0.00034	0	0	0	0	0	0.1	0	0	0	0	0	0	0	0	
10	0	0	0	0	0.00196	0.06107	0	0	0	0	0	0.1	0	0	0	0	0	0	0	0	
11	0	0	0	0	0.00322	0.01713	0	0	0	0	0	0.1	0	0	0	0	0	0	0	0	
12	0	0	0	0	0.00208	0.01415	0	0	0	0	0	0.1	0	0	0	0	0	0	0	0	
13	0	0	0	0	0.00307	0.0528	0	0	0	0	0	0.1	0	0	0	0	0	0	0	0	
14	0	0	0.12	0.1	0	0	0	0	0	0	0	0	0	0	0	0	0	0	0	0	
15	0	0	0.04	0	8.00E-05	0.00136	0	0	0	0	0	0	0	0	0	0	0	0	0	0	
16	0	0	0	0	0.00197	0.01368	0	0	0	0	0	0.1	0	0	0	0	0	0	0	0	
17	0	0.02	0.09	0	0.01032	0	0	0	0	0	0	0	0	0	0	0	0	0	0	0	
18	0	0	0.12	0.05	0	0	0	0	0	0	0	0	0	0	0	0	0	0	0	0	
19	0	0	0.03	0	9.00E-05	0.00035	0	0	0	0	0	0.1	0	0	0	0	0	0	0	0	
20	0	0	0.01	0	0.01073	0.00333	0	0	0	0	0	0.1	0	0	0	0	0	0	0	0	
21	0	0	0	0	0.00239	0.02195	0	0	0	0	0	0.1	0	0	0	0	0	0	0	0	
22	0	0	0	0	0.00228	0.08595	0	0	0	0	0	0.1	0	0	0	0	0	0	0	0	
23	0	0	0	0	0.00213	0.30725	0	0	0	0	0	0.1	0	0	0	0	0	0	0	0	

Dataset_Anomaly

Figure 3. Misuse dataset

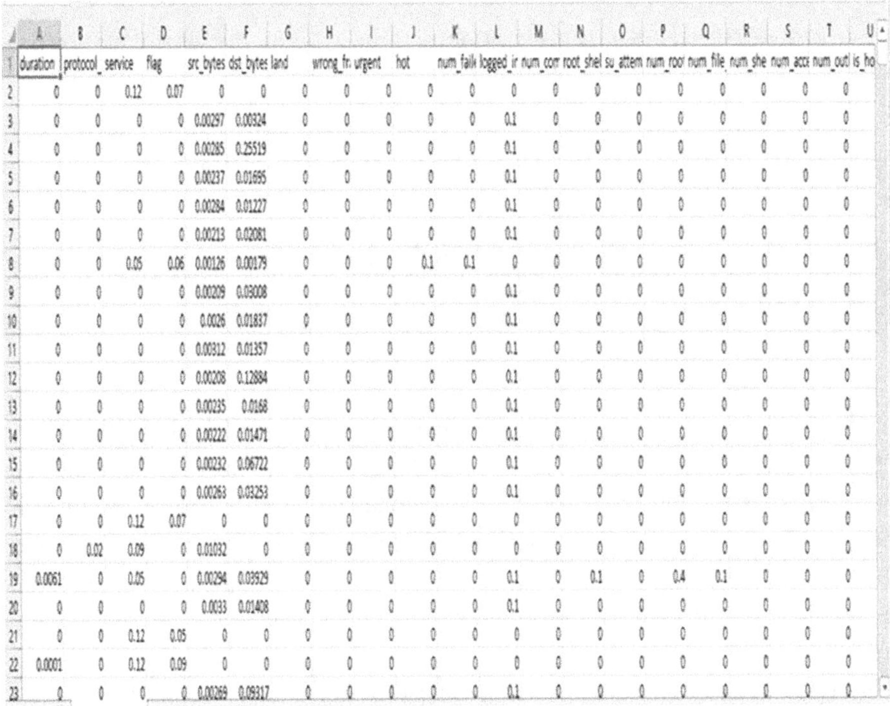

	A	B	C	D	E	F	G	H	I	J	K	L	M	N	O	P	Q	R	S	T	U
1	duration	protocol_	service	flag	src_bytes	dst_bytes	land	wrong_fr	urgent	hot	num_faile	logged_ir	num_corr	root_shel	su_attem	num_roo	num_file	num_she	num_accs	num_outl	is_ho
2	0	0	0.12	0.07	0	0	0	0	0	0	0	0	0	0	0	0	0	0	0	0	
3	0	0	0	0	0.00297	0.00324	0	0	0	0	0	0.1	0	0	0	0	0	0	0	0	
4	0	0	0	0	0.00285	0.25519	0	0	0	0	0	0.1	0	0	0	0	0	0	0	0	
5	0	0	0	0	0.00237	0.01695	0	0	0	0	0	0.1	0	0	0	0	0	0	0	0	
6	0	0	0	0	0.00284	0.01227	0	0	0	0	0	0.1	0	0	0	0	0	0	0	0	
7	0	0	0	0	0.00213	0.02081	0	0	0	0	0	0.1	0	0	0	0	0	0	0	0	
8	0	0	0.05	0.06	0.00126	0.00179	0	0	0	0.1	0.1	0	0	0	0	0	0	0	0	0	
9	0	0	0	0	0.00209	0.03008	0	0	0	0	0	0.1	0	0	0	0	0	0	0	0	
10	0	0	0	0	0.0026	0.01837	0	0	0	0	0	0.1	0	0	0	0	0	0	0	0	
11	0	0	0	0	0.00312	0.01357	0	0	0	0	0	0.1	0	0	0	0	0	0	0	0	
12	0	0	0	0	0.00208	0.12884	0	0	0	0	0	0.1	0	0	0	0	0	0	0	0	
13	0	0	0	0	0.00235	0.0168	0	0	0	0	0	0.1	0	0	0	0	0	0	0	0	
14	0	0	0	0	0.00222	0.01471	0	0	0	0	0	0.1	0	0	0	0	0	0	0	0	
15	0	0	0	0	0.00232	0.06722	0	0	0	0	0	0.1	0	0	0	0	0	0	0	0	
16	0	0	0	0	0.00263	0.03253	0	0	0	0	0	0.1	0	0	0	0	0	0	0	0	
17	0	0	0.12	0.07	0	0	0	0	0	0	0	0	0	0	0	0	0	0	0	0	
18	0	0.02	0.09	0	0.01032	0	0	0	0	0	0	0	0	0	0	0	0	0	0	0	
19	0.0061	0	0.05	0	0.00294	0.03929	0	0	0	0	0	0.1	0	0.1	0	0.4	0.1	0	0	0	
20	0	0	0	0	0.0033	0.01408	0	0	0	0	0	0.1	0	0	0	0	0	0	0	0	
21	0	0	0.12	0.05	0	0	0	0	0	0	0	0	0	0	0	0	0	0	0	0	
22	0.0001	0	0.12	0.09	0	0	0	0	0	0	0	0	0	0	0	0	0	0	0	0	
23	0	0	0	0	0.00269	0.09317	0	0	0	0	0	0.1	0	0	0	0	0	0	0	0	

Attributes Selection Using Weka Tool

This subsection is used for optimal attributes selection for the proposed method using weka tool. Waikato Environment for Knowledge Analysis (Weka), developed at the University of Waikato, New Zealand. It is free software licensed under the GNU General Public License, and the companion software to the book "Data Mining: Practical Machine Learning Tools and Techniques". The weka tool contains some attributes that shown in Figure 4. The steps of attribute selection using Weka tool is given below.

Step 1: Open WEKA and select the Explorer Mode.

Step 2: Open Dataset_Anomaly.csv and Dataset_Misuse.csv which are our datasets in Weka.

Step 3: Under Preprocess data tab select: Fil-ter>Unsupervised>Attribute>RemoveUseless ().

Step 4: We generate new datasets after saving from WEKA as: Dataset_Anomaly_AttributeSelection.csv and Da-taset_Misuse_AttributeSelection.csv

Figure 4. Attribute before filter used

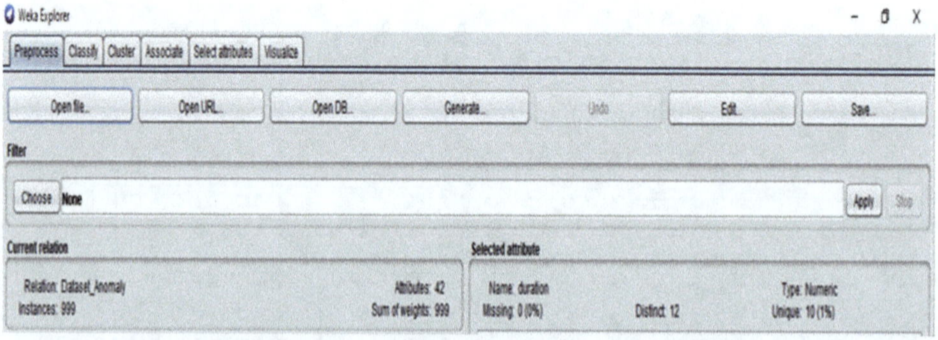

In Figure 4, shows the working environment of the weka tool. In the explorer mode we have a vast number of functionalities which can be seen. Those functionalities are Preprocess, classify, cluster, associate, selected attribute, visualize etc. So, to select necessary attribute we used a predefined filter function known as RemoveUseless () function. So, in the above figure the photo taken before the uses of the filter. We can see that it has 1000 records and 42 attributes. There are total of 12 distinct attributes present.

In Figure 5, it is a working environment of the weka tool. In the explorer mode we have a vast number of functionalities which can be seen. Those functionalities are Preprocess, classify, cluster, associate, selected attribute, visualize etc. So, to select necessary attribute we used a predefined filter function known as RemoveUseless () function. So, in the above figure the photo taken after the uses of the filter. We can see that it has 1000 records and 34 attributes. There are total of 12 distinct attributes present.

Figure 5. Attributes after filter used

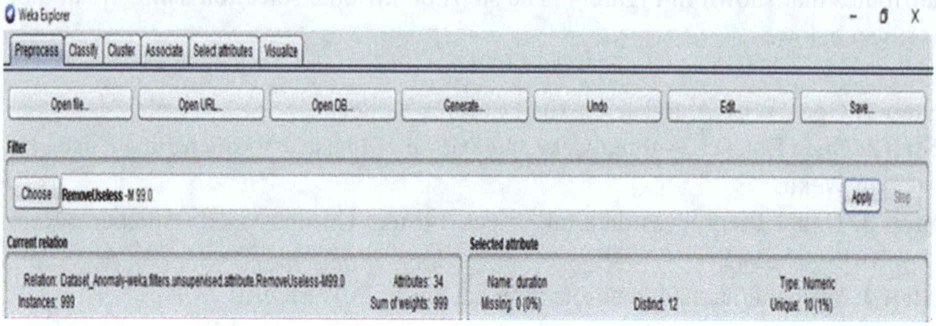

In this subsection, classifier is design using SVM with the help of R programming in R studio. Table 1 shows that how kernel SVM classifier is design using R language. It related captions shown in Table 2.

Table 1. Kernel SVM classifier is design

Data_Classifier<-ksvm(AttackType~.,data= DataSet_TrainingSet,type = 'C-svc', kernel = 'rbfdot')

Table 2. Related captions of classifier design

Caption	Description
Dataset_Classifier	Name of ther svm model
ksvm()	Kernel-SVM.
AttackType	Target attribute or class attribute
rdfdot	Redial basis function kernel

Simulation and Analysis

The proposed method is validated in R language for predicting and analysis of intrusion detection system accuracy. The system configuration shown in Table 3. In this work, a confusion matrix is designed for analysis and predicting results, estimating the model accuracy and errors.

Table 3. Simulation parameters

Software/Hardware	Specification
Windows OS	10
System Type	64 bit
MS Office	MS Office 2016
R Studio	3
Processor	Intel i5
Speed	2.2GHz
RAM	4 GB
Hard Disk	600 GB

In the problem of statistical classification, a confusion matrix, also known as an error matrix used, which is a table that is often used to describe the performance of a classification model on a set of test data for which the true values are known. It allows the visualization of the performance of an algorithm. R studio is used to compare the propose model prediction based on actual test dataset. Here is the following R code shown in Figure. 6 which is used for implementation of confusion matrix. Prediction of the proposed system model shown in Figure 7.

Figure 6. The proposed confusion matrix for system model

```
library(kernlab)
library(caret)
library(e1071)

anomaly<-read.csv("Dataset_Anomaly.csv", na.strings=c(".", "NA", "", "?"), strip.white=TRUE, encoding="UTF-8")
aRow<-nrow(anomaly)
aCol<-ncol(anomaly)
summary(anomaly$AttackType)

sub<-sample(1:aRow,floor(0.66*aRow))
anomalyTrainingSet<- anomaly[sub,]
anomalyTestSet<- anomaly[-sub,]

anomalyClassifier<- ksvm(AttackType~.,data=anomalyTrainingSet,type = 'C-svc', kernel = 'rbfdot')
anomalyPrediction<-predict(anomalyClassifier, anomalyTestSet[,-aCol])
confusionMatrix(anomalyPrediction,anomalyTestSet[,aCol])
```

CONCLUSION

The IDS will give a detailed report of attacks to its administrator. IDS will detect the vulnerabilities and try to fix them all with or without administrator. IDS will order the level of risks (High, Moderate, Low etc.) in a hierarchy structure. Prompt possible preventive steps needed to overcome that particular situation. The results that we have above are drawn from running tests for 3 approaches, the summary and snapshots are attached above. The original dataset that we had with 42 attributes which were reduced to 36 using Weka's RemoveUseless () and then eliminating other redundant attributes with low variance. If we ignore the difference in the number of these attributes everything else in the dataset is just same which is why we have similar scripts for these different data sets.

Figure 7. Prediction of system model

```
C:/Users/Sagar Samal/Desktop/ProjectWork/
> confusionMatrix(anomalyPrediction,anomalyTe
Confusion Matrix and Statistics

          Reference
Prediction Attack Normal
    Attack    120      1
    Normal      0    219

               Accuracy : 0.9971
                 95% CI : (0.9837, 0.9999)
    No Information Rate : 0.6471
    P-Value [Acc > NIR] : <2e-16

                  Kappa : 0.9936

 Mcnemar's Test P-Value : 1

            Sensitivity : 1.0000
            Specificity : 0.9955
         Pos Pred Value : 0.9917
         Neg Pred Value : 1.0000
             Prevalence : 0.3529
         Detection Rate : 0.3529
   Detection Prevalence : 0.3559
      Balanced Accuracy : 0.9977

       'Positive' Class : Attack
```

REFERENCES

Borkar, G. M., Patil, L. H., Dalgade, D., & Hutke, A. (2019). A novel clustering approach and adaptive SVM classifier for intrusion detection in WSN: A data mining concept. *Sustainable Computing: Informatics and Systems*, *23*, 120–135. doi:10.1016/j.suscom.2019.06.002

Bosch, P., De Schepper, T., Zeljković, E., Famaey, J., & Latré, S. (2020). Orchestration of heterogeneous wireless networks: State of the art and remaining challenges. *Computer Communications*, *149*, 62–77. doi:10.1016/j.comcom.2019.10.008

Chatterjee, S., Hore, S., Dey, N., Chakraborty, S., & Ashour, A. S. (2017). Dengue fever classification using gene expression data: a PSO based artificial neural network approach. In *Proceedings of the 5th international conference on frontiers in intelligent computing: theory and applications* (pp. 331-341). Springer.

Das, S. K., Das, B., & Burnawal, A. P. (2014). Intelligent energy competency routing scheme for wireless sensor network. *International Journal of Research in Computer Applications and Robotics*, *2*(3), 79–84.

Das, S. K., & Giacalone, M. (Eds.). (2022). *Fuzzy Optimization Techniques in the Areas of Science and Management*. CRC Press. doi:10.1201/b23138

Das, S. K., Kumar, A., Das, B., & Burnwal, A. P. (2013). Ethics of reducing power consumption in wireless sensor networks using soft computing techniques. *International Journal of Advanced Computer Research*, *3*(1), 301.

Das, S. K., Samanta, S., Dey, N., & Kumar, R. (Eds.). (2020). *Design frameworks for wireless networks*. Springer Singapore. doi:10.1007/978-981-13-9574-1

Das, S. K., & Tripathi, S. (2017). Energy efficient routing formation technique for hybrid ad hoc network using fusion of artificial intelligence techniques. *International Journal of Communication Systems*, *30*(16), e3340. doi:10.1002/dac.3340

Das, S. K., & Tripathi, S. (2018). Adaptive and intelligent energy efficient routing for transparent heterogeneous ad-hoc network by fusion of game theory and linear programming. *Applied Intelligence*, *48*(7), 1825–1845. doi:10.100710489-017-1061-6

Das, S. K., & Tripathi, S. (2018). Intelligent energy-aware efficient routing for MANET. *Wireless Networks*, *24*(4), 1139–1159. doi:10.100711276-016-1388-7

Das, S. K., & Tripathi, S. (2019). Energy efficient routing formation algorithm for hybrid ad-hoc network: A geometric programming approach. *Peer-to-Peer Networking and Applications*, *12*(1), 102–128. doi:10.100712083-018-0643-3

De, D., Mukherjee, A., Das, S. K., & Dey, N. (Eds.). (2020). *Nature inspired computing for wireless sensor networks*. Springer. doi:10.1007/978-981-15-2125-6

Devi, R., Jha, R. K., Gupta, A., Jain, S., & Kumar, P. (2017). Implementation of intrusion detection system using adaptive neuro-fuzzy inference system for 5G wireless communication network. *AEÜ. International Journal of Electronics and Communications*, *74*, 94–106. doi:10.1016/j.aeue.2017.01.025

Dey, N., Ashour, A. S., Beagum, S., Sifaki Pistola, D., Gospodinov, M., Peneva Gospodinova, E., Manuel, R. S., & Tavares, J. (2015). Parameter optimization for local polynomial approximation based intersection confidence interval filter using genetic algorithm: An application for brain MRI image de-noising. *Journal of Imaging*, *1*(1), 60–84. doi:10.3390/jimaging1010060

Dey, N., Ashour, A. S., Shi, F., Fong, S. J., & Sherratt, R. S. (2017). Developing residential wireless sensor networks for ECG healthcare monitoring. *IEEE Transactions on Consumer Electronics*, *63*(4), 442–449. doi:10.1109/TCE.2017.015063

Duque, S., & Omar, M. N. (2015). Using data mining algorithms for developing a model for intrusion detection system (IDS). *Procedia Computer Science*, *61*, 46–51. doi:10.1016/j.procs.2015.09.145

El Bekri, M., & Diouri, O. (2019). Pso based intrusion detection: A pre-implementation discussion. *Procedia Computer Science*, *160*, 837–842. doi:10.1016/j.procs.2019.11.002

Elbasiony, R. M., Sallam, E. A., Eltobely, T. E., & Fahmy, M. M. (2013). A hybrid network intrusion detection framework based on random forests and weighted k-means. *Ain Shams Engineering Journal*, *4*(4), 753–762. doi:10.1016/j.asej.2013.01.003

Fang, W., Tan, X., & Wilbur, D. (2020). Application of intrusion detection technology in network safety based on machine learning. *Safety Science*, *124*, 104604. doi:10.1016/j.ssci.2020.104604

Hidayanto, B. C., Muhammad, R. F., Kusumawardani, R. P., & Syafaat, A. (2017). Network intrusion detection systems analysis using frequent item set mining algorithm FP-max and apriori. *Procedia Computer Science*, *124*, 751–758. doi:10.1016/j.procs.2017.12.214

Idhammad, M., Afdel, K., & Belouch, M. (2018). Distributed intrusion detection system for cloud environments based on data mining techniques. *Procedia Computer Science*, *127*, 35–41. doi:10.1016/j.procs.2018.01.095

Khalili, A., & Sami, A. (2015). SysDetect: A systematic approach to critical state determination for Industrial Intrusion Detection Systems using Apriori algorithm. *Journal of Process Control*, *32*, 154–160. doi:10.1016/j.jprocont.2015.04.005

Khan, K., Mehmood, A., Khan, S., Khan, M. A., Iqbal, Z., & Mashwani, W. K. (2020). A survey on intrusion detection and prevention in wireless ad-hoc networks. *Journal of Systems Architecture*, *105*, 101701. doi:10.1016/j.sysarc.2019.101701

Koucham, O., Mocanu, S., Hiet, G., Thiriet, J. M., & Majorczyk, F. (2018). Efficient Mining of Temporal Safety Properties for Intrusion Detection in Industrial Control Systems. *IFAC-PapersOnLine*, *51*(24), 1043–1050. doi:10.1016/j.ifacol.2018.09.719

Krzysztoń, M., & Marks, M. (2020). Simulation of watchdog placement for cooperative anomaly detection in bluetooth mesh intrusion detection system. *Simulation Modelling Practice and Theory*, *101*, 102041. doi:10.1016/j.simpat.2019.102041

Mahdavi, E., Fanian, A., & Amini, F. (2020). A real-time alert correlation method based on code-books for intrusion detection systems. *Computers & Security*, *89*, 101661. doi:10.1016/j.cose.2019.101661

Manan, J., Ahmed, A., Ullah, I., Merghem-Boulahia, L., & Gaïti, D. (2019). Distributed intrusion detection scheme for next generation networks. *Journal of Network and Computer Applications*, *147*, 102422. doi:10.1016/j.jnca.2019.102422

Mazini, M., Shirazi, B., & Mahdavi, I. (2019). Anomaly network-based intrusion detection system using a reliable hybrid artificial bee colony and AdaBoost algorithms. *Journal of King Saud University-Computer and Information Sciences*, *31*(4), 541–553. doi:10.1016/j.jksuci.2018.03.011

Mohammad, M. N., Sulaiman, N., & Muhsin, O. A. (2011). A novel intrusion detection system by using intelligent data mining in weka environment. *Procedia Computer Science*, *3*, 1237–1242. doi:10.1016/j.procs.2010.12.198

Mohammad, M. N., Sulaiman, N., & Muhsin, O. A. (2011). A novel intrusion detection system by using intelligent data mining in weka environment. *Procedia Computer Science*, *3*, 1237–1242. doi:10.1016/j.procs.2010.12.198

Mukkamala, S., Sung, A. H., & Abraham, A. (2005). Intrusion detection using an ensemble of intelligent paradigms. *Journal of Network and Computer Applications*, *28*(2), 167–182. doi:10.1016/j.jnca.2004.01.003

Nadiammai, G. V., & Hemalatha, M. J. E. I. J. (2014). Effective approach toward Intrusion Detection System using data mining techniques. *Egyptian Informatics Journal*, *15*(1), 37–50. doi:10.1016/j.eij.2013.10.003

Özyer, T., Alhajj, R., & Barker, K. (2007). Intrusion detection by integrating boosting genetic fuzzy classifier and data mining criteria for rule pre-screening. *Journal of Network and Computer Applications*, *30*(1), 99–113. doi:10.1016/j.jnca.2005.06.002

Qu, H., Qiu, Z., Tang, X., Xiang, M., & Wang, P. (2018). Incorporating unsupervised learning into intrusion detection for wireless sensor networks with structural co-evolvability. *Applied Soft Computing*, *71*, 939–951. doi:10.1016/j.asoc.2018.07.044

Raschellà, A., Bouhafs, F., Mackay, M., Shi, Q., Ortín, J., Gállego, J. R., & Canales, M. (2020). A dynamic access point allocation algorithm for dense wireless LANs using potential game. *Computer Networks*, *167*, 106991. doi:10.1016/j.comnet.2019.106991

Roy, S., Karjee, J., Rawat, U. S., & Dey, N. (2016). Symmetric key encryption technique: A cellular automata based approach in wireless sensor networks. *Procedia Computer Science*, *78*, 408–414. doi:10.1016/j.procs.2016.02.082

Thaseen, I. S., & Kumar, C. A. (2017). Intrusion detection model using fusion of chi-square feature selection and multi class SVM. *Journal of King Saud University-Computer and Information Sciences*, *29*(4), 462–472. doi:10.1016/j.jksuci.2015.12.004

Chapter 2
Wireless Sensor Networks With the Context of Knowledge-Based Management

Sunil Gautam
Institute of Technology, Nirma University, Ahmedabad, India

Azriel Henry
Institute of Advanced Research, Gandhinagar, India

ABSTRACT

Wireless sensor networks (WSN) play in vital role in different research area technology, electronics, and telecommunication, but these wireless sensor networks have limited computational capability and memory. The basic functions of WSN are analysis, processing, storage, and mining of data. Due to the flexibility of WSN in solving problems of various fields, it has gained more popularity. WSNs have been successfully applied in a variety of applications. In these days, knowledge management systems are attracting research that emerge with wireless sensor networks to reduce drawback of sensors. This chapter presents the review literature of knowledge management and demonstrates creating, managing, sharing, and utilizing in organizations.

INTRODUCTION

In these days, cyber world collects the information for enhancing the business. Therefore, researchers have developed an information system to supporting the business using different process such as analysis, design and implementing the

DOI: 10.4018/978-1-6684-7343-6.ch002

information. A knowledge based management system (KMS) support these process to developing the knowledge creation, storage, retrieval, transfer and application.

Knowledge based Management System can be described as an action of consistent managing, something which can be for service to prosper new window of opportunities in the social, technological, academic or business setups. It can be applied to be aware of on-going arising requirements and recognize for making full use of the obtained knowledge. It seeks to push up distinct and shared efficiency by easing knowledge flows and sharing. It changes itself dynamically in an application which are reflective in nature, and hence results in assistance to form the shared comprehension, makes informative knowledge easier, and makes an appearance of the regulations that regulates the application of tasks (Luque, 2016).

It's a well-known reality that one of the main methods of accomplishing an ambitious gain is Knowledge Management. For furthering the progress of the plan of action, each and every organization requires to exercise knowledge. The procedure accountable for such exercising is Knowledge Management (Domínguez Gonzalez & Martins, 2014). Knowledge management system is helpful for monitoring the things using wireless sensor network.

A wireless network containing special diffusible, self-governing gadgets that make use of devices which detect or measure a physical property and natural states are defined as a Wireless Sensor Network (WSN). These self-governing gadgets, or junctions, united with a gateway and routers produce a Wireless Sensor Network having distinctive qualities (Prasanna & Rao, 2012). It is a cluster of sensors which observes checks and keeps continuous record of some exceptional states comprising of material or nature.

The figure 1 has shown basic units of wireless sensor network. It is a self-arranged form and framework fewer wireless networks. Zillions of sensor junctions are held within these junctions and they exchange information amongst each other by utilizing radio waves (Rizvi, 2017).

Through radio waves, the sensor junctions share or exchange information amongst each other. A wireless sensor junction contains necessary items like detecting and calculating gadgets, electrical power units, radio transmitters and receivers. In wireless sensor networks, the independent junctions have finite storage scope, operating momentum, and transmission bandwidth. The functioning means of the sensor junction might be uninterrupted or interrupted based on event. For acquiring position and location based knowledge, Global Positioning System (GPS) can be utilized.

Because of many limitations, WSN needs unconventional patterns for designing new protocols such as MAC Protocols, Routing Protocols and IEEE 802.15.4 Ménage Protocols etc. (Matin & Islam, 2012).

Figure 1. Fundamental Units of WSN

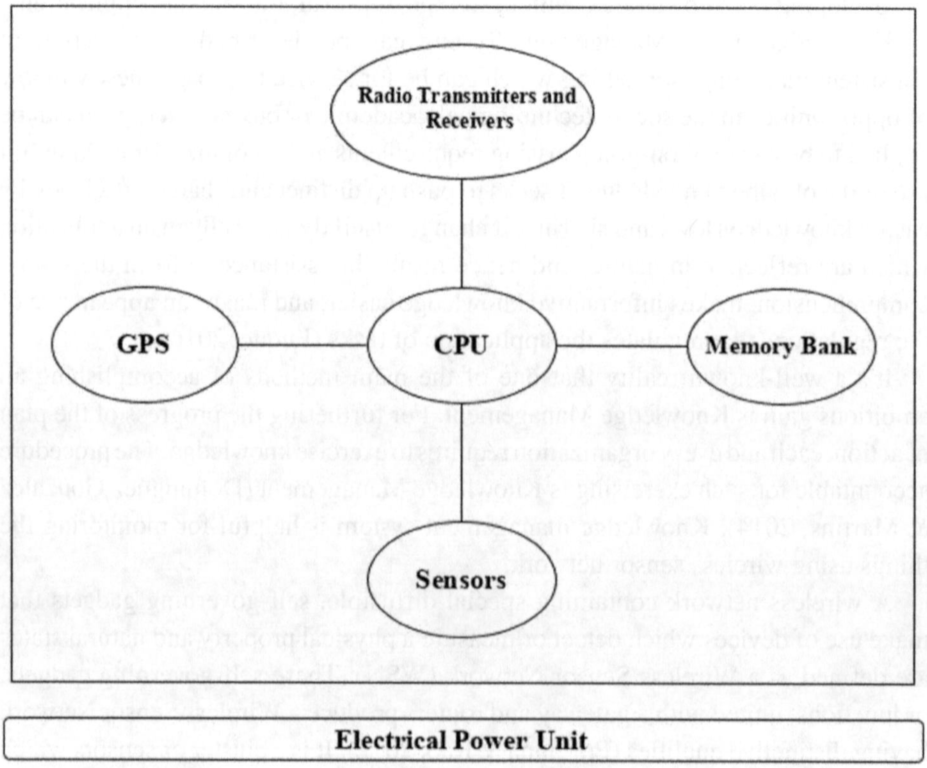

ARCHITECTURE OF KNOWLEDGE MANAGEMENT SYSTEM

In spite of the tendency for extremely limited devices in sensors and networks of sensors, there is a continuous mode to add more and more functions in the sensors to decrease the requirement for transmission in the sensor and to lengthen accumulator lifespan (Canada-Bago et al., 2010).

Sensors are really tiny, computerized gadgets which accomplishes the aim of enclosed sensors in any location because of the advancement in minimized automations as their dimension will be decreased (Martínez et al., 2010). A sensor contains various fundamental constituents as shown in the given figure 2:

1. Power Unit
2. Transmission Unit
3. Centre component with CPU, cache, etc.
4. Other types of sensors

Figure 2. Units of a Junction

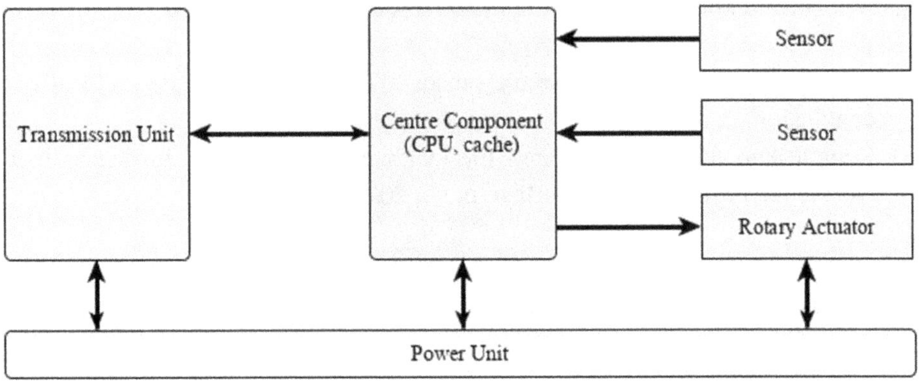

It is requisite to grow a newly discovered plan of action through which sensors can carry out a small-scale knowledge based method of functionalities, evacuating alternate functionalities to be carried out in an organization with prominent data handling ability. The present data handling ability of sensors is moderate because of the microcontroller unit and little amount of the memory bank that such gadgets generally accommodate.

To authorize end users for upgrading sensor information or to observe the development of a specific entity variable, sensors do not contain human-machine interaction. Nevertheless, sensors are allowed to distribute information and knowledge by the sensor networks. Knowledge amends and association between sensors can assist the accomplishment of worldwide purposes (Canada-Bago et al., 2010).

The knowledge that might vary for each and every sensor permits every single sensor to deduce or conclude its localized out-turns. Thus, it might be contemplated as localized knowledge. Under other conditions, knowledge based organizations and WSNs are skillfully composed and self-assured to take additional benefit of cooperation, knowledge and information distributing among components of the structure to assist worldwide purpose.

Each and every sensor is capable to distribute its information and knowledge with other sensors, with a collection of sensors or with every component of the structure. Hence, it is imaginable that a sensor might have localized and remote information.

1. The information which is acquired direct from the sensor as a result of its own assets is known as localized information. For example, assessment of climate conditions.
2. The information which is acquired by different sensors and later on passing on to a particular sensor is known as remote information.

The very same distribution setting takes place with knowledge. Sensor might contain localized knowledge or it might contain isolated knowledge.

1. Localized knowledge permits the sensor to deduce or conclude its localized conclusion.
2. Remote knowledge supplements the localized knowledge to further assist a worldwide purpose (Canada-Bago et al., 2010).

COMPONENTS OF KNOWLEDGE MANAGEMENT

Knowledge Management

It is an outcome of procedures which includes amalgamation, penetration, correlation and survey of accessible data by the field specialists. It involves accession and repository of knowledge, formation and presentation of knowledge, disclosure and implementation of knowledge, deduction and distribution of knowledge. It is assisted by field knowledge, observed knowledge and field specialists.

Figure 3. Design of KM

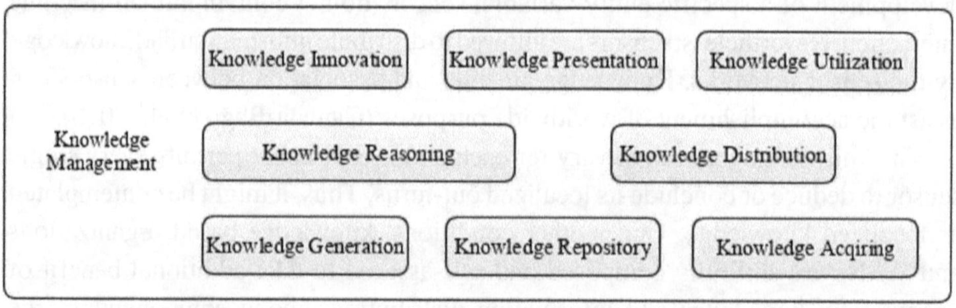

Data Handling

It is the procedure of basic information into significant data. Data Handling might contain different factors. Here, we have taken into consideration the two factors. The first factor is the data handling for present WSNs, and another factor is data handling for ancient anticipating information. For the present WSNs, there are abundant interrelated functionalities of network process, like chain transmission, junction locating, information integration, phase synchrony, surveillance administration, geography administration, situation observation, data accumulation, depository

Figure 4. Data Handling Layout

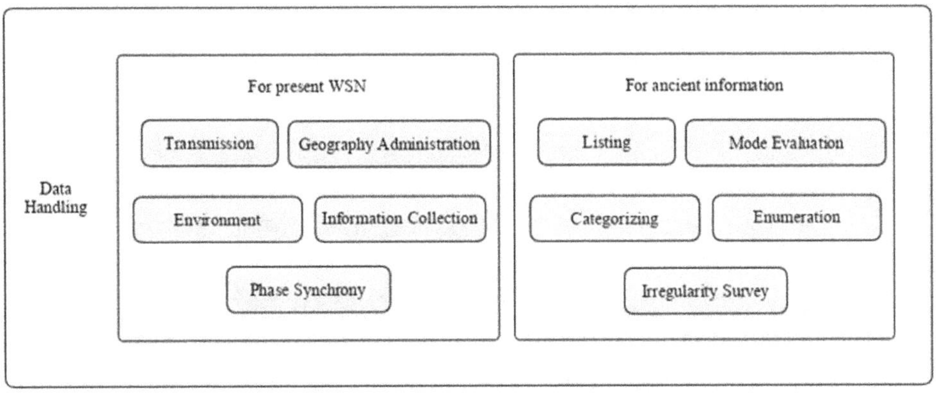

administration, analysis supervision, information transmitting, information inquiry, accomplishments and reasoning, and so on. For ancient anticipating information, there are data listing, information categorizing, mode evaluation, information enumeration, irregularity disclosure, information excerpt and penetrating, etc.

Information Processing

It accumulates and supplies all types of basic information, designs and knowledge. It contains perceiving information, network data, functioning operations, executed design, meta-data, theoretical knowledge, contented knowledge, rationalization regulations, etc. (Zhang et al., 2014).

Figure 5. Information Processing Layout

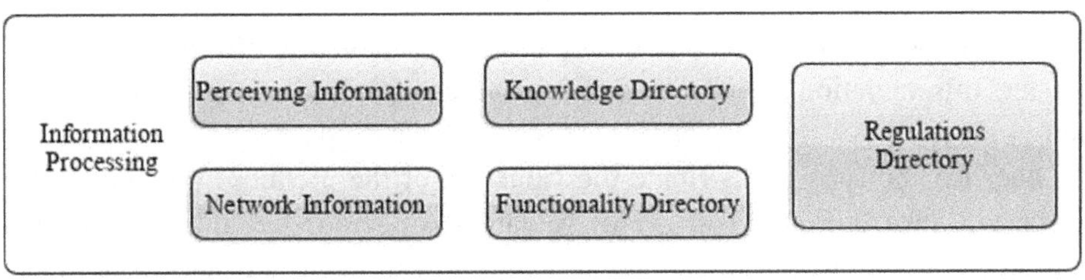

Data Distribution Using Sensors

Each and every knowledge based sensor owns a collection of inserted parameter, produced parameter and a knowledge base. From these, practicality is gained through

Figure 6. Distribution of inserted and produced parameters among the two sensors

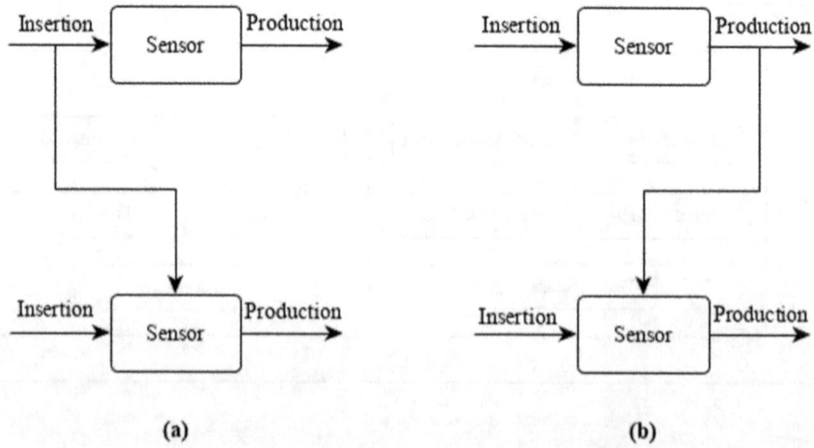

its reasoning mechanism. Because of the network system, it is feasible to distribute the below mentioned data among the other structure in a systematic way to gain worldwide purposes:

1. Distributed information might contain the valuation of inserted parameter.
2. The valuation of the produced parameter, parameter clarities and regulations.

Sensors might distribute the valuation of parameters that are specifically acquired with different sensors by dispatching it through the network. Thus, the valuation of the inserted parameters for a sensor junction might be provincially or distantly assessed and regarded as the valuation of the precursor of its knowledge (Canada-Bago et al, 2010).

A sensor distributing an inserted parameter is shown in the Figure 6 (a) which is given below. In this scheme sensor assesses the valuation of a parameter and transfers this valuation to different sensors that can use it to conclude and deduce their product.

Other feasibility is to distribute the valuation of the produced parameter of a sensor as shown in Figure 6 (b). This valuation is decided through the localized knowledge of the sensor and that data is deduced by the sensor. In such conditions, the distributed valuation reckons with the transferring sensor's knowledge. Hence, knowledge can be distributed by using this mode.

Knowledge Extraction from Information

The below mentioned approaches are recommended to combine localized and remote data (information and knowledge):

1. Use just a knowledge-base that assimilates localized and remote data. Remote parameters and regulations are combined to the knowledge-base and hence the regulations are capable to utilize peripheral information in their interpretations. The interpretation mechanism will then allow for those abstracted regulations if the remote data is accessible.
2. Assimilate a finalized cooperative knowledge-base as shown in the figure below. In this way, the sensor uses a knowledge-based with localized data and other finalized knowledge-based with cooperative data. The interpreted mechanism will then allow the cooperative knowledge-base that prepares each peripheral information to be accessible after deducing the result established on the localized knowledge base. This process displays supplementary benefits:
   a. The statistics of regulations is lower.
   b. The regulations are much more decipherable by the cause of two knowledge bases.
3. Lastly, the third approach is to construct a framework of an ordered scheme that can fuse knowledge-bases and localized, remote information (Canada-Bago et al, 2010).

Knowledge Management in Practice

An example of an existent mankind use is knowledge based wireless sensor network which conclusively designs a method of olive tree pests. The evolution of the olive tree flies is actively connected to the weather conditions and atmospheric moistures of the environs. If the menace of the infestation arriving in a region exceeds a margin level, the adequacy of pesticide nursing utilizations must be assessed. Systems based on evidence and reasoning makes resolution based on the organization's knowledge of the identified circumstance (Canada-Bago et al, 2010).

Applications of WSNs range from front line transmission organization to meteorological observing structures; such as environment tracking, alchemical noticing, framework safety, and backlog and traffic administration. For instance, sensors might be disseminated over an apartment so as to discover and determine an individual and alter meteorological variables to adapt a single figure (Martínez et al., 2010).

Figure 7. A sensor containing cooperative and localized knowledge

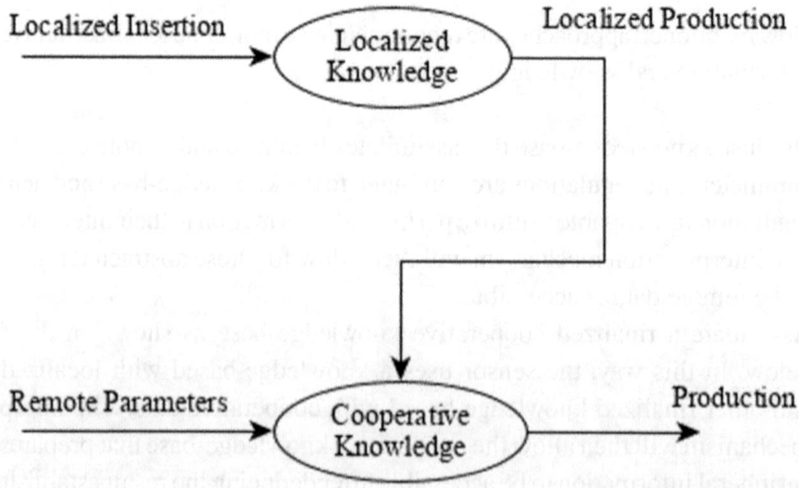

Accumulation of varied information from distinct surroundings is a valuable property of WSNs distributed in countryside environments. Through this method, organizations can acquire information about surroundings, making it feasible to collect acceptable accords and ease knowledge analysis. The aim is to be able to supply a bright habitat to make resolutions and accurately administer assets. The procedure of gathering varied data is associated to the terminology "Big Data" that manages with the amount, range, speed, and complicatedness of information created day-to-day. The Big Data not just imports conjointly abundant quantity of information, the specimen also makes it feasible to operate different kinds of information. These information flow form always requests considerable handling rates and so far, their repository should be monetary. The foremost aim of the Big Data in experimentation and automations is to administer and transfer accessible personal and corporeal information into knowledge and to construct well organized conclusions established on associational prerequisites and requirements (Zhang et al., 2014).

MOTIVATION

The novel advancement in maneuvering, transmission and networking have influenced recent sensor layouts, data sciences and wireless structures. Aforementioned leading sensors could be utilized as a link among the substantial environment and the cybernetic environment. At present, wireless sensor network has evolved into the extreme inspiring networking automations to propose the anticipated gathered information with limited potential capability. A meaningful number of industries have already acknowledged the likelihood of enhancement which prevails in a sensible business with knowledge and WSN. In accordance to this belief, different activities including knowledge management and WSN has been proposed amid previous senescence, nearly all of them originating with resolutions established on data automations formed to reinforce company methods and also an effective commutation of knowledge (Finke & Will, 2003; Canada-Bago et al., 2010). In this work, a cooperative knowledge based resolution for wireless sensor network has been discussed as knowledge based sensors are useful for a broad area of applications.

RELATED WORK

The two main constituents of this work are: the knowledge management (KM) and wireless sensor networks (WSN). These constituents are portrayed here which are related to the nearby works.

The basic purpose of sensors in WSN is to sense physical or environmental settings such as temperature, pressure, sound, humidity, etc. Education system, entertainment system, industrial applications, industrial process monitoring and control commercial application, civil structure monitoring, animal movement tracking, etc. are some of the applications of WSN (De et al., 2020).

For many years, WSN have been a main topic for analysis. Wireless Sensor Network authorizes innovative and engaging answers for data accumulation over the range of goal containing transition, enterprise, medical-management, and ecological supervision. Regardless of these approaches, the epidemically growing information obtained from wireless sensor network is not assimilating enough service due to the absence of knowledge, time and wealth through which the information may be enhanced and reserved for forthcoming utilization. The later era of wireless sensor network would be profited when sensor information is combined to sites, implicit areas, and communal web functions. This conversion of information obtained from sensor networks in an important material for data starving appliances would be profitable from methods being refined for the arising cloud computing automations. Early large Potential Computing ways might be replenished or discover a part in information stratagem before the information is transported in the Cloud. An innovative outline is projected to accommodate the Cloud Computing design with wireless sensor network. Redistributed wireless sensor network would be interconnected to the projected framework (Ahmed & Gregory, 2011). Knowledge Management has lately emanated as distinct field in the analysis of institutions and repeatedly mentioned as the prior of institutional practices. Its goal is at inspecting the collision of Knowledge Management procedures on institutional procedures of small and medium-scaled enterprises (SME) in Serviceability Corporation. Well-known machine learning methods along with numerical determinant studies are utilized to establish prognostic and analytical designs. The information for this is gathered from SMEs functioning in the Serviceability Corporation within the larger urbane part. The study determined that there is a powerful and definite relation among the application stage of Knowledge Management procedures and institutional practices associated to Knowledge Management. It gives a rundown in searching of the analysis and gives administrative association to enhance the managerial procedure of SMEs by productive application of Knowledge Management processes. Moreover, there are various architectural solutions for wireless network security (Delen et al., 2013; Das et al., 2021).

WSNs survey active environs which can modify promptly over pace of time. This active demeanour is either induced through extrinsic determinants or proposed by the method creators themselves. To acclimate to aforementioned settings, sensor networks generally select machine learning methods to remove the requirement of avoidable remodel. Machine learning also encourages numerous realistic answers which enlarge source usage and extend the lifetime of the network. A far-reaching related work over the time pace 2002-2013 of machine learning techniques which were utilized to focus on simple problems in wireless sensor networks were addressed. The benefits and detriments of each and every projected innovation are assessed in contrast to the reciprocal issues. A provisional catalogue to help wireless sensor network architects in constructing appropriate machine learning answers for particular ultimatums is also supplied (Alsheikh et al., 2014).

An astounding improvement is seen in surpassing knowledge in the Indian academic precinct for as much as last decennium. In current generation of knowledge, IT performs a main part to acquire prosperity to the academic establishment. Although efficient knowledge management is a main key to assure that education is capable to prosper the academic possibility in entirety. In the generation of worldwide integration and knowledge based environment, knowledge is been treated as the principal operator of this environs. Knowledge distribution is a recent window of opportunity which has been surfaced within the scope of data science, data administration and newly Cloud Computing. In modern sequence of events, various academic organizations are interfacing the difficulties with the rising requirements of IT and framework. In the recent economic critical situation and planet of frugality more advanced education machinery abutment for knowledge management exercises advancement in the organizations. It explains meta-surveillance through a cloud environs over the knowledge fanlight. And also made clear a perception on recent awareness in gaining knowledge management over cloud computing in more advanced academics as rising requirements of framework have to expend a huge volume to accomplish the requirements and claims of educational collaborators (Bimol et al., 2014).

Big Data is a well-known terminology in the area of data and transmission automation. WSNs is one of the famous subscribers of big data. Wireless sensor network consists various sensor junctions which collectively observe the surroundings. Each and every network contains of sensor junction, cache, and transmission gadgets. Information produced by a sensor junction is limited although information produced by shared sensor network is naturally huge and it is known as big data. The analytical problem in wireless sensor network is power usage and information collection. Its focal point is on Expectation Management (EM) established assembled strategy performance and the implementation study of wireless sensor network by utilizing an ambulatory fall. From the study, a relation among the figure of ambulatory falls needed for a specific network with an accustomed figure of sensor junctions. Experiential outputs displays that the figure of ambulatory falls is also a main variable to effectively collect data in wireless sensor network. Moreover, there is a range of intelligent algorithms that can be used to filter useful information and efficiently solve particular problems (Babu et al., 2015; Das et al., 2021).

The rumble phrasing these days is Big Data (BD). Everybody is speaking regarding its ability, its capacity, and its speed. KM has been all over from a long period of time. The aim of knowledge management has been to gather, classify, accumulate, extract and refine information into knowledge. There are various ways of knowledge gaining from institutional civilization to the later. Conventional procedures change information into data by long-established collection of data and then practiced enterprise intellect and information extracting methods to mine knowledge. Both big data and knowledge management could support each other enormously. Knowledge management when used properly, have aided administrators to make accords speedy and excelling, obviated rediscovering the circle, maintained some intelligent procedures by preserving trail of elite processes, and indicated novelty because of knowledge distribution and propagation. Big Data handles large number of information and doesn't need a long-established collection to be efficient. It has its devices and need which could be improved by knowledge management. The conclusive goal is to create a design wherein both the KM and BD exist together. Moreover, bio-inspired optimization algorithms, such as genetic algorithms (GA), artificial neural networks (ANN) and artificial immune systems (AIS) can be used in the contexts of fault analysis and diagnosis, and traffic management (Hijazi, 2017; Das et al., 2021).

Development in appliances and wireless sensor network devices has settled us at the threshold of novel generation wherein small wireless gadgets would bestow with approach to data all the time, everywhere and also promptly provide in developing intelligent environment. The sensors in WSNs congregate data concerning the elements they are used to feel. However, these sensors are prohibited in their accomplishment by prohibitions of energy and frequency range. Machine Learning techniques could assist them in overpowering such prohibitions. All along the former year, wireless sensor networks have observed more and more accurate application of extremely advanced machine learning innovations for data managing and developing network execution. Machine learning affect uncountable practical descriptions along which produce higher presence of network. Particularly, wireless sensor network architectures have efficiently complied to machine learning models to handle with extensive deliberated issues related to local, information collection, error identification, safety, junction gathering, forecasting designs and power attention subdue (Chander et al., 2018).

As computerized protection hazards grow in commonness and culture, institutions are understanding that one of their powerful sources to battle computerized invasions exists in the developing amount of information at their transfer. Although, historic knowledge management automations have definite capacity to efficiently operate and evaluate these greater information amounts which could access administrators with relevant data to make improved cognizant protective choices. Through the usage of analysed information from enterprises and data protection administrations and designing from the source established observation, it can be seen that the institutions which redistributed big data analysis stated meaningful advances in their computerized knowledge management abilities. Alternatively, the discovery also affirmed a necessarily clear cooperation among computerized deftness and big data analysis. Particularly, institutions which redistributed big data analysis resolutions stated meaningful advances in their capabilities to discover and acknowledge speedily to computerized hazard. In the end, in comparison to institutions without big data analysis, the enterprises ` has redistributed big data analysis automations deliberated their computerized protection abilities enough and potent much to preserve analytical data resources from computerized hazards. Generally, the output displayed in that institutions requires improved information abilities to efficiently influence IT sources, permitting them to acknowledge computerized occurrences with velocity and deftness that eventually accumulates in the higher computerized security of important data resources. The output also gives some valuable associations for examining and managing procedures. Artificial intelligence and soft computing techniques play a vital role in developing a more robust algorithm suitable for the dynamic and heterogeneous environment, making the network self-managed, self-operational, and self-configurational, and efficiently reducing uncertainties and imprecise information (Obitade, 2019; Das et al., 2020).

Bera S. et al. proposed an intelligent routing algorithm in WSN based on African Buffalo Optimization (ABO). They claim that the proposed technique efficiently optimizes various parameters in WSN. Also, it reduces the energy consumption of the nodes and the link of the WSN.

Some of the challenges that sensor networks suffer from are Energy, node deployment, congestion, wireless medium, hardware constraints, and security. Renuka K. et al. have proposed a method to address these challenges by developing two algorithms namely Delay Controlled Energy Efficient Route Matrix Detection (DCEERMD) and Optimized Routing and Congestion Control Algorithm. Evaluation of their proposed system is carried out using parameters such as loss rate of packet, energy consumption, and average of the routing hops (Das et al., 2021).

As there is a remarkable growth in WSNs, there is also a growth in the chances of attacks in the network. Thus, developing a robust system to detect and prevent intrusions is a necessity. A comparative study carried by Chander P. et al. (2018) shows how Machine Learning and Deep Learning proves to be an effective method to develop such systems for WSN. They have used certain machine learning algorithms such as Random Forest, Decision Tree, Naïve Bayes, etc. as well as a Deep Learning algorithm called Convolution Neural Network (CNN). They have reported that deep learning technique proved to be better than machine learning algorithms in detecting intrusions in WSN (Das et al., 2021). Yang W. et al. (2018) shows the equipment and capability provided by the advanced wireless communication system. Moreover, the study also analyses system architecture including function, user, information, and communication to establish the system for interconnection, sensing, and intelligent decision making among all transportation elements and systems. Jayakumar et al. (2020) proposed a mode selection method for D2D communication using Multi-Criteria Decision Making (MCDM) AHP-TOPSIS. They have reported that their proposed system allows the network operator to choose the best possible mode for D2D users based on their type. Moreover, the proposed system outperforms in all three different services operated between D2D pairs in simulation (Jayakumar et al., 2020). There are several approaches to optimize the quality of the clusters, however, they are difficult to implement due to complexity and resource requirements. To address this issue, Bera S. et al. proposed a method called centre ordering of vectors on ring (COVR). They claim that the model is easy to implement to any size of problem. Moreover, it can provide the opportunity to observe intermediate stages and analyse them (Bera et al., 2017). Das and Giacalone (2022) proposed a book for the purpose of fuzzy logic and its related usage in the areas of engineering and management. It contains several related works that helps to deals with several types of issues and problems that helps in the real-life applications.

Need of Knowledge Management

Institutional continuity, competing distinction, worldwide integration accoutrements and developing personnel are the main aspects that are urging the requirement for Knowledge Management. All in all, the management motions at present day, the duty of managing knowledge needs extreme concentration as most of the effort is established by data. It is an accepted reality that institutions emulate on the foundation of knowledge, as goods created and aids have been progressively complicated (Omotayo, 2015).

Long established knowledge management methods are concentrated on the capability to occupy knowledge in integrated methods and make it accessible at the next point in time. A new time period in history was unlocked for knowledge management by knowledge distribution and cooperation activities. There are many important reasons behind the requirement of Knowledge Management. Some of them are as follows:

1. It is simple to discover the data or the person who has the data needed through Knowledge Management. It expands adeptness and output and concedes to perform more excellently, decreasing the impulse to recreate the circle.
2. The later generally known case for administrating Knowledge Management is to assist in competing distinction. All institutions, either for benefit or for obstruction, vie inside an area. Knowledge Management is an analytical operator of competing benefits on account of institutions to introduce so that to discern oneself from its opponents. Institutions which are not able to introduce at a reasonable time would need the capability to constantly draw attention of new clients, which in proper sequence would guide to their downfall. Although institutions which are capable to introduce would be capable to guard, and also hold their competing ranks in the rialto (Omotayo, 2015).
3. All the data is not treated to be knowledge. The consideration of the association of the accustomed data is the knowledge. The knowledge management procedure recognizes, examines and makes the data better and this data is transmitted as litigable knowledge and observations. When a knowledge management method is utilized, it guarantees that this knowledge is properly recorded and is ready for use to everyone who needs it.
4. Glancing at the manner the planet is advancing in AI (Artificial Intelligence), it is apparent that Artificial Intelligence will take the place of a lot of the efforts made by persons in the coming time. We can guess that they can have same approach to the knowledge of the planet. Their exclusive distinction can be in their knowledge management method that decides in what manner an Artificial Intelligence utilizes this knowledge.

5. Generating a knowledge stand that is finely adapted for the aim and achieving required procedure and processes assists to alter the education and hence undergoes first-rate outputs.
6. The count of affiliates also evolves when an institution evolves. An orderly establishment and transmission of knowledge that makes usage of the knowledge consumed by all the affiliate can generate the novelty procedure. Generally, activities and actions generate repeated knowledge inside an institution and knowledge might not be correctly distributed among distinct groups of the aforementioned institution. This can point to recreating the circle. Knowledge Management expands adeptness and capacity and grants for better functioning, decreasing the direction of movement to recreating the circle.
7. Knowledge Management supplies a method to examine the lack of success for the sake of recognizing the weak points and logic for the lack of success in institutions. It assists to distribute the favourable outcomes and logics in the institutions to generate education of shared belief and faith. For example, an educating institution known as the (AARs) After Action Reviews was developed by the United States Army to determine their system constantly and to observe for methods for advancing. After all the necessary action or act, army crews analyse tasks, determines favourable outcomes and the lack of success, and explores methods to carry out well later on.
8. The arrival of worldwide integration has also compelled the requirement of Knowledge Management, as institutions pursuit to discover efficient devices and ways for obtaining and distributing knowledge across numerous fundamental and educational obstructions. Hence, worldwide integration has generated an immediate requirement for institutions to be capable to manage knowledge beyond nations and continents. The academic central requires to be apprehended so the forthcoming era in these task atmospheres doesn't have to rerun the errors and recreate knowledge.

The greatest vital power urging the advancement of any association is knowledge. The planet, actually, is learning the data or knowledge innovation. Knowledge is a basic ability which can supply ambitious border to singletons, institutions and communities. Knowledge eras, managing knowledge and communicating and distributing knowledge are, accordingly, of analytical significance. Knowledge is an essential source in smart resolution establishing, anticipating, constructing, preparation, examination, reasoning, assessment, and spontaneous acumen. It is built in and distributed among singleton and cumulative intelligence. It doesn't evolve with practice, favorable outcomes, lack of success and educating over pace. Knowledge permits for establishing forecasting, informal cooperatives, or anticipative resolutions in relation to what should be done like distinct data, that naturally provides us with the actuality. Briefly, knowledge grants the formation of ability that concludes the capability to do stuffs.

Application of Knowledge based Management System

The area of application of Knowledge Based Systems is increasing consistently as recent exploration subject matters arise. In the 90's, the anticipated utilization instructions of KM Systems were layout, analysis, direction deciphering noticed information, observing, forecasting, instruction of medications for breakdowns. The application of knowledge based systems increased widely since 1996, at present escalating in voice identification, CPU image, intellectual methods, and many more. At present, knowledge based systems are entrenched in browsers that commemorate last explorations, permissible operating system, public operating system associating, computerized aviators, medicinal analysis, contact centers, Computer-Aided Design functions, error removing devices.

The range of exploration on knowledge application is very large and increasing very quickly. It amounts to exploration, distribution and application. It concerns to the application of the knowledge produced, suited, and distributed by the analysts, and acknowledged and approved by the end users for accomplishing particular aims and purposes. It is a convoluted position including governmental, institutional, social, economic and behavioral determinants in inclusion to knowledge. It is also the range of extraction involved with determinants which makes clear about the usage of experimental and high-tech knowledge by resolution-establishers and those in experienced proceedings. Today analysts identify knowledge created as a convoluted procedure comprising singletons, institutions, and civilizations and also governmental, social, economic, behavioral, and other located determinants (Mohajan, 2017).

The accumulation of appropriate knowledge which is gathered in the CPU and is systematized in such a way, it could be utilized for conjectures. Resolution establishing is fundamentally fulfilled through the resolution assistance method. In the resolution establishing procedure differing stages are concerned that are as follows:

The knowledge is utilized in various distinct methods, for example, to evaluate expenses and essential sources for recent productions or assistances, to discover likeness in distinct application circumstances, or to escalate the rehash of preceding works.

Information Communications Technology (ICT) and society has depicted the application scheme for the projected Knowledge based Management System (Ribino et al., 2011). Their prime aim is the advancement of management organizations. The acceptance of an assimilated structure that effectively assists the system employees

Figure 8. Different stages of Resolution Establishing Procedure

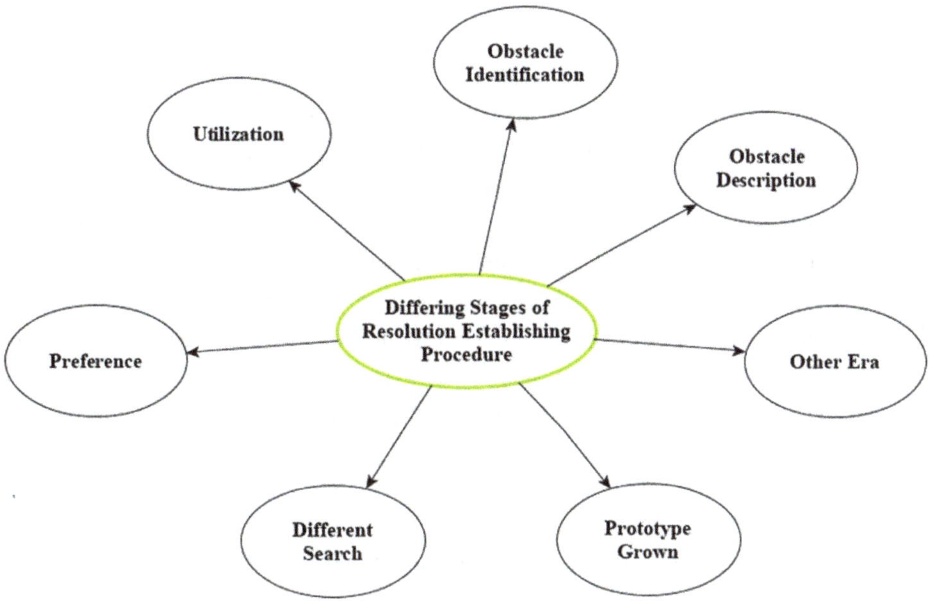

in their field actions, and the usage of smart appliances for resolution making, could enhance administrational accomplishment.

Unfortunately, establishing the administrational knowledge approachable to all those who could be assisted from its application is a crucial project because of the distinct types of the knowledge stocked. The fundamental aim of the projected core is to handle distinct forms of knowledge so as to supply a valuable device for field resolution reinforce and to enhance the administrational output. So as to accomplish this product, a new amalgam of AI (Artificial Intelligence) methods able of managing with implicit and accurate knowledge is projected (Ribino et al., 2011).

Knowledge Management proposes a management system for the industry to make secure that their knowledge estates when produced are correctly recorded and that the knowledge in distinct area holders would be distributed among the administration. When knowledge estates are recorded and distributed, knowledge application would be expedited. This is the platform in Knowledge Management where amount generation is deposited. Through exploiting knowledge from distinct knowledge areas and abilities over the administrations, directly effects to the aim and targets of the industry could be reached.

Illustrations of categorization techniques utilized as a portion of knowledge management contains the categorizing of the outpatients against basic wellness program hubs to professionals; the application of inclination technique to aid

Knowledge Management and resolution making procedures; and the combo of the data mining and resolution assistance accessing in preparing of the territorial wellness program organization. Cheng, Lu & Sheu started a philosophy occupied way of Knowledge Management and knowledge distributing in economic knowledge management system in the economical organization and practiced the Self-Organizing Feature Map (SOFM) / Learning Vector Quantization (LVQ) classifier of composite assembling and categorizing data mining methods to categorize allied relations (Silwattananusarn & Tuamsuk, 2012). Edible material community department, information extraction could enhance by knowledge nurturing technique especially Ebstein's and Extension data mining (EDM) for minuscule and intermediate company (Silwattananusarn & Tuamsuk, 2012). This technique was the unification of data mining and knowledge management to enhance a resolution reinforce method stage for excelling resolutions. As knowledge estates performed the main piece in knowledge business, Cantu & Chen engaged data mining operatives for obtaining beneficial design establishers in producing advantages from the knowledge estates and utilized KIN (Knowledge Information Network) stage for directing the knowledge estates (Silwattananusarn & Tuamsuk, 2012). In the industry administrations among the huge number of tasks, the industries desired to advance their understanding about what the secret designs amongst the Knowledge Management and its accomplishment, utilizing the combo of data mining methods. Rough Set Theory (RST) and Bayesian Network (BN) in their industries can assist associations generating the Knowledge Management to be accomplished efficiently and carry out greater adeptness outcome (Silwattananusarn & Tuamsuk, 2012).

In consideration of the advantages of ontology and the disadvantage of classical method of accumulating and recapturing knowledge, an ontology based knowledge management would assist the IT service management. Furthermore, IT service management which comprises of most of the fields for Information Technology business in India at present intensifies this requirement.

A linguistic internet ontology based layout for management of knowledge for Information Technology assistances. This layout provides to the requirement of knowledge end users who examine monotonous kind of task. The computerized era of ontology established knowledge base would assist the end users by contributing recyclable result to the diversity of complications encountered by them while damage assessment. This would allow the Information Technology assistance providers to examine the health of different framework factors being examined, take suitable conducts based on the preceding data accessible in the knowledge base and supervise the development of health of framework factors.

CONCLUSION

This chapter discusses Wireless Sensor Networks with the context of Knowledge based management approach to obtain organizational objectives. The chapter discusses the fundamental units of WSN and Knowledge Management System along with its components. Further in the chapter there are various related works that showcases the applications of WSN in different fields. The last segment of the chapter focuses on the need of the Knowledge based management system and its applications. The discussion show how effective knowledge management system proves in general when used with WSN.

REFERENCES

Ahmed, K., & Gregory, M. (2011, December). Integrating wireless sensor networks with cloud computing. In *2011 seventh international conference on mobile ad-hoc and sensor networks* (pp. 364-366). IEEE.

Alsheikh, M. A., Lin, S., Niyato, D., & Tan, H. P. (2014). Machine learning in wireless sensor networks: Algorithms, strategies, and applications. *IEEE Communications Surveys and Tutorials*, *16*(4), 1996–2018.

Babu, K. R., Suja, G. J., Samuel, P., & Jos, S. (2015, September). Performance Analysis of Big Data Gathering in Wireless Sensor Network Using an EM Based Clustering Scheme. In *2015 Fifth International Conference on Advances in Computing and Communications (ICACC)* (pp. 109-113). IEEE.

Bera, S., Chattopadhyay, M., & Dan, P. K. (2017). A two-stage novel approach using centre ordering of vectors on agglomerative hierarchical clustering for manufacturing cell formation. *Proceedings of the Institution of Mechanical Engineers. Part B, Journal of Engineering Manufacture*, *232*(14), 2651–2662.

Bimol, S., Saikia, M., & Devi, L. P. (2014, March). Achieving knowledge management through Cloud Computing: A case in higher education. In *2014 International Conference on Computing for Sustainable Global Development (INDIACom)* (pp. 222-227). IEEE.

Canada-Bago, J., Fernandez-Prieto, J. A., Gadeo-Martos, M. A., & Velasco, J. R. (2010). A new collaborative knowledge-based approach for wireless sensor networks. *Sensors (Basel)*, *10*(6), 6044–6062.

Canada-Bago, J., Fernandez-Prieto, J. A., Gadeo-Martos, M. A., & Velasco, J. R. (2010). A new collaborative knowledge-based approach for wireless sensor networks. *Sensors (Basel), 10*(6), 6044–6062.

Chander, B., Kumar, B., & Kumaravelan, P. (2018). Analysis of Machine Learning in Wireless Sensor Network. *IACSIT International Journal of Engineering and Technology, 7*, 185–192.

Das, S. K., Das, A., Sinha, H. K., Das, S. P., & Nayak, S. (2021). Game theory based optimal decision-making system. In *Intelligent and Cloud Computing* (pp. 761–769). Springer.

Das, S. K., Das, S. P., Dey, N., & Hassanien, A. E. (Eds.). (2021). *Machine learning algorithms for industrial applications*. Springer.

Das, S. K., & Giacalone, M. (Eds.). (2022). *Fuzzy Optimization Techniques in the Areas of Science and Management*. CRC Press.

Das, S. K., Patra, N., Das, B. R., Das, S. P., & Nayak, S. (2021). Fuzzification-Based Routing Protocol for Wireless Ad Hoc Network. In *Intelligent and Cloud Computing* (pp. 237–245). Springer.

Das, S. K., Samanta, S., Dey, N., & Kumar, R. (Eds.). (2020). Design frameworks for wireless networks. Springer Singapore.

De, D., Mukherjee, A., Kumar Das, S., & Dey, N. (2020). *Nature inspired computing for wireless sensor networks*. Springer.

Delen, D., Zaim, H., Kuzey, C., & Zaim, S. (2013). A comparative analysis of machine learning systems for measuring the impact of knowledge management practices. *Decision Support Systems, 54*(2), 1150–1160.

Domínguez Gonzalez, R. V., & Martins, M. F. (2014). Knowledge management: An analysis from the organizational development. *Journal of Technology Management & Innovation, 9*(1), 131–147. doi:10.4067/S0718-27242014000100011

Finke, I., & Will, M. (2003). Motivation for knowledge management. In *Knowledge Management* (pp. 66–91). Springer.

González-Briones, A., Castellanos-Garzón, J. A., Mezquita Martín, Y., Prieto, J., & Corchado, J. M. (2018). A framework for knowledge discovery from wireless sensor networks in rural environments: A crop irrigation systems case study. *Wireless Communications and Mobile Computing*, 2018.

Hassanien, A. E. (Ed.). (2021). *Architectural Wireless Networks Solutions and Security Issues*. Springer Singapore Pte. Limited.

Hijazi, S. (2017). *Big Data and Knowledge Management: A Possible Course to Combine Them Together*. Association Supporting Computer Users in Education.

Jayakumar, L., & Dumka, A., & Janakiraman, S. (2020). Energy Aware Dynamic Mode Decision for Cellular D2D Communications by Using Integrated Multi-Criteria Decision Making Model. *International Journal of Ambient Computing and Intelligence*, *11*(3), 131–151.

Luque, A. M. A. (2016). *Brief Introduction to Knowledge Management in Organizations*. Academic Press.

Martínez, J. F., Castillejo, P., Zuazua, M., García, A. B., López, L., DaSilva, A., & Hernández, V. (2010). Wireless sensor networks in knowledge management. *Procedia Computer Science*, *1*(1), 2291–2300.

Matin, M. A., & Islam, M. M. (2012). Overview of wireless sensor network. *Wireless Sensor Networks-Technology and Protocols, 1*(3).

Mohajan, H. (2017). *The roles of knowledge management for the development of organizations*. Academic Press.

Obitade, P. O. (2019). Big data analytics: A link between knowledge management capabilities and superior cyber protection. *Journal of Big Data*, *6*(1), 1–28.

Omotayo, F. O. (2015). Knowledge Management as an important tool in Organisational Management: A Review of Literature. *Library Philosophy and Practice (e-journal)*.

Prasanna, S., & Rao, S. (2012). An overview of wireless sensor networks applications and security. *International Journal of Soft Computing and Engineering*, *2*(2), 2231–2307.

Ribino, P., Augello, A., Lo Re, G., & Gaglio, S. (2011). A Knowledge Management and Decision Support Model for Enterprises. *Advances in Decision Sciences*.

Rizvi, D. R., Amjad, M., & Doja, M. N. (2017). A Study of Wireless Sensor Networks. *Int. Res. J. Eng. Technol*, *4*(7), 2078–2081.

Silwattananusarn, T., & Tuamsuk, K. (2012). *Data mining and its applications for knowledge management: a literature review from 2007 to 2012*. arXiv preprint arXiv:1210.2872.

Yang, W., Wang, X., Song, X., Yang, Y., & Patnaik, S. (2018). Design of Intelligent Transportation System Supported by New Generation Wireless Communication Technology. *Intelligent Systems*, 715-732.

Zhang, S., Liu, X. D., & Song, W. W. (2014). *Using Knowledge-based Information Systems to Support Management of Wireless Sensor Networking Systems*. Academic Press.

Chapter 3
Water Monitoring System in Agriculture Through Wireless Devices

Javed Miya
Galgotias College of Engineering and Technology, Greater Noida, India

M. A. Ansari
Gautam Buddha University, Greater Noida, India

Ranjit Kumar
Galgotias College of Engineering and Technology, Greater Noida, India

Shamimul Qamar
https://orcid.org/0000-0002-1091-7715
King Khalid University, Abha, Saudi Arabia

Sanjay Kumar
KCC Institute of Technology and management, Greater Noida, India

ABSTRACT

Irrigation of farmlands is a tough task to farmers, and due to electricity issues in villages, the overhead increases. Breaking of bund and overflow of water to other farmlands is a big issue to a farmer in spite of his/her loss due to drowning of crops. India being the economy of farmlands, farmers are contributing 74% to it. Therefore, any problem of farmers is the problem of the country. To protect this, farmers have to watch over tube well and bunds the whole time until the irrigation is done. Due to electricity issues in village areas, farmers many times need to spend their complete night in the farmlands, and especially in winters when it's chilly out there. So, to help farmers in this, the authors propose a device that helps farmers do so in an easy way. The visual surveillance and machine learning will now do this job for the farmer.

DOI: 10.4018/978-1-6684-7343-6.ch003

INTRODUCTION

Crop yield is very crucial for farmers and water overflow in the fields being a problem while irrigation. Due to the overflow of water, the crops drown and leads to yield destruction. Water overflow also causes breakage of bunds of the main field and water many-times flows out to the other's field causing their crops to drown getting farmer into a bigger problem. To protect this, farmers have to watch over tube well and bunds whole time till the irrigation is done. Due to electricity issues in village areas, farmers many times need to spend their complete night in the farmlands, and especially in winters when it's chilly out there. So, to help farmers in this, we are proposing a device which helps farmers do so in some easy way. The visual surveillance and Machine Learning will now do this job for the farmer by having a watch over farmlands with irrigation facility using tube well (Soma et al., 2019). The farmer will be able to switch-On/Off remotely using one-click alert systems.

Precision Farming is defined as 'management of farm practices that uses computers, satellite positioning systems and remote sensing devices to provide information on which enhanced decisions can be made. Precision agriculture is the marriage of advanced agronomics with big data and predictive analytics. Under pressure to produce more with fewer resources, today's farmers are harnessing sensor technologies, GIS data, and expert decision-support algorithms in order to increase yields and lower costs. Sometimes called connected agriculture because of the major role played by always-connected IoT devices, precision agriculture has spawned an entire ecosystem of innovative solutions — from veteran companies seeking to tap into new opportunities, and from exciting new start-ups. Precision Agriculture (PA) is comprised of near and remote sensing techniques using IoT sensors, which help to monitor crop states at multiple growth levels. PA involves the acquisition and processing of a large amount of data related to crop health. Multiple parameters are involved in plants health, including water level, temperature and others. PA enables a farmer to know precisely what parameters are needed for healthy crop, where these parameters are needed and in what amount at a particular instance of time (Soma et al., 2019).

With the pressure of growing population and increasing water shortages, there is increasing awareness that water resources have to be stewarded more carefully. One of the responses has been to raise the price of water to levels that are closer to the actual cost of extracting the water, cleaning it and then distributing that water will be helpful for the use. It is not surprising, therefore, that farmers today are highly motivated to use as little irrigation water as possible. One way to conserve is through the "hardware", i.e., to migrate to more efficient irrigation methods, such as drip or spray irrigation. These methods reduce unnecessary surface water runoff as well as loss of water through evaporation. In order to make irrigation even more precise, farmers are adopting centralized command and control solutions that analyze crop, soil and weather data in order to optimize when and how much any given field should be irrigated (Konstantinos et at, 2018; James & Zhang, 2018).

Throughout the world, surveillance cameras or Closed-Circuit Television (CCTV) are gradually being used to surveil public as well as personal spaces. Video surveillance are increasingly being used by government and law enforcement authorities in diverse scenarios extending from criminal investigations, the safeguarding of government buildings and urban living, traffic management, the tracking of demonstrators along with the context of investigation of several other criminal activities. Digital video monitoring is therefore being used nearly in all types of environments. By means of electronic equipment, video monitoring lets individuals to surveil from a distance such as closed-circuit television cameras (CCTVs). Often, these cameras are linked via a recording aid or an IP network, enabling it to be tracked locally.

Along with security purposes, video surveillance systems have other unique applications as well. Home to IP video specialist VideoSurveillance.com takes the necessary measures to listen to every obligation when it is about the surveillance over the internet. In indoor as well as outdoor settings, internet security cameras can be used alike. Online security cameras are available in various designs mainly because of the evolution in IP video monitoring. Internet security cameras ranging from bullet like cameras and domes to PTZ based models and box shaped cameras are not fixed to any specific form feature. Be it discreet, demanding to spot camera for a retail market or a small sized camera to monitor a backyard. Often, the scientists who monitor the conduct and actions of animals due to lack of resources are unable to do so on a regular basis. Due to such limitation, scientists, interns, and specialists depend on live video monitoring cameras to broadcast the conduct of animals that are often strenuous to see and monitor (James & Zhang, 2018). A key role was played by an IP video surveillance to monitor Pandas in China, a species with differing breeding tendencies in some of our reports. For retailers, an excellent way to monitor customer traffic and understand shopper behavior is through operations management. We will be able to monitor the frequency as well as the shopping patterns of the shoppers once inside the store as well locate aisles and sections that need to be restocked. By connecting through a network recording device or a set up port forwarding, the installed internet security cameras via a remote device like a smartphone or laptop can be viewed. The network ports are sent through post forwarding by an IP camera by way of World Wide Web for online viewing. Forwarded through a router these network ports, are configurable by IP video experts at VideoSurveillance.com. By setting up port-forwarding, the IP address must be obtained by the user. Contemplate moving with an NVR producer like Smartvue if port-forwarding is not an option. By allowing users to surveil their live feed through camera and view videos recorded by Internet-ready machines, Smartvue proposes distinctive plug-and-play surveillance solutions based upon cloud. In order to view the camera feed, the user has to sign in to the Smartvue cloud service via their user account number and password (Lin et al., 2017). The user then gets automatically secure to the internet surveillance video system.

Motivation

The agriculture is a field which has undergone constant and rapid changes. These changes are often attributed to newer farming techniques, impact of technology and the way farmers view farming (Bhattacharya et al., 2008). The study explores the different motivating factors which influence the farmers to switch from conventional farming to organic farming methods which are sustainable. Factors such as concern for the environment, moral and social responsibility, and economic considerations are studied through a qualitative research method. Often changes in the outlook of the society also influence farmers to adopt new and improved farming methods which are sustainable.

PROBLEM FORMULATION

The author have read various research papers regarding the problem and found out few methodologies and techniques that can use for the project. The description of these problems and technologies are as given below:

Environmental Land Management: Agricultural water management has for many years been a key topic in numerous Member States of both the FAO and the IAEA. It has also been one of the major topics for the SWMCN Subprogram (Joint FAO/IAEA). Water management in agricultural production requires significant efforts and plays a significant role in hydrological, climatological, and agronomical balance (Dhawan & Sihalath, 2017).

Canal Network Flow Monitoring System (CNFMS): Good water management practices will increase yields, improve crop quality, conserve water, save energy, decrease fertilizer requirements, and reduce non-point source pollution (Reddy et al., 2016). Information is most critical to decide on exact amount of water required by a crop in a given climatic condition and for effective design and management of irrigation system, irrigation scheduling, etc. Proper scheduling can avoid over watering and excessive runoff. There are many advantages of using smart technologies over the conventional methods (Theseken et al., 2018).

ICT-based monitoring systems are widely applied in a number of agricultural areas. Such systems include the remote monitoring of farm conditions and the remote control of farm equipment, such Environments 2017, 4, 6 3 of 12 as irrigation

Table1. Literature review

Paper Name	Method Name	Founded Problem Description	Author	Date
Precision Agriculture in India challenges and Opportunities	Satellite farming or Site-Specific (SSCM) and Decision Support System (DSS)	Reducing soil degradation., Reducing the chemicals used crop production, Utilizing water resources in an economical and effective manner, Propagating the adoption of modern farm practices across the country in order to increase the output.	Soma et al.	June 2019
Machine Learning in Agriculture: A Review	ELM	Estimation of weekly evapotranspiration based on data from two meteorological weather stations	Konstantinos et at	August 2018
Crop Row Detection Using Python and OpenCV	Python and OpenCV	Color-Shifting Canny edge detection skeletonization and Hough line transformation	James Theseken	May, 2018
Dark Line detection with Line width extraction	Python and OpenCV	First order derivative of Gaussion filter by finding the centre of a line, the direction of a line and the width of a line	Bhattacharya et al.	January, 2018
Sensors for water monitoring for improved on farm water management	Canal network flow monitoring system (CNFMS)	It enables canal network flow management through creation of stages/ nodes in a canal network. The system will facilitate receiving daily water availability, flows and released information	Reddy et al.	November, 2016

systems, through smartphone applications. In fact, there is a large body of work in this field, and a number of associated tools have been developed (Vladimir & Malabanan, 2010).

For a long time, most programmers have struggled with digital image processing. Mainly because of the following reasons: Firstly, late appearance of digital image processing in the making of computers and hence, the demand for computers that assist graphical operating systems came to light. Secondly, an alert optimization for real-time implementation (Patin, 2003) is required. Along the years, the operation of digital image processing became more acknowledged with the advent of different algorithms in the growth of contemporary technologies. Surpassing its simple practice; these implementations can now be employed in television, security, movies, medical imagery, etc.

Monitoring water level through digital image processing is another application. As stated by Ken Chapman (2009)," A comprehensive solution to a set of problems may be very advantageous but is often not achievable." Due to many issues that may come forward while implementing digital image processing, programmers implementing such technology face some kind of hardship.

Canal Network Flow Monitoring System (CNFMS): The existence of waves dispensing the uniformity of water extremity is a major problem that may arise in determining the water extremity. Techniques such as threshold and hyperbolic tangent function are used to detect waves. According to Hwung, Kuo, and Chien (2009), algorithm for hyperbolic tangent function is ineffective in comparison to threshold-based algorithm by analyzing complete outcome with the computation of the wave gauge. This is beneficial in examining repository or artificial vessel in an open area where waves are existing. In a study conducted by Park (2009), similar algorithm examines integrated histogram where it determines fluctuations, between previous and current images whether small or large. Due to this, the algorithm becomes competent enough to separate water from land. Through the conversion of image water value by the use of algorithm integrated histogram along band pass filter the actual water level was achieved.

Using sequence of images water level detection is more efficient: In Varonen's (2008) conference, the viability of details is examined to acknowledge real time while water level statistics is transferred. This figure is then sent via a simple data message. The other techniques involved in calculating water level can be generalized into two parts: contact and noncontact procedures. The procedures that require contact with water belong to contact technique such as electric depth gauge or electric sounder, wetted tape and air line (Trimmer, 1991). The procedures involve difficult deployment but is more specific and have less possibility of fault and mistakes.

Unlike the methods in which contact is required, non-contact process such as processing of digital image are easy to utilize using a simple camera. However, it may be vulnerable to fault in accuracy and processing. A classical problem in digital images is automated line detection with various implementations such as detection of road in remote sensing images, detection of vessels in medical analysis of image. Various algorithms based on line detection have been proposed like local contrast analysis-based methods, edges based parallel seeking techniques, and ridge based seeking methods. Nonetheless, line detectors are capable to respond to both lines as well as edges. Example, when dark lines are only detected, they achieve huge contribution to the rims of bright lines/blobs. In support, large number of the contemporary line detectors are ineffective at measuring line width. Gabor filter was implemented to improve ridges in the real part of an image. Nonetheless, like other band-pass filters like the Gaussian derivative of second order, its feedback include both lines as well as edges. This creates error detection in various images. Scale selection of the filter is another problem. To achieve detection of lines having arbitrary widths, it is beneficial to examine the detection steps in the scale space.

Conventionally, a single optimal scale or by the maximum response of all scales the lines in an image are detected. Nonetheless this leads to overestimation of the line width of small lines. The Hough transforms and Radon transform based procedures are ineffective at either improving lines or guessing the line width. It requires external assistance in line detection through preprocessing and those preprocessing generally causes errors. Various techniques were proposed, to approximate line width by examining the response filter. These techniques can achieve precise line width. Nonetheless, various error detections were still reported (Somsak & Sihalath, 2010).

Detection of Dark Line along Line Width Extraction: To find the parallel edges a set of line detectors was used. By integrating the data of a few line detectors, it will erase little wrong detections. Unfortunately, this procedure is ineffective to erase wrong data because of the requirement of scales (small bright lines will be unknowingly detected by large scale filters when dark lines are needed) and it also under values line widths of large sized lines. The following chapter presents a line detection method based on the first order Gaussian derivative, to pivot on dark line detection. The detection of bright lines is achieved by reversing the image gray level. (Design Frameworks, Springer, 2020) The enlisted approach can examine dark lines without much wrong examination on blobs or bright lines along with estimation of the width of the line simultaneously.

Problem Statement

According to the sources, over the past few years, a number of farmer families have suffered huge crop losses due to over irrigation and resulting infertility of the crop-field soil. The most often heard excuse is the carelessness and oblivion of the farmers. In this part, author will be giving a short succinct explanation of the issues like breaking of bunds in the farmlands and the oblivion resulting to over-irrigation. In light of the above issue, we will be discussing about proposed ideas, their requisites, etc.

Breaking of bunds is the main problem of farmers because, during irrigation, the water flows out to other fields making their crops drown. Crops in the over-filled farm will also drown causing a big loss to the farmer. To cop it up, farmers have to keep a constant eye on tube well and fields during irrigation (Soma et al., 2019). Due to electricity issues, a many times farmer has to do that same work during the night. Also, seeing the insufficiency of water, we shall not waste any water. Overflow causes plenty of water wastage (Uferah et al., 2019).

Proposed Idea

A camera that keeps an eye on farms and monitors water flow and the bunds. The camera will detect the crops and farm boundary when the farm gets filled with water, it will alert the farmer. Also, when there will be any breakage of bund and leakage of water, it will also alert farmer (Dhawan & Sihalath, 2017). After the breakage of bund or filling of farms followed by an alert, it will also ask him to switch off the device. If the farmer is using the river for irrigation, then it will simply alert the farmer.

Requirement Analysis

This chapter is based on water monitoring and precision agriculture and our stakeholders are farmers. The main purpose of this chapter is to monitor water during irrigation (Nature Inspired Computing, Springer, 2020). So, the requirements of this chapter divided into two parts, one is hardware and another is the software part:

Hardware Requirements are Described As:

1. For water monitoring, there is a need for a visual surveillance camera and for operating to this camera there is need of a network, electricity and power backup in the absence of electricity.
2. Every stakeholder or farmer should have a mobile phone so that he/she can get alert from the visual camera and operate tube well.
3. For the bund – boundary detection algorithm there is a need for python and OpenCV.

Software Requirements are Described As:

1. An application for operating tube well and visual camera.

Context Diagram

Requirement analysis of this chapter has done with the help of a context diagram and use case diagram (See Fig. 1 and Fig. 2). The Context Diagram shows the system under consideration as a single high-level process and then shows the relationship that the system has with other external entities (systems, organizational groups, external data stores, etc.).

Use-Case Diagram

A use case diagram is a dynamic or behavior diagram in UML. Use case diagrams model the functionality of a system using actors and use cases. Use cases are a set of

Figure 1. Context diagram of water monitoring system in agriculture

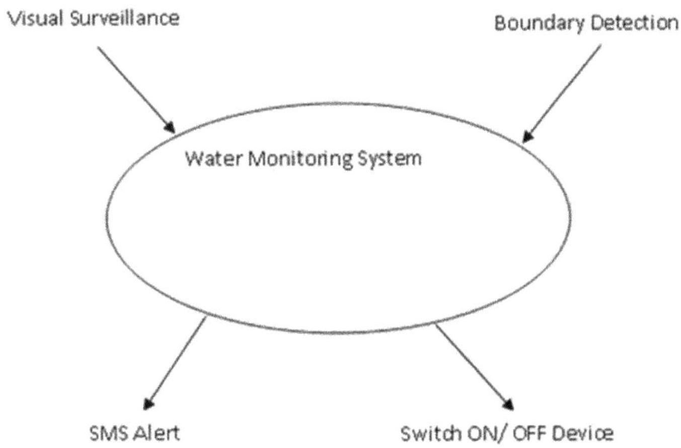

actions, services, and functions that the system needs to perform. In this context, a "system" is something being developed or operated, such as a web site. The "actors" are people or entities operating under defined roles within the system.

PROBLEM SOLUTION

In this section, we will discuss about plan, scope and goals of having this chapter, few tools and technologies we intend to work with, a discussion of used references and our proposed methodology for the work model. Few enhancements in the previous technologies have also been discussed. The methodology contains description of methods, area of planting the system and workflow diagrams in the form of Data flow and ER diagrams.

Planning to this chapter has done according to the Software Development Life Cycle (SDLC) planning phase. The planning is including following steps:

1. Vision
2. Goals
3. Scope
4. Tasks
5. Tools & Technology

Figure 2. Use case diagram of water monitoring system in agriculture

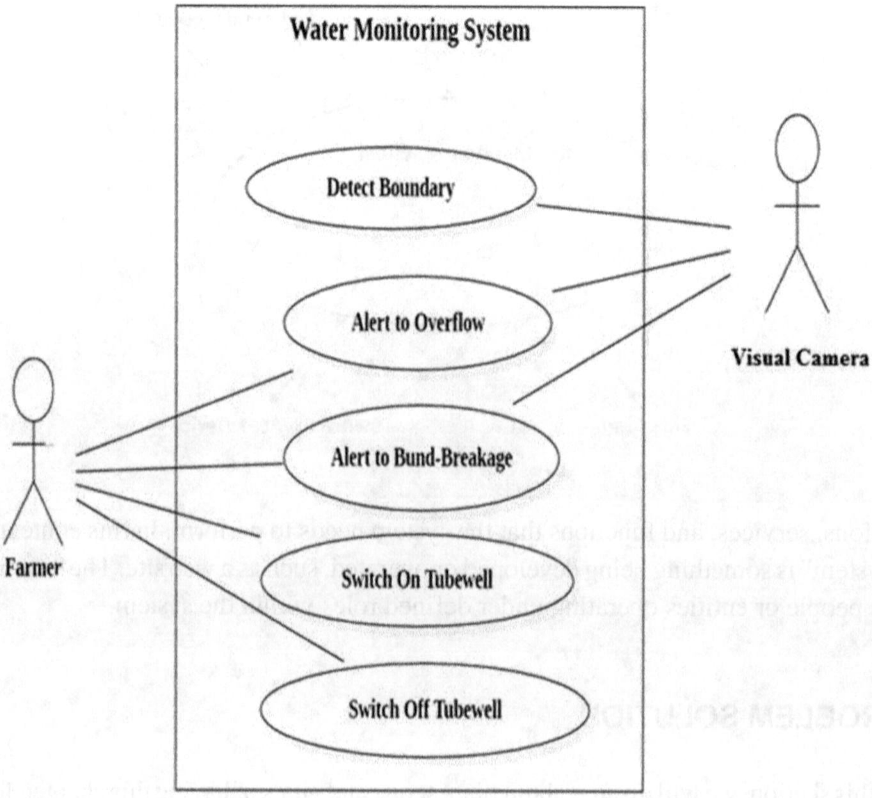

Vision

The Vision of this idea is to helping farmers by converting their hard work into smart work using visual surveillance technology for water monitoring in agriculture.

Goals

This Chapter will help to reduce working hour for farmers.

1. Enhancing Precision Agriculture in India.
2. Reducing Farmers effort.
3. To stop water wastage.
4. To Provide facility to farmer monitoring water from home.
5. To convert hard work into smart work.

Scope

India is an agrarian country where agriculture is the backbone of the economy and precision farming can be quite useful, so scope of this chapter is whole Indian agriculture market. India ranks 2nd worldwide in farm output. Agriculture and allied sectors like forestry and fisheries accounted 13.7% of the GDP (Gross Domestic Production) in 2013, and employed 50% of the workforce. The irrigation infrastructure includes a network of canals from rivers, ground water, well based systems, tanks and other rainwater harvesting products for agriculture activities. Today ground system is the largest, covering – 160 million ha of cultivated land in India with 39 million ha irrigated by ground water (Vladimir & Malabanan, 2010). The average size of the around 138 million farms was around 1.15 ha in 2010/11 and average size of large-scale farmers' farms (170.000) is around 37 ha in 2016 (BEML India country report 2016) (James, 2018).

Agricultural extension has only one extension worker per 800-1000 farmers and degree of mechanization reaches less than 50% (BEML India country report 2016). Indicators of water stress and scarcity are generally used to reflect the overall water availability in a country or a region (James, 2018).

Tools & Technology

Tools and Technologies of Water Monitoring system are given as:

1. Python.
2. Machine Learning.
3. OpenCV.
4. Visual Surveillance Camera.

METHODOLOGY

The complete automation of farming is inevitable. Farmers have been adopting each new wave of technology since the invention of the wheel. By learning how to take advantage of this, using the advancements in computer vision and machine learning to provide farmers with new precision agriculture systems. Our method uses an advanced version of the line detector algorithm. Bunds around the farmlands are just lines when seen from a far distance. Therefore, the author will be changing the line detector into the bund detector by applying a few feature enhancements and linear data coefficient integrations.

The proposed technique of line detection by (Design Frameworks-Springer, 2020). The underlined method is inspired by the line detection method proposed in (Design Frameworks-Springer, 2020), and the distinction is that we aim at detecting the thick lines (bunds) in the image. Through the Gaussian second-order derivative filter we will be able to enhance the thickness of the line. It may be possible to diagnose dark colored lines with less false detection on blobs or bright colored lines (Dey at el., 2017, Binh at el. 2016). Meanwhile, the proposed approach can also approximate the line width simultaneously. The next step is to enhance water level detection in a more effective way through a sequence of images. Live tracking of water level measurement can be visually verified by visual surveillance of the actual event.

The processing of image operation procedure used before measuring the depth of water, which incorporates color conversion to grayscale, image smoothing and detection of edge. By using the one-dimensional kernel and threshold the rim points of the water are regulated. Then, by applying linear regression an equation for the top line model that matches the set of rim points is found. Calibration through fiducials or markers determines the real-world coordinates of the water level. First, the images are transformed toward grayscale image which is exclusively contains of grey colored shade (Miya et al., 2020; Miya et al., 2020; Dandapat et al., 2005). All pixel image consists of Red, Green, Blue (RGB) portions that are constituted by an integer value ranging from 0 to 255. Each pixel integer value is calculated through the average RGB constituents of every source pixels of image in a grayscale image. The noise of the calculated grayscale image is blurred, and the noise is reduced. The program makes use of Gaussian technique of smoothing, achieved through image convolving. The weighted average value of every pixel's adjacent neighbor with the weighted average value toward the value of the centric pixels' kernel is computed. Now, using a line detector the level can be easily detected and we can use cut-off value for the alert system.

Rest we can use the further program to calculate values of cut-off accordingly after a few tests on the farm. The alerts and other methods are well known and can be used externally via advanced electronics.

1. For the detection of bunds, author will be using a modified version of line detector.
2. The line detector can also detect the water clogging in the water bunds.
3. It can also detect when the bunds get destroyed by overflowing of water and reaching other fields.
4. We'll begin the pre-processing phase to extract the information author want from the picture.
5. For the use-case, author want to be sure the computer is focusing on this information:

Figure 3. A demonstration of workspace and process

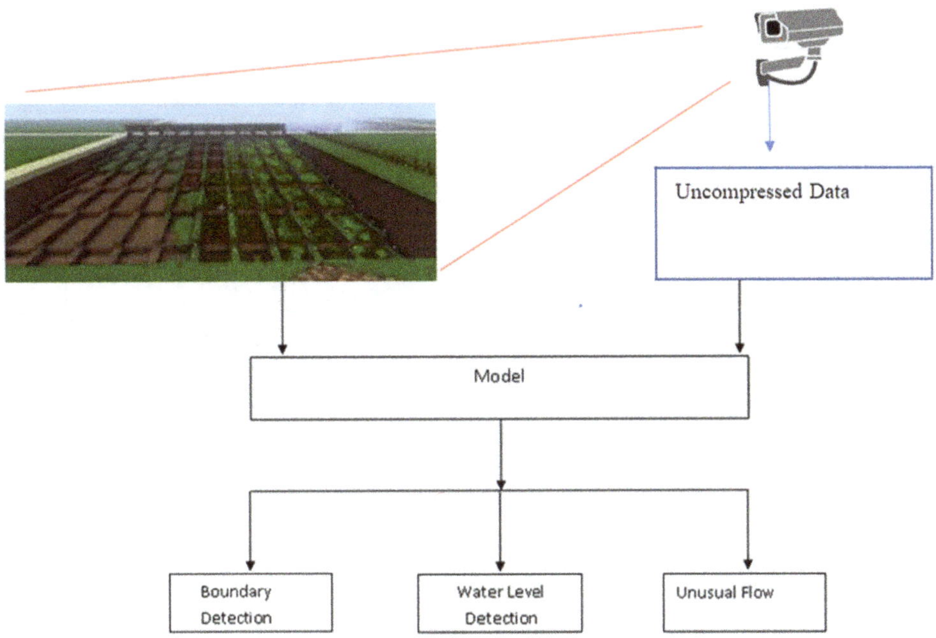

a. The bund lining that estimates farm boundary.
b. The drowning of kyaries to estimate the irrigation is done.
c. Any unusual flow of water out the field.

The process of detection and alert system is featured in the figure below:

DFD Diagram

A data-flow diagram (DFD) is a way of representing a flow of a data of a process or a system (usually an information system). The DFD also provides information about the outputs and inputs of each entity and the process itself. A data-flow diagram has no control flow, there are no decision rules and no loops.

ANALYSIS OF RESULT

This chapter is providing a solution for the problems faced by farmers and saves their time and efforts that they have to put while irrigation. Breaking of bund and overflow of water to other farmlands is a big issue to a farmer in spite of his/her

Figure 4. DFD for Water Monitoring System

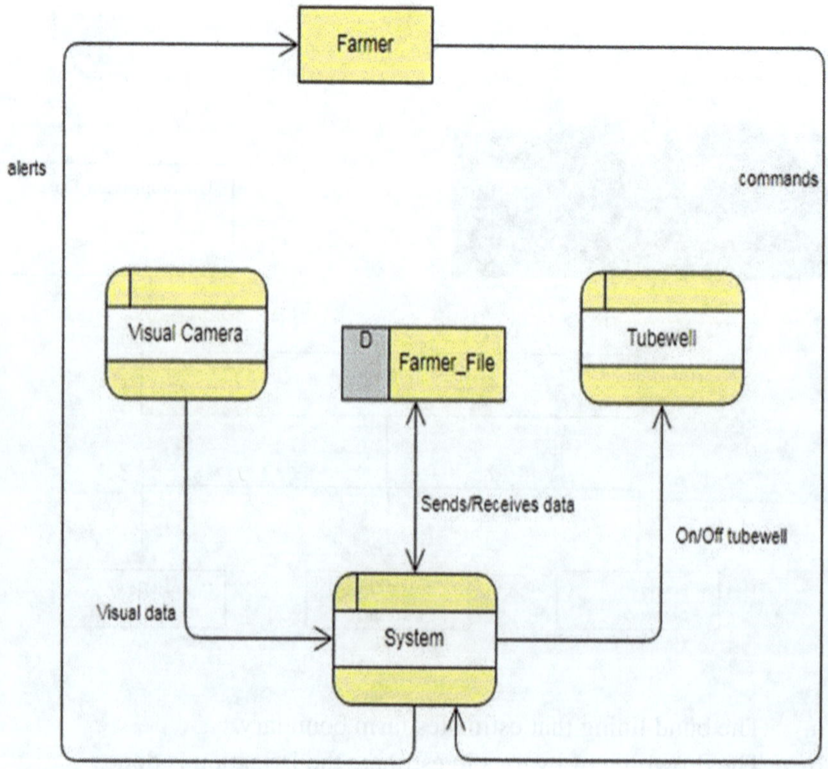

loss due to drowning of crops. To protect this, farmers have to watch over tube well and bunds whole time till the irrigation is done. Due to electricity issues in village areas, farmers many times need to spend their complete night in the farmlands, and especially in winters when it's chilly out there. So, to help farmers in this, we are proposing a device which helps farmers do so in some easy way. The visual surveillance and Machine Learning will now do this job for the farmer. The system with Machine learning algorithm will do surveillance of the farmland and notifies if any break occurs in the bund during irrigation or water overflow occurs. Water contributes 70% of body but contributes 100% of life. So, it our duty to save water. This chapter also saves water from being wasted.

As discussed in the proposed approach, every pixel image consists of Red, Green, Blue (RGB) portions that are constituted by an integer value ranging from 0 to 255. Each pixel integer value is calculated through the average RGB constituents of every source pixels of image in a grayscale image. The noise of the calculated grayscale image is blurred, and the noise is reduced. The program makes use of Gaussian technique of smoothing, achieved through image convolving. The weighted average

Figure 5. Filter description

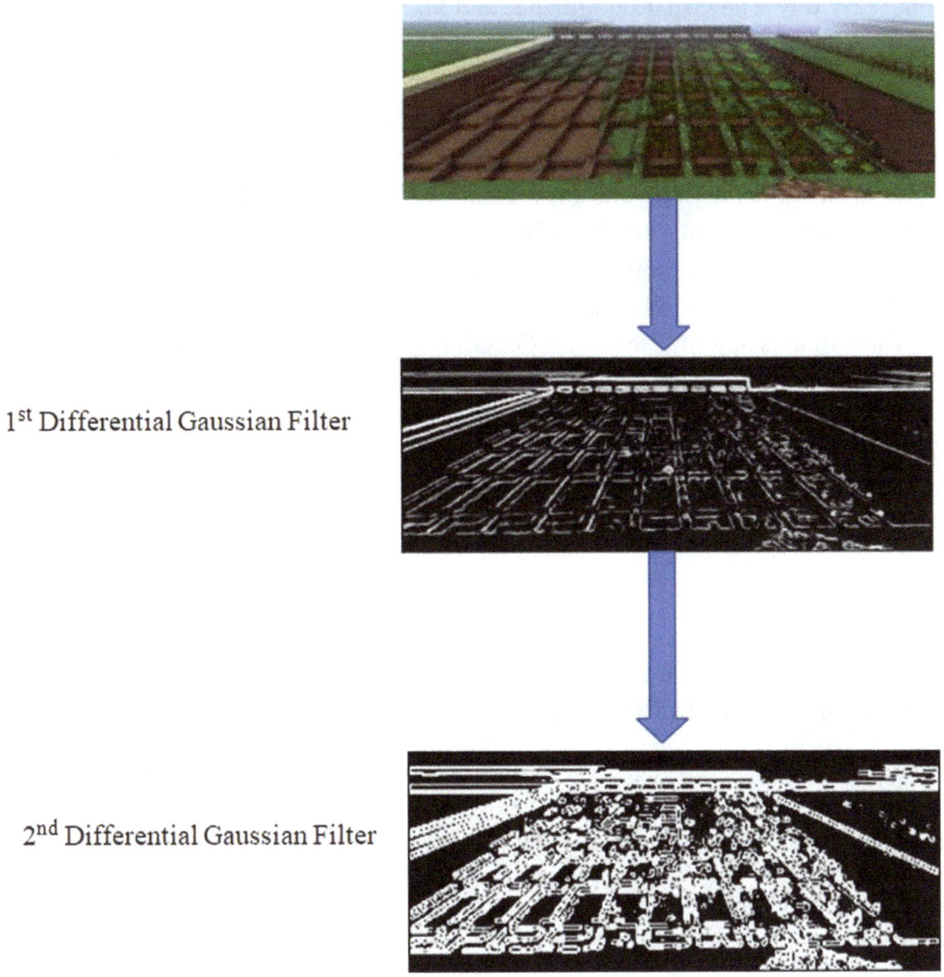

value of every pixel's adjacent neighbor with the weighted average value toward the value of the centric pixels' kernel is computed. Now, using a line detector the level can be easily detected and we can use cut-off value for the alert system. The below diagram (Fig. 5) is showing the implementation of the above text.

CONCLUSION

Due to electricity issues in village areas, farmers many times need to spend their complete night in the farmlands, and especially in winters when it's chilly out there.

So, to help farmers in this, we are proposing a device which helps farmers do so in some easy way. The visual surveillance and Machine Learning will now do this job for the farmer. The system with Machine learning algorithm will do surveillance of the farmland and notifies if any break occurs in the bund during irrigation or water overflow occurs. Water contributes 70% of body but contributes 100% of life. So, it our duty to save water. This chapter also saves water from being wasted. The water overflow in the fields and any breaking of bunds due to that overflow has also been taken care of using the Bund Detector. The visual surveillance and Machine Learning algorithm are now doing this job for the farmer by having a watch over farmlands with irrigation facility using a tube well.

The farmer is now able to switch-On/Off remotely using one-click alert systems. For the time being, we are focusing on the bund detection and water monitoring algorithms that help farmers and saves their time and efforts that they have to put while irrigation, but for future enhancements, we can suggest proper insecticides, animal intrusion alert and fire alert too.

REFERENCES

Bhattacharya, Zhang, Li, Zhang, & You. (2008). *Dark Line Detection with Line Width Extraction*. IEEE. DOI: doi:10.1109/ICIP.2008.4711831

Binh. (2016). Improved cuckoo search and chaotic flower pollination optimization algorithm for maximizing area coverage in wireless sensor networks. *Neural Computing & Applications*, *30*(7), 2305–2317. doi:10.100700521-016-2823-5

Choomchuay & Sihalath. (2010). *An Application of Second Derivative of Gaussian Filters in Fingerprint Image Enhancement*. IEEE.

Dandapat, & Satyarthi, & Kumar. (2005). Gaussian Intensity Distribution Modelling of Blood Vessels in Fundus Images. *IEEE Indicon Conference, Chennai*, 228-232.

Design Frameworks for Wireless Networks. (2020). Springer.

Dey. (2017). Developing residential wireless sensor networks for ECG healthcare monitoring. *IEEE Transactions on Consumer Electronics*, *63*(4), 442–449.

Dhawan & Sihalath. (2017). *Water and Agriculture in India.* Background paper for the South Asia expert panel during the Global Forum for Food and Agriculture.

Liakos, K. G., Busato, P., Moshou, D., Pearson, S., & Bochtis, D. (2018). Machine Learning in Agriculture: A Review. *Sensors (Basel)*, *2018*(18), 2674. doi:10.339018082674

Mariano & Malabanan. (2010). *Real-time Water Level Monitoring Using Image Sequence Processing*. Academic Press.

Miya, J., & Ansari, M. A. (2020). Medical images performance analysis and observations with SPIHT and wavelet techniques. *Journal of Information and Optimization Sciences*, *41*(1), 273–282. doi:10.1080/02522667.2020.1721616

Miya, J., & Ansari, M. A. (2020). Wavelet Techniques for Medical Image Performance Analysis and Observations with EZW and Underwater Image Processing. In *Wireless Personal Communications*. Springer. https://rdcu.be/b3eSe

Nature Inspired Computing for Wireless Sensor Networks. (2020). Intelligent Routing in Wireless Sensor Network Based on African Buffalo Optimization. In Nature Inspired Computing for Wireless Sensor Networks (pp. 119-142). Springer.

Reddy, Reddy, Nagothu, & Reddy. (2016). *Sensors for water monitoring for improved on farm water management*. Academic Press.

Shafi, U., Mumtaz, R., García-Nieto, J., Hassan, S. A., Syed, A. R. Z., & Iqbal, N. (2019). Precision Agriculture Techniques and Practices: From Considerations to Applications. *Precision Agriculture Techniques and Practices: From Considerations to Applications.*, *3796*(17), 3796. Advance online publication. doi:10.339019173796 PMID:31480709

Soma, M. K., Shaheen, M., & Zeba, F. (2019). Precision Agriculture in India-Challenges and Opportunities. Academic Press.

Theseken. (2018). *Crop Row Detection Using Python and OpenCV*. Academic Press.

Theseken & Zhang. (2018). Crop Row Detection using Python and OpenCV. In *Applications of Information and Communication Technology for Improvements of Water and Soil Monitoring and Assessments in Agricultural Areas- A Case Study in the Taoyuan Irrigation District*. DOI: doi:10.3390/environments4010006

Theseken, J., Singh, I., & Bansal, M. (2018). *Monitoring Water Level in Agriculture Using Sensor Networks*. Academic Press.

Chapter 4
Theoretical Analysis of a Microwave Antenna for Optically Detected Magnetic Resonance (ODMR) in NV Centre of Diamond

Arka Ghosh
University of Siegen, Germany

ABSTRACT

The nitrogen vacancy (NV) centre is one of the most popular stable point defect centres in diamond. It acts as a single photon source even at room temperature. In order to analyse the optically detected magnetic resonance (ODMR) with the help of NV colour centre in diamond, an external magnetic field is needed as well as a microwave (MW) antenna. In this chapter, the authors present a simulated model of a microwave antenna and propose further modifications in order to increase the intensity of the microwave field.

INTRODUCTION

Colour centres are basically one or several point defects centres. The origin of these defect centres are vacant lattice sites or impurity atoms, which can be perfectly identified by individual absorption and emission spectra. Colour centres in diamond are responsible for the typical colouration of diamond. At the starting of this kind of research, natural diamonds were used intensively, but now a days industrially

DOI: 10.4018/978-1-6684-7343-6.ch004

grown diamonds are used for this purpose. Diamond nanoparticles have promising fluorescence characteristics with many potential applications in the domain of biolabeling. The electronic characteristics of such impurities are the field of interest for physicists. The nitrogen Vacancy defect (Gruber et al., 1997) is one of such kind of impurities and along with some exceptional electronic characteristics it shows quantum as well as magnetic behaviour even at room temperature. The proposed work is based on constraint decision making system that contains several types of decision making. In the last decade, several types of decision making works are done such as different architecture design and its methodologies of wireless network and its components (Das et al., 2021; Das et al., 2020; Das et al., 2021), metaheuristic and soft computing based work (Binh & Dey, 2018; Binh et al., 2020), machine learning based work (Das et al., 2021), smart application design based work (Das, 2021; Das & Giacalone, 2022), etc. A constraint of a system is responsible for slowing down of its performance or preventing it from the desired goal. In order to keep the flow of a project going identification, exploitation and elevation of constraints are very important.

This report focuses on both physical as well as simulation groundwork with the goal of optical sensing of local magnetic field as well as microwave. Finite difference time domain software is used to determine and vary the intensity of the microwave field over a specific region on which the diamond sample would be kept on. A paramagnetic iron is also used to increase the strength of magnetic field.

THEORETICAL BACKGROUNDS

Nitrogen Vacancy Centre

The Nitrogen Vacancy (NV) centre in diamond is one of the numerous point defects (Schirhagl et al., 2014) in diamond. Photoluminescence is the one of the most discussed as well as useful properties of NV centres. The NV centre`s electron spin, localized at atomic scale, can be manipulated at room temperature by external factors such as magnetic or electrical fields, microwave radiation, or light, resulting in sharp resonance in the intensity of photoluminescence. An individual NV centre can be used as a basic unit for a quantum computer, a qubit, used for quantum cryptography.

Crystal and Electronic Structure

As nitrogen is the most abundant impurity in diamond, the NV centre (Schirhagl et al., 2014) was already discovered in the 1970s. Due to diamond`s outstanding properties, which includes mechanical hardness, optical transparency, and high Debye

Figure 1. Atomic structure of the NV centre in diamond. The arrow indicates its dipole orientation orthogonal to its (1 1 1) symmetry axis (Reuschel).

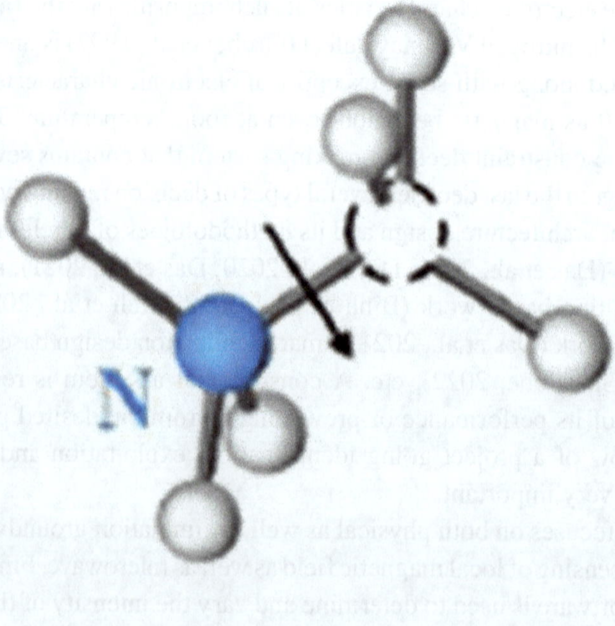

temperature, no extensive cooling or vacuum is necessary to observe luminescence from the NV centre. One disadvantage, however, is that the emission is around 100 nm wide because of strong coupling with phonons of local oscillation modes. Therefore, only 4% of the emitted light lies within the zero-phonon line (ZPL). Over the years, many alternative colour centres have been found, which show most of the emission in their ZPL (Wolters et al., 2010). Those include the Silicon-Vacancy (SiV), the Germanium Vacancy (GeV). Because of its special sensing application in optically detected magnetic resonance (ODMR), this report will focus on the increasing of strength of magnetic field by using a paramagnetic iron disc as a function of its position, radius, and thickness.

Optical Excitation and Spin Polarization

Because of its unique spin-optical properties, the negatively charged nitrogen vacancy centre (NV^-) has many different applications, while neutral nitrogen vacancy centre (NV^0) does not. That is why the following section will focus on the NV^- centre. The NV^- (Schirhagl, 2014) centre can be excited by energy larger than the difference between its ground and excited state of 1.945 eV. This energy can be resonantly delivered by a laser with wavelength λ of 637 nm according to

Figure 2. Emission spectra of NV0 and NV- in nano diamonds. The ZPL of NV0 and NV- are visible at 575 nm and 637 nm, respectively
(Karaveli et al., 2016).

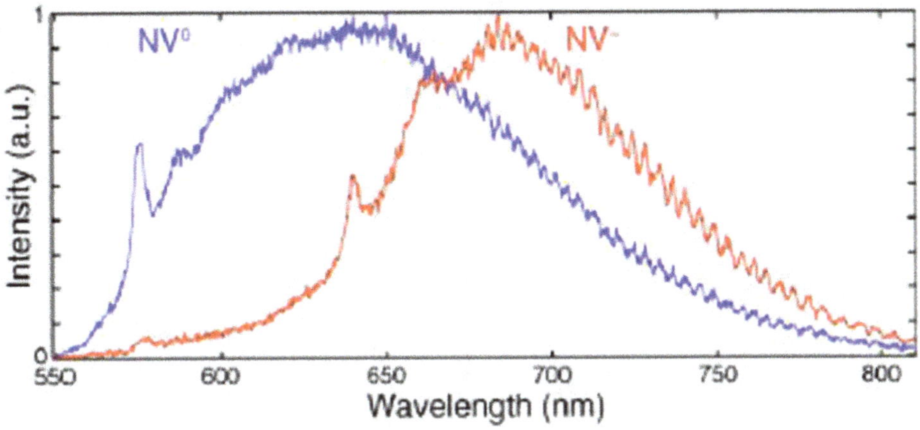

$$E = hf = h\frac{c}{\lambda}$$

OPTICALLY DETECTED MAGNETIC RESONANCE

To create a nitrogen vacancy inside diamond, carbon atom is replaced by a nitrogen atom and subsequently create an adjacent vacancy. The three possible charge state during creation of nitrogen vacancy centres (Doherty et al., 2014) are positively charged NV centres (NV^+), neutrally charged NV centres (NV^0), negatively charged NV centre (NV^-). Among these three types, only NV^- is suitable for optically detected magnetic resonance (ODMR) activity. ODMR is a sensing technique using NV^- centre. Here the centre is being pumped optically during the detection of fluorescence signal. Microwave (MW) radiation with a frequency close to zero-field splitting (D_g = 2.87 GHz) is applied at the centre by using desired MW antenna.

There are three types of energy state for NV^- centre namely a triplet ground state *g*, a triplet excited state *e*, shelving state *s*. Without the application of any external magnetic field spin sub levels are splitted (m_s= 0, m_s= ±1) due to the application of zero field splitting (D) MW frequency. For the ground state D_g= 2.87 GHz and for the excited state D_e= 1.42 GHz. On the other hand, with an external magnetic field, (B^1 0) due to the Zeeman effect (Schiff & Snyder, 1939), the spin sub-levels

Figure 3. (a) Three different energy state of NV- centre, (b) ODMR spectrum and effect of external magnetic field on it
(Sum, 2012).

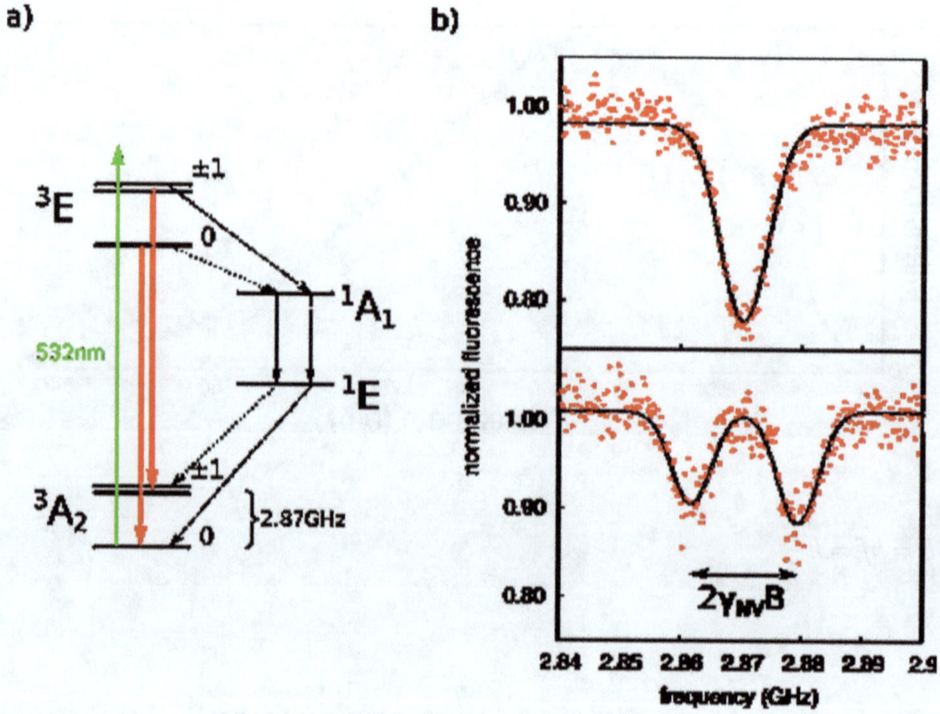

energy split is determined by a formula: $\Delta E = 2\gamma e_B$. Where, B is the Applied external magnetic field.

During the absence of external magnetic field, sweeping of MW frequency as well as collection of fluorescence signal leads to single intensity dip visibility. Due to MW pumping the centre transfers from m_s=0 state to degenerated $m_s = \pm1$ state, which leads to an increment of inter-system crossing (ISC), which results into less fluorescence. Due to the application of external magnetic field along the main axis of NV^-, this dip splits into two dips with an energy separation of: $\Delta E = 2\gamma e_B$.

Experimental Setup

The required tool for this experiment is a laser beam which can excite NV centre, collection of fluorescence shift, apart from this, in order to drive the spin sub-levels application of magnetic field is necessary along with microwave field.

The diamond sample used here is having (2.6 mm × 2.6mm × 1mm) dimension which is rich with nitrogen vacancy (NV). In order to generate microwave (MW) signal, a voltage-controlled oscillator (VCO) must be used. With the help of solenoid and current source, a static magnetic field is created. To collect the fluorescence signal, a spectrometer is used. By integrating the overall NV spectrum along with the influence of varying MW frequency, the spectrometer is used to measure ODMR.

DESIGN OF ANTENNA

The Finite Difference Time Domain Method

FDTD stand for Finite-difference time-domain which is implemented in Lumerical. FDTD is used to determine the impulse response of an electromagnetic system. In case of Lumerical simulation, FDTD is used for solving Maxwell`s equation. FDTD is basically a direct time and space solution, which provides detailed solution for photonics as well as electromagnetic. Though FDTD works directly with time domain, but it can also give solution in frequency domain by expanding time domain function with the help of Fourier transform. As a result of it, quantities like reflection or transmission of light and complex pointing vector can be calculated.

Apart from this, in case of non-magnetic materials, Maxwell`s curl equations are solved by FDTD. On a discrete spatial as well as temporal grid, these equations are solved.

Generally, FDTD is solved in a rectangular mesh in a cartesian coordinate. As the mesh size becoming smaller, simulation time as well as memory requirements will increase.

Geometry of the Structure

As referred to the figure we can evaluate that our structure is basically a Sub Miniature version A (SMA) connector with an additional helix and two wires. These additional parts are basically our deserving antenna which we are looking for. Using Lumerical this structure is made of four structural groups namely- Connector Group, TL, helix, FDTD, Mesh Group, Directivity. Connector group specially make the foundation of the SMA connector which consists of one base, four posts. All these components are rectangular in shape and made up of perfectly electrical conductor (PEC). The structural group TL is made up of one circular pin, one coaxial PEC and one coaxial dielectric both having ring structure among which pin and coaxial PEC are made up of perfectly electrical conductor (PEC), but coaxial dielectric is made up of dielectric having dielectric constant 1.48. The pin has a span of 55 mm and radius is 0.8 mm, the CoaxPEC has a span of 50 mm with an outer radius of 3.75 mm and an inner radius of 2.9 mm. Apart from this, Coaxdielectric has a z span of 50.8 mm

Figure 4. (a)Sketch of experimental setup. (BS) Beamsplitter, (CMOS) Complementary metal- oxide semiconductor, (LP) Longpass Filter, (VCO) Voltage controlled oscillator. (b) Planar microwave antenna with SMA connector. Red circle indicates the position of diamond sample
(Reuschel et al.).

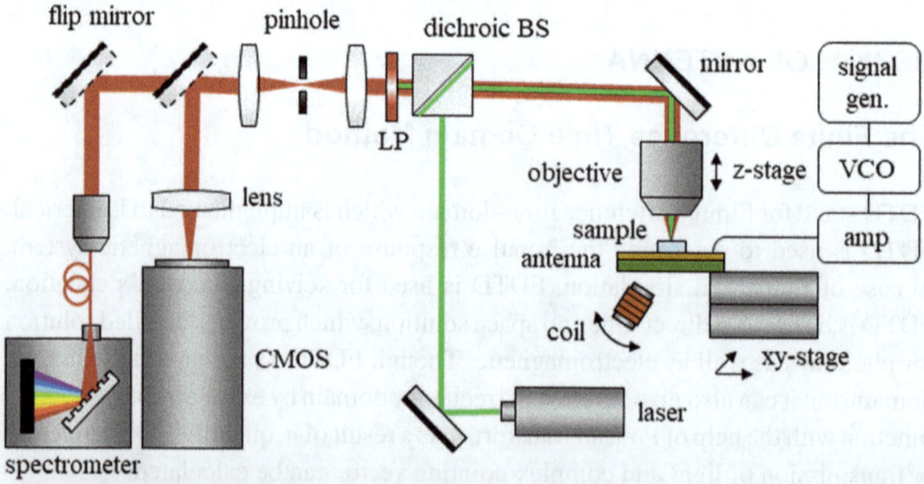

with an outer radius of 2.9 mm and an inner radius of 0.8 mm. The next structural group is helix which itself is the circular part of our desired antenna apart from this, two circular wires are there to complete the structure of antenna and all these are made up of PEC. These two circular wires are having radius of 0.5 mm. Due to the unavailability of ferromagnetic material in the material database, paramagnetic material with high magnetic permeability (~1000) is used as a magnetic iron. The permittivity of this iron is kept as 1. FDTD structural group consists of port. Port can act as a mode of source as well as field monitor and field expansion monitor. In the mesh group it consists of three meshes namely-mesh cable, mesh connector, mesh helix. Directivity group consists of six monitors from whom we can get the profile for electric field, magnetic field, far field, power etc. All those six monitors in the geometry are as follows- y1, y2, y3 these three monitors are parallel to each other along with the helix structure group. Y1 per monitor is perpendicular to all these three monitors and it monitors the parametric variation along the perpendicular direction of all those three monitors. Apart from these, monitor pin and monitor pin 2, which monitor the parametric variation of pin of the structure.

Monitors

Monitors record data from simulations. There are 5 types of monitors namely- refractive index monitor, field time monitor, movie monitor, frequency domain field profile, frequency domain field and power monitor. Different types of monitors are used to measure different types of parameters. For example, an index monitor can be used to measure the refractive index of the structure, a field time monitor can be used to measure the electric fields over time, and a frequency domain power monitor can be used to record transmission and reflection. Frequency-domain field (FDT) monitors works with the data in frequency domain within the simulation region of FDTD, var FDTD solvers. The two types of frequency domain monitors are- ′Frequency domain field profile monitor′ and ′Frequency domain field and power monitor′. In most situations, using of the 'field and power' monitor is recommended. This monitor delivers more accurate result than the other one by taking snaps to the nearest mesh cell, which eventually leads to minimization of required interpolation. The ′profile monitor′ records the data of its located place. Though it can be useful in some specific occasion, but requirement of extra interpolation reduces the accuracy of data. The port object can act as both a monitor and a source and is available characterizing the scattering parameters or S- parameters of a component element of a photonic integrated circuit. The port object can act as both a monitor and a source and is available characterizing the scattering parameters, or S-parameters of a component element of a photonic integrated circuit.

Mesh and Mesh Order

The mesh order is a property depending on which the simulation mesh is being determined for overlapping objects. So, basically it is useless for the objects those do not overlap. Mesh order needs to be set manually at material level as well as at object level. At material level it can be set from material database and at object level it can be set from object properties.

Material having lower mesh order will take priority over material having higher mesh order. So, from the figure below it can be seen that material having mesh order

Figure 5. (a) Views of geometry of structure from different coordinate. (b) Perspective view of the geometry of structure (with coordinate).

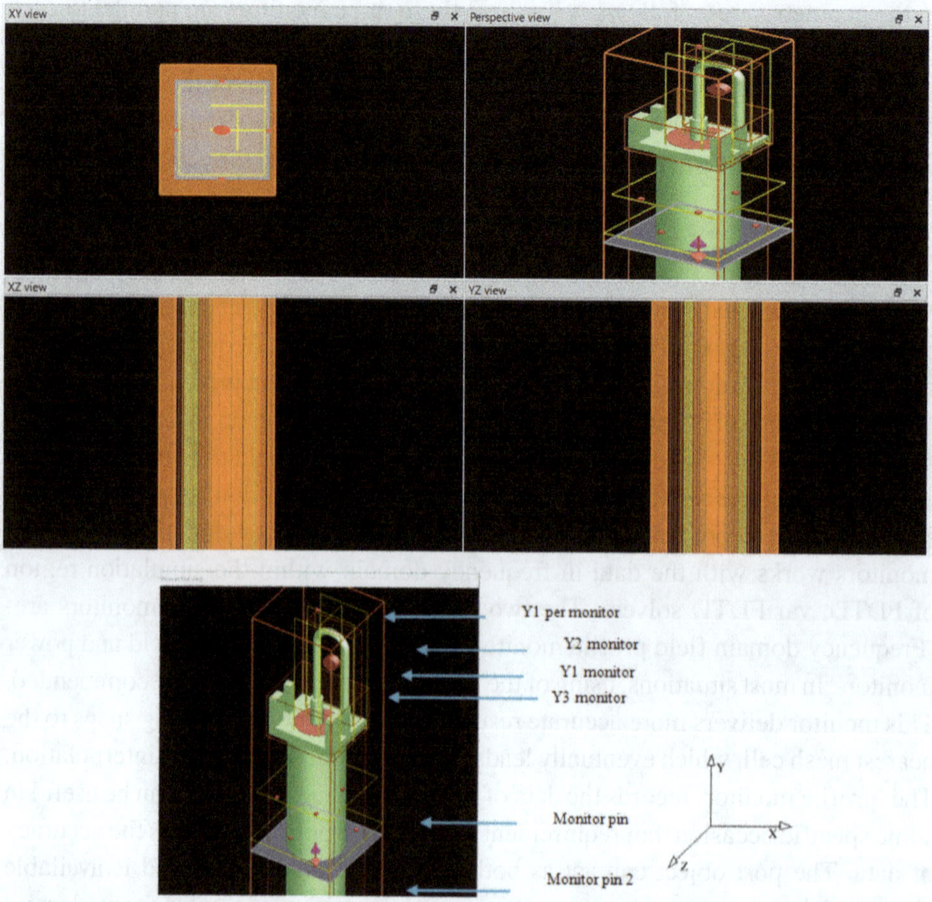

1 will take priority over material having mesh order 2. In the figure, the overlapping objects are being simulated depending on their mesh orders. If overlapping materials are having same mesh order, then priority will be determined from the object tree. Object at the bottom of the tree will have the highest priority while object at the top of the tree will have the least priority. 2D object is having more priority than 3D objects regardless of their mesh order.

Usage of Paramagnetic Iron

In between the two wires of desired antenna structure an experimental disc of paramagnetic iron is used with an aim to increase the strength of magnetic field.

Figure 6. Mesh and mesh order.

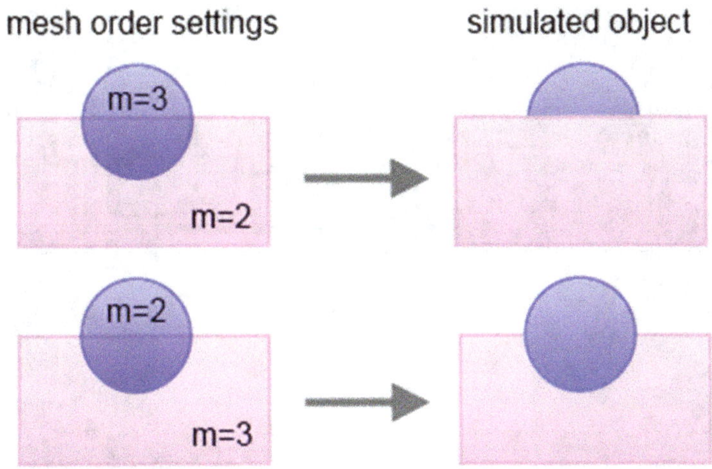

There are three types of magnetic materials namely- diamagnetic, paramagnetic, ferromagnetic. Ferromagnetic materials are having very high permeability and are used for the purpose of increasing magnetic strength (Fescenko et al., 2020). Ferromagnetic materials are generally used in electromagnet, core of transformer, other electrical machines. Due to unavailability of ferromagnetic material in the material database of Lumerical, a disc of paramagnetic iron has been used here with a high permeability (μ). This will increase the magnetic field in the region between two wires of the antenna. The three main variable parameters for this disc are ´r´, `y´, and `t´. where, `r´is the radius of the magnetic iron disc. `y´ is the position of the disc in three dimensional (3D) cartesian coordinate system and `t´ is the thickness of the disc.

Disadvantage of Using Sandwich of Irons

With an intension of increasing the strength of magnetic field two paramagnetic iron discs are used together by keeping both at a certain distance. But keeping the feasibility of this structure in mind during the laboratory experiment this type of structure does not get further approval. Because if this structure is supposed to get implemented, then the diamond sample should be kept in between this two iron rings and very little displacement of that sample leads to major change in the result. Apart from this there will be some air gap in between those two iron rings. So due to the discontinuity of magnetic material and presence of air gap desired magnetic field will also be discontinuous as seen from the figure. As a result of it this structure

Figure 7. Uses of paramagnetic iron (Red Disc).

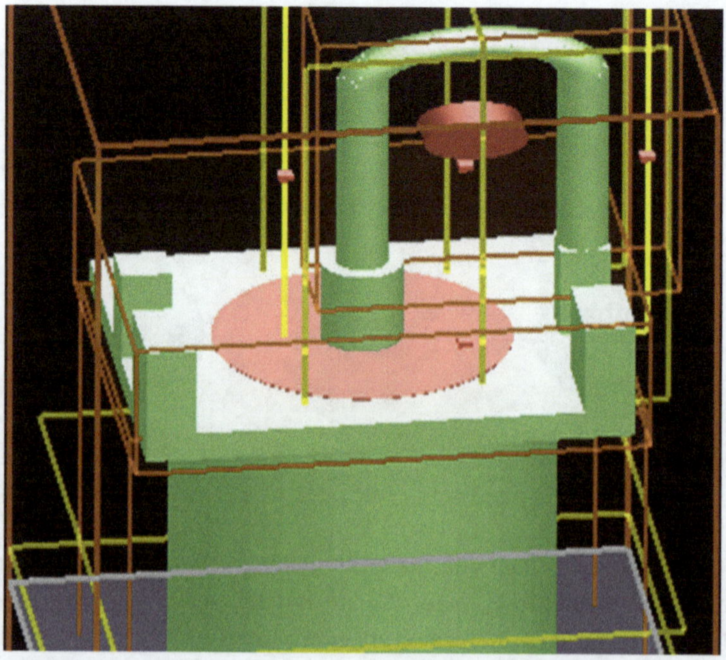

does not necessarily serve our purpose of incorporating magnetic material in the experiment. That is why it should be rejected.

Variation of Magnetic Field as a Function of Radius (r)

In order to optimize the strength of magnetic field in between the helical structure of the antenna, some manipulations have been done with the radius of the paramagnetic iron which is presented over there. Four different radii have been considered for checking the variation of magnetic field. The parametric variations for these four cases are (a) without iron, (b) y=0, r=0.5, t=1, (c) y=0, r=0.25, t=1, and (d) y=0, r=1, t=1. The reading has been taken from y1 monitor as well as y1 per monitor in all the four cases. In all the individual cases thickness of the disc as well as position of the disc are kept constant. In the fig: 9 the observation taken from y1 monitor for all the four cases and plotted by MATLAB for better and comparative understanding. Whereas fig: 10 correspond with the y1 per monitor. Fig: 11(a) and 11(b) are 2D view of two different cases from monitor y1 where the parameters are (a) y=0, r=1.5, t=1, (b) y=0, r=0.5, t=1 respectively and on the other hand fig: 11(c) and11(d) correspond to y1 per monitor where the parametric variations are (c) y=0, r=1.5, t=1, (d) y=0, r=0.5, t=1 respectively.

Figure 8. (a) Uses of sandwich of two paramagnetic irons, (b) Corresponding 2D plot from y1 monitor.

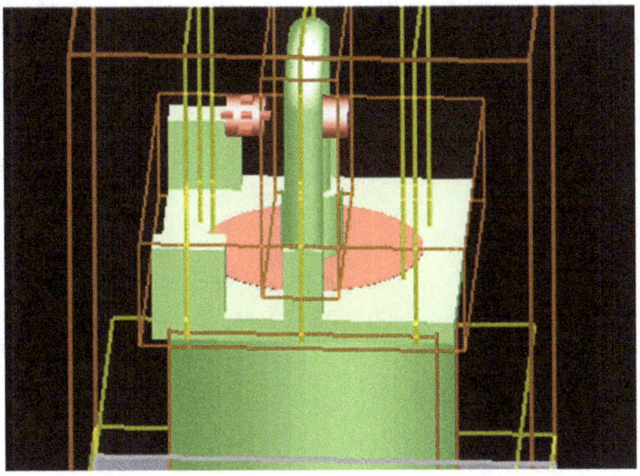

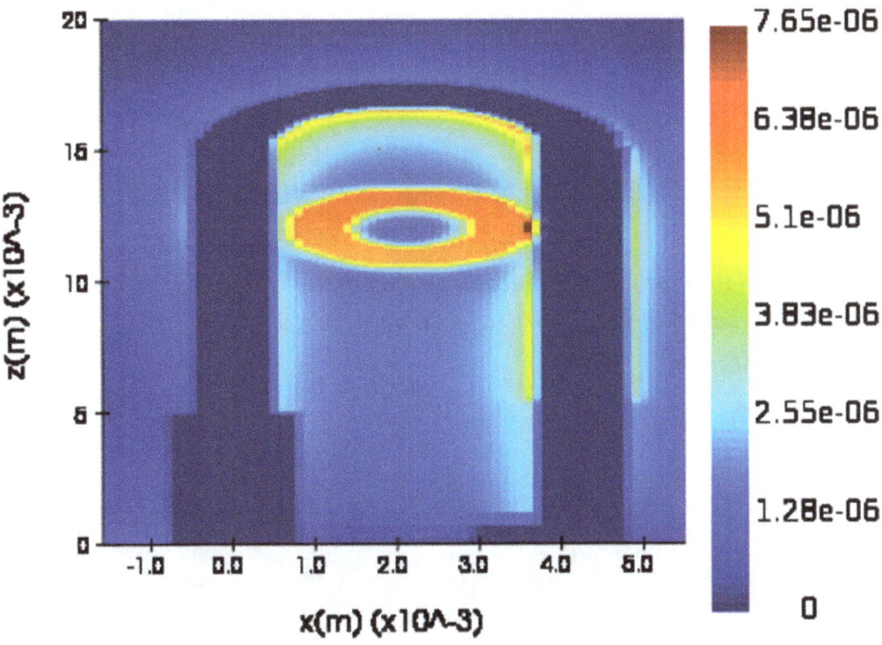

Variation of Magnetic Field as a Function of Thickness (t)

In order to optimize the strength of magnetic field in between the helical structure of the antenna, some manipulations have been done with the thickness of the paramagnetic iron which is presented over there. Six different thicknesses have been

Figure 9. Variation of magnetic field as a function of radius for y1 monitor where the parameters of graphs are (a) without iron, (b) y=0, r=0.5, t=1, (c) y=0, r=0.25, t=1, (d) y=0, r=1, t=1.

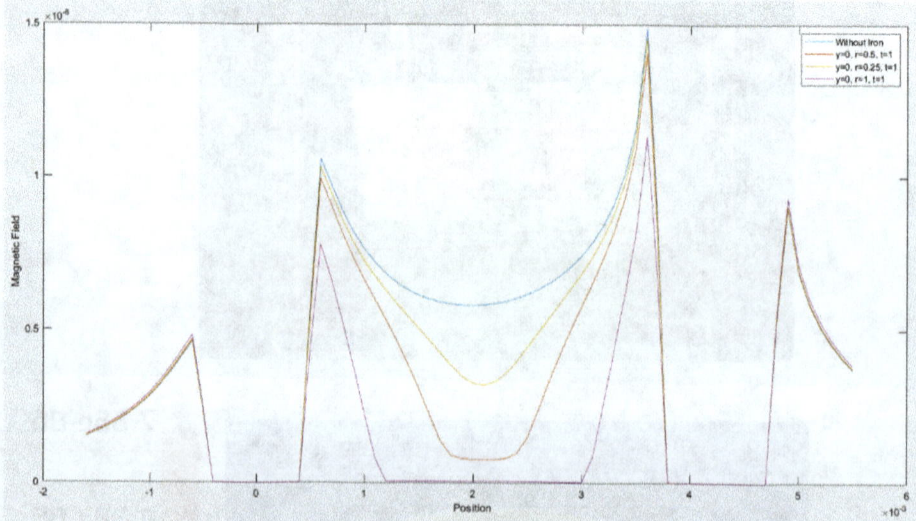

Figure 10. Variation of magnetic field as a function of radius for y1 per monitor where the parameters of graphs are (a) without iron, (b) y=0, r=0.5, t=1, (c) y=0, r=0.25, t=1, (d) y=0, r=1, t=1.

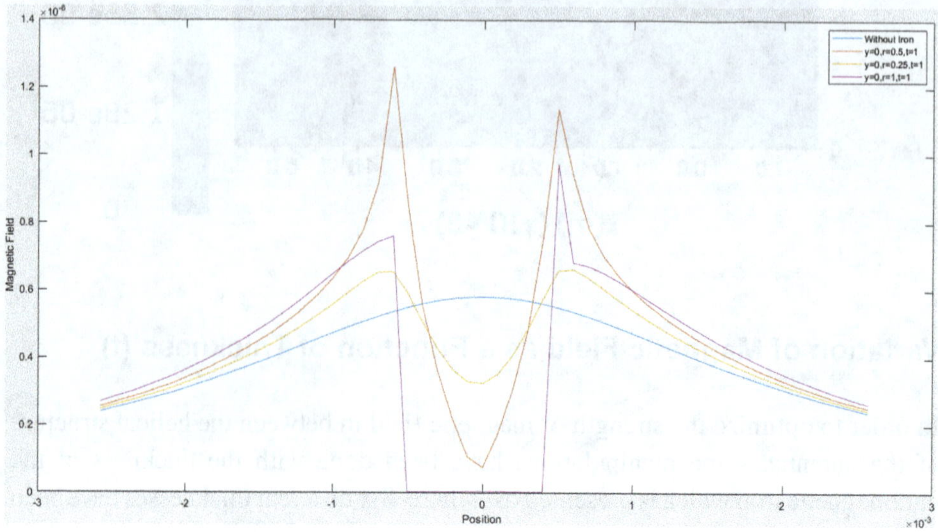

Figure 11. Variation of magnetic field for y1 monitor where the parameters of graphs are (a) y=0, r=1.5, t=1, (b) y=0, r=0.5, t=1, (c) y=0, r=1.5, t=1, (d) y=0, r=0.5, t=1.

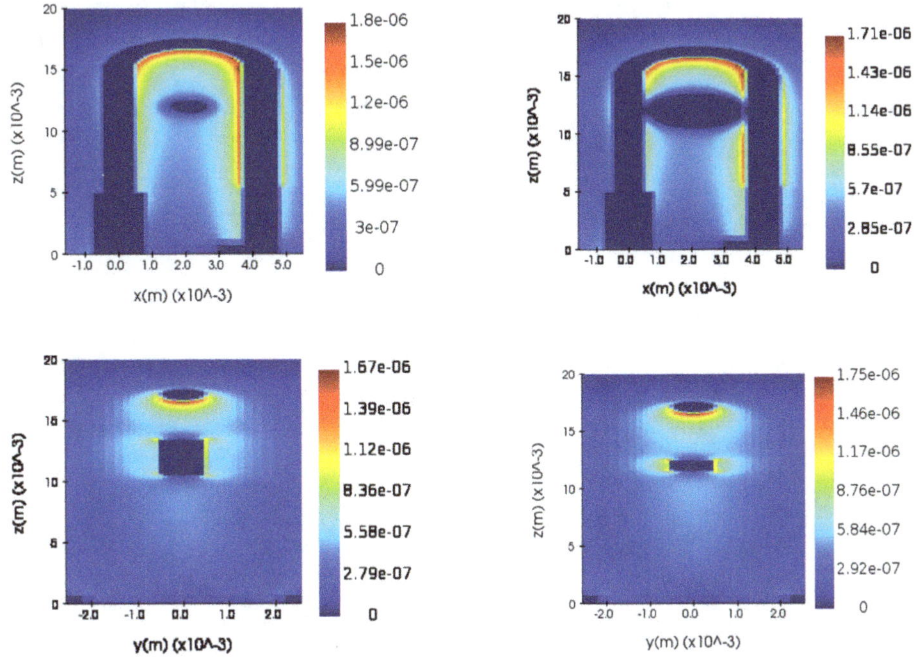

considered for checking the variation of magnetic field. The parametric variations for these six cases are (a) without iron, (b) y=0, r=1.5, t=0.5, (c) y=0, r=1.5, t=1.5, and (d) y=0, r=1.5, t=1. (e) y=0, r=1.5, t=2.5, (f) y=0, r=1.5, t=2. The reading has been taken from y1 monitor as well as y1 per monitor in all the six cases. In all the individual cases radius of the disc as well as position of the disc are kept constant. In the fig: 12 the observation taken from y1 monitor for all the six cases and plotted by MATLAB for better and comparative understanding. Whereas fig: 13 correspond with the y1 per monitor. Fig: 14(a) and 14(b) are 2D view of two different cases from monitor y1 where the parameters are (a) y=0, r=1.5, t=0.5, (b) y=0, r=1.5, t=2.5 respectively and on the other hand fig: 14 (c) and 14(d) correspond to y1 per monitor where the parametric variations are (c) y=0, r=1.5, t=0.5, (d) y=0, r=1.5, t=2.5, respectively.

Variation of Magnetic Field as a Function of Position (y)

In order to optimize the strength of magnetic field in between the helical structure of the antenna, some manipulations have been done with the thickness of the paramagnetic iron which is presented over there. Six different positions have been

Figure 12. Variation of magnetic field as a function of thickness for y1 monitor where the parameters of graphs are (a) without iron, (b) y=0, r=1.5, t=0.5, (c) y=0, r=1.5, t=1.5, (d) y=0, r=1.5, t=1, (e) y=0, r=1.5, t=2.5, (f) y=0, r=1.5, t=2.

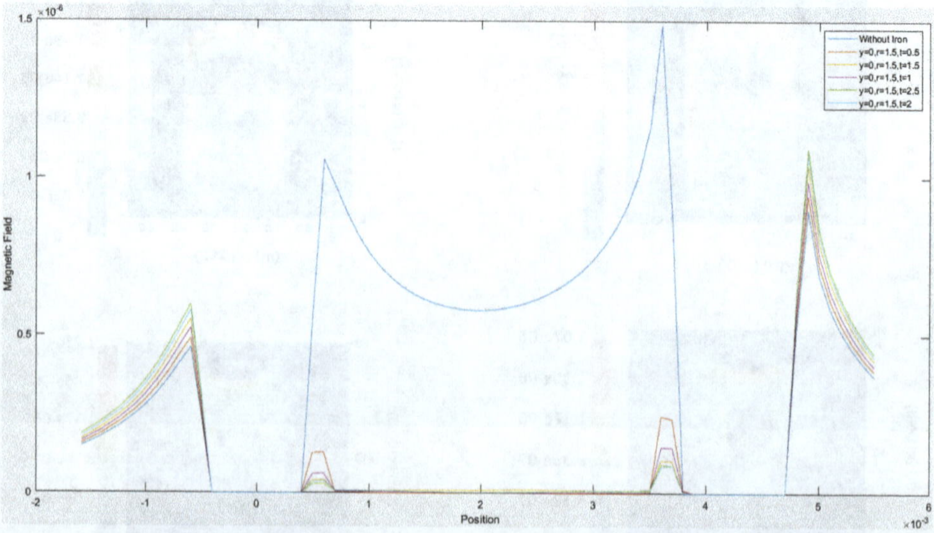

Figure 13. Variation of magnetic field as a function of thickness for y1 monitor where the parameters of graphs are (a) without iron, (b) y=0, r=1.5, t=0.5, (c) y=0, r=1.5, t=1.5, (d) y=0, r=1.5, t=1, (e) y=0, r=1.5, t=2.5, (f) y=0, r=1.5, t=2.

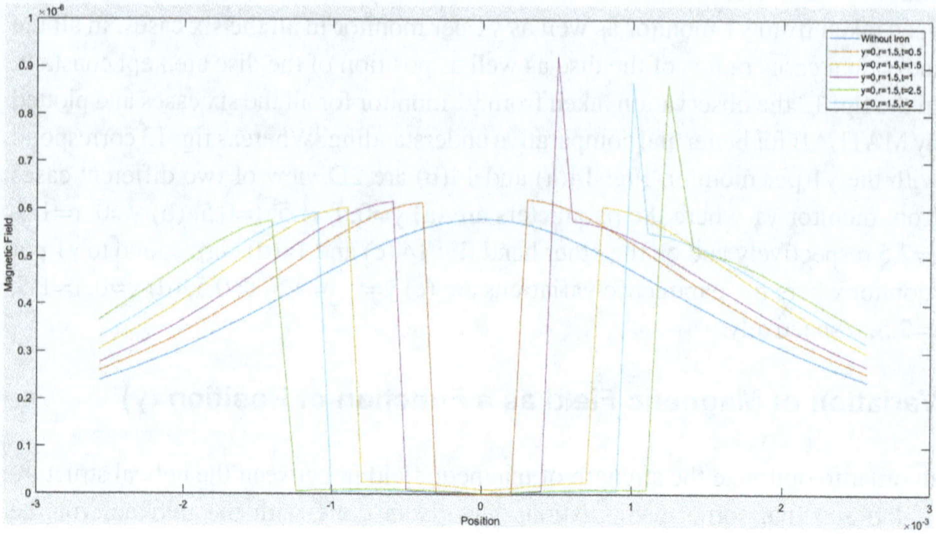

Figure 14. Variation of magnetic field for y1 monitor where the parameters of graphs are (a) y=0, r=1.5, t=0.5, (b) y=0, r=1.5, t=2.5, (c) y=0, r=1.5, t=0.5, (d) y=0, r=1.5, t=2.5.

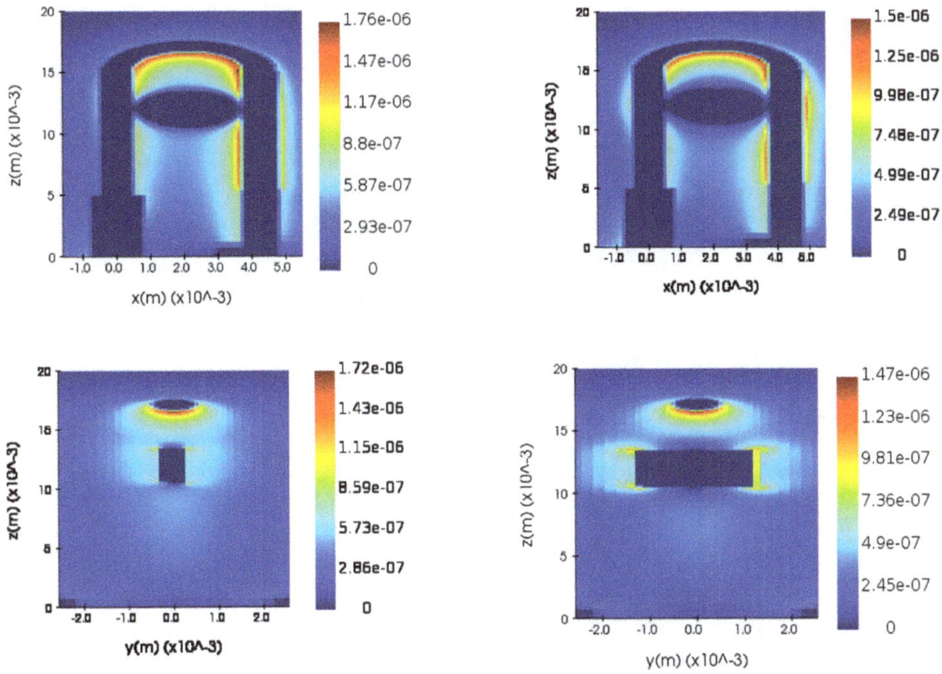

considered for checking the variation of magnetic field. The parametric variations for these six cases are (a) without iron, (b) y=-0.5, r=1.5, t=1, (c) y=-1, r=1.5, t=1, and (d) y=0, r=1.5, t=1. (e) y=0.5, r=1.5, t=1, (f) y=1, r=1.5, t=1. The reading has been taken from y1 monitor as well as y1 per monitor in all the six cases. In all the individual cases thickness of the disc as well as radius of the disc are kept constant. In the fig: 15 the observation taken from y1 monitor for all the six cases and plotted by MATLAB for better and comparative understanding. Whereas Fig. 16 correspond with the y1 per monitor. Fig: 17(a) and 17(b) are 2D view of two different cases from monitor y1 where the parameters are (a) y=-0.5, r=1.5, t=1, (b) y=1, r=1.5, t=1 respectively and on the other hand fig: 17(c) and 17(d) correspond to y1 per monitor where the parametric variations are (c) y=-0.5, r=1.5, t=1, (d) y=1, r=1.5, t=1 respectively.

Figure 15. Variation of magnetic field as a function of position for y1 monitor where the parameters of graphs are (a) without iron, (b) y=-0.5, r=1.5, t=1, (c) y=-1, r=1.5, t=1, (d) y=0, r=1.5, t=1, (e) y=0.5, r=1.5, t=1, (f) y=1, r=1.5, t=1

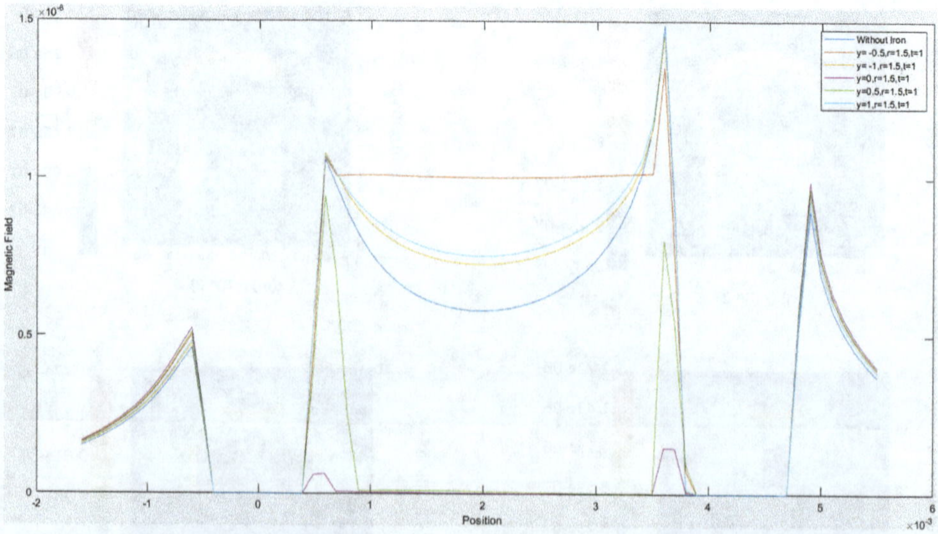

Figure 16. Variation of magnetic field as a function of position for y1 per monitor where the parameters of graphs are (a) without iron, (b) y=-0.5, r=1.5, t=1, (c) y=-1, r=1.5, t=1, (d) y=0, r=1.5, t=1, (e) y=0.5, r=1.5, t=1, (f) y=1, r=1.5, t=1

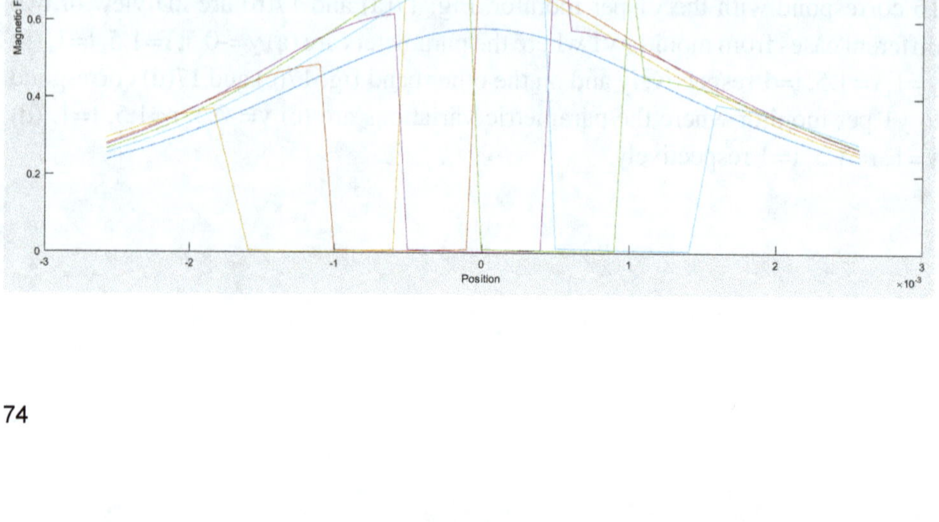

Figure 17. Variation of magnetic field for y1 monitor where the parameters of graphs are (a) y=-0.5, r=1.5, t=1, (b) y=1, r=1.5, t=1, (c) y=-0.5, r=1.5, t=1, (d) y=1, r=1.5, t=1.

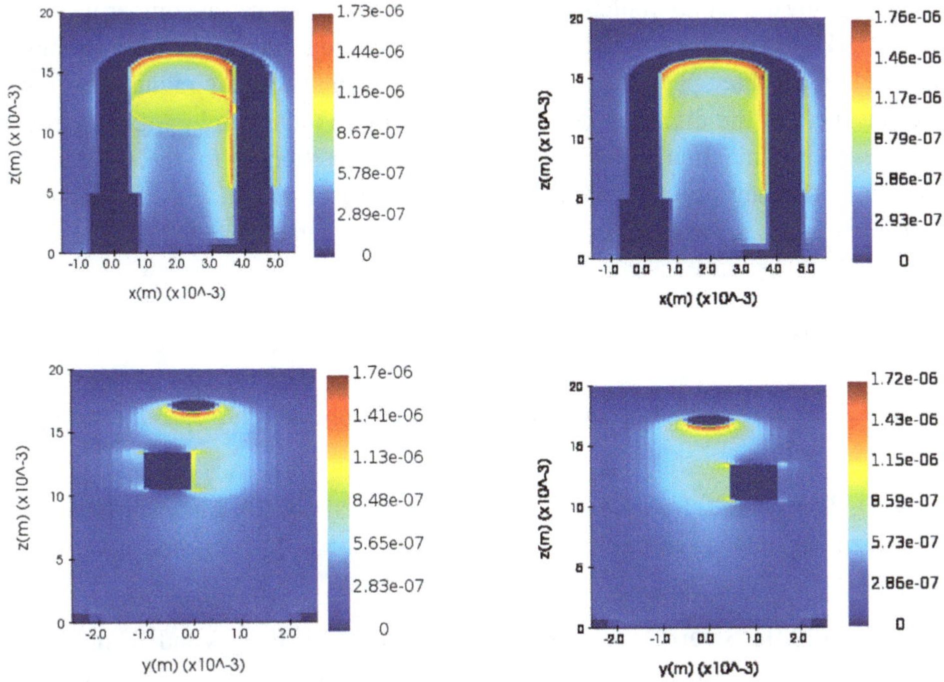

CONCLUSION

The main objective of this report was to simulate a microwave (MW) antenna by using Lumerical for optically detected magnetic resonance (ODMR). With the help of scanning confocal microscope only the fluorescence of each and individual nitrogen vacancy point defect centre can be detected in diamond lattice. These defect centres are photostable, photo sensitive and there is no change in the rate of fluorescence emission spectra. Magnetic resonance is also possible for individual point defect centre even in room temperature. On the other hand, even if this magnetic resonance spectra stays almost unchanged with respect to time, but it shows significant changes with respect to the existence and variation of magnetic field.

In order to have the maximum intensity of magnetic field for the sample of nano diamonds having a nitrogen vacancy in it, several simulation possibilities are discussed for the modelling of a microwave antenna. A circular disc of paramagnetic iron with relative permeability (μ= 1000) has been incorporated in between two wires of the antenna to maximize the magnetic field. This iron disc has three parameters to vary in order to maximize the magnetic field and these parameters are- position of the disc, radius of the disc, and thickness of the disc. All these parameters have been taken into consideration and have been used for individual simulation.

Depending on all the simulation results from lumerical as well as comparing plots from MATLAB it can be concluded that the generating microwave field strength depends on the three parameters of the magnetic iron. These are radius, position, and thickness of the disc. An ideal condition of these three combining properties will results into maximum outcome of magnetic field strength for the same input voltage. After analysing all these results and comparisons it can be concluded that y=1, r= 0.5, t=1 is the ideal for having the maximum microwave field strength.

REFERENCES

Binh, H. T. T., & Dey, N. (Eds.). (2018). *Soft computing in wireless sensor networks*. CRC Press.

Binh, H. T. T., Hanh, N. T., Nghia, N. D., & Dey, N. (2020). Metaheuristics for maximization of obstacles constrained area coverage in heterogeneous wireless sensor networks. *Applied Soft Computing*, *86*, 105939.

Das, S. K. (2021). Smart Design and Its Applications: Challenges and Techniques. In *Nature-Inspired Computing for Smart Application Design* (pp. 1–6). Springer.

Das, S. K., Das, S. P., Dey, N., & Hassanien, A. E. (Eds.). (2021). *Machine learning algorithms for industrial applications*. Springer.

Das, S. K., & Giacalone, M. (Eds.). (2022). *Fuzzy Optimization Techniques in the Areas of Science and Management*. CRC Press.

Das, S. K., Maheswari, V., & Sharma, A. (2021). Wireless Networks: Applications, Challenges, and Security Issues. In *Architectural Wireless Networks Solutions and Security Issues* (pp. 1–10). Springer.

Das, S. K., Samanta, S., Dey, N., & Kumar, R. (Eds.). (2020). *Design frameworks for wireless networks*. Springer Singapore.

Das, S. K., Samanta, S., Dey, N., Patel, B. S., & Hassanien, A. E. (Eds.). (2021). *Architectural Wireless Networks Solutions and Security Issues*. Springer Singapore Pte. Limited.

Doherty, M. W., Michl, J., Dolde, F., Jakobi, I., Neumann, P., Manson, N. B., & Wrachtrup, J. (2014). Measuring the defect structure orientation of a single NV-centre in diamond. *New Journal of Physics, 16*. doi:10.1088/1367-2630/16/6/063067

Fescenko, I., Jarmola, A., Savukov, I., Kehayias, P., Smits, J., Damron, J., & Acosta, V. M. (2020). Diamond magnetometer enhanced by ferrite flux concentrators. *Physical Review Research, 2*(2), 023394.

Gruber, A., Drabenstedt, A., Tietz, C., Fleury, L., Wrachtrup, J., & Borczyskowski, C. V. (1997). Scanning confocal optical microscopy and magnetic resonance on single defect centers. *Science, 276*(5321), 2012–2014. doi:10.1126cience.276.5321.2012

Karaveli, S., Gaathon, O., Wolcott, A., Sakakibara, R., Shemesh, O. A., Peterka, D. S., & Englund, D. (2016). Modulation of nitrogen vacancy charge state and fluorescence in nanodiamonds using electrochemical potential. *Proceedings of the National Academy of Sciences of the United States of America, 113*(15), 3938–3943.

Schiff, L. I., & Snyder, H. (1939). Theory of the quadratic zeeman effect. *Physical Review, 55*(1), 59–63. https://doi.org/10.1103/PhysRev.55.59

Schirhagl, R., Chang, K., Loretz, M., & Degen, C. L. (2014). Nitrogen-vacancy centers in diamond: Nanoscale sensors for physics and biology. *Annual Review of Physical Chemistry, 65*(1), 83–105.

Sum, W. C. Y. (2012). *Towards magnetometry with nitrogen-vacancy center in diamond* [Undergraduate Thesis]. Department of Physics, National University of Singapore.

Wolters, J., Schell, A. W., Kewes, G., Nüsse, N., Schoengen, M., Döscher, H., & Benson, O. (2010). Enhancement of the zero phonon line emission from a single nitrogen vacancy center in a nanodiamond via coupling to a photonic crystal cavity. *Applied Physics Letters, 97*(14). https://doi.org/10.1063/1.3499300

Chapter 5 Impact of Hall Effect on the Electromagnetohydrodynamic (EMHD) Flow of Nanofluid Through a Porous Microchannel

Amalendu Rana
https://orcid.org/0000-0003-1457-7284
Jadavpur University, Kolkata, India

Motahar Reza
GITAM University (Deemed), India

ABSTRACT

A theoretical investigation on EMHD flow through a microchannel with porous medium is considered in this chapter. Impact of Hall Effect on the nanofluid flow transport is analyzed by considering the Brinkman momentum equation to govern the flow performance through porous microchannel. Constant pressure gradient and interfacial slip boundary condition at the walls of the microchannel has been considered. The analytical solution for the nanofluid flow has been expressed by considering blood and ferromagnetic nanoparticles. Variations in velocity and skin friction coefficient associated with the combined EMHD flow in porous microchannel has been evaluated in this study. The present investigation reveals that the velocity and the skin friction both are enhanced by the influence of Hall current. Further, it is also observed from the study that the velocity becomes maximum near to the walls of the microchannel due to interfacial slip boundary conditions. To validate the result an excellent agreement is shown by making a comparison with the results of some available scientific literature.

DOI: 10.4018/978-1-6684-7343-6.ch005

INTRODUCTION

For the wide applications in bio-medical areas, physiology and medicine, microfluidics is taking attentions of researchers and engineers, not only that to design more advanced microfluidic device, microfluidic technology have being undertaken in so many fields such as bio chemical analysis instruments, micro-heat exchanger, lab-on-chip devices, micro-pumps, micro-turbines, etc. (Stone et al., 2004; Laster & Santiago, 2004). Moreover, in current years, various technological applications, including analysis and separation of various biological and chemical samples, material processing, and micro-electro-mechanical systems (MEMS) (West et al., 2002; Weston et al., 2010; Jang and Lee, 2000) are performing significant role in the relevant newly emerging research. MEMS is also very popularly used in modern application of computer science (Das et al., 2015; Das et al., 2021a; Das et al., 2021b; Das et al., 2021c; De et al., 2020; Das et al., 2020. Many real life problems can be modeled using decision constraints. It has immense significance in linear programming problems, optimization theory etc. This notion may also be useful in the study of flow transport to explore some of its important aspects. The flow performance through fractal-like heat sinks, subject to both geometrical and flow constraints are assessed which are crucial in order to satisfy physical requirements of a design of microfluidic devices.

Microfluidic transport is generated by the influence of electrokinetic effect and electromagnetohydrodynamic (EMHD) effects. Recently, researchers are investigated various theoretical models of different flow problems to analyze the flow transportation phenomenon of different fluids. EMHD flow in microchannel has been widely studied by so many researchers and various theoretical and experimental investigations are done by them (Liu et al., 2018; Liu and Jian, 2019; Sarkar et al., 2017; Wang et al., 2016; Sarkar et al., 2017; Vargas et al., 2017; Reza et al., 2021; Rana et al., 2022). The two phase flow transport of bio-fluid is analysed by Liu and Yang (Liu & Yang, 2009). Chakraborty and Paul (Chakraborty & Paul, 2006) performed the flow control through parallel plate microchannel under the EMHD effects. The heat transfer performance is analyzed by Chakraborty et al. (Chakraborty et al., 2013) under the influence of the applied magnetic field. An analytical investigation on the thermal characteristics of interfacial slip flow through a porous microchannel is done by Reza et al. (Reza et al., 2020a). They (Reza et al., 2020b) also studied the similar investigations through porous microtube. Reza and Rana (Reza & Rana, 2020) reported an analytical exact solution of EMHD flow control of casson fluid through microchannel. Buren et al. (Buren et al., 2014) studied the corrugated channel flow through a microparallel wall with EMHD effect by applying Perturbation technique. Azari et al. (Azari et al., 2020) performed Electroosmotic flow transport and thermal performance through a microtube with heterogeneous surface. Very recently, EMHD flow of non-Newtonian second grade fluid through a porous microchannel with corrugated wall investigated by Rashid et al. (Rashid et al., 2018) by applying the perturbation technique. A comprehensive analysis of heat transfer of nano-fluid through a microtube under the streaming potential effects is done by Zhao et al (Zhao et al., 2016a). Zhao et al. (Zhao et al., 2016b) represented the thermal transport of a nanofluid flow through microchannel which is influenced by pressure gradient effect and streaming potential.

It can be observed from the above references that in most of the investigations, there may be ignored the effect of Hall current. However, it is well known to all that when the system is considered in a strong applied magnetic field, there may be generated a complicated phenomenon, which is dominated by Hall current in the flow field. Due to the ionization of the electrolyte solution, the Hall current effect is generated inside the electrically conducting liquid.

In our study, different parameters, variables and symbols are used which are described on the following nomenclature:

NOMENCLATURE

B_y Applied magnetic field along y-axis
Da Darcy Number
E_x External applied electric field along x-axis
E_z External applied electric field along z-axis
e Charge of the electron
H Half channel height
Ha Hartmann Number
k_B Boltzmann constant
L Length of the microchannel
m Hall parameter
n_e Number density of the free electron
Re Reynolds number
S Transverse electric field
$\vec{U}$ Flow velocity vector
u Flow velocity along x-axis
u_{HS} Helmholtz-Smoluchowski velocity
W Width of the channel
β_s Non-dimensional slip parameter
ε_f Permittivity of the base fluid
ε_n Permittivity of the nanoparticle

ε_{nf} Effective permittivity of the nanofluid
ζ Wall zeta potential
κ Dybye-Hückle parameter
μ_f Viscosity of the base fluid.
μ_{nf} Viscosity of the nano-fluid
ϕ Nanoparticle volume friction parameter
ρ_f Density of the base fluid
ρ_n Density of the nanoparticles
ρ_{nf} Density of the nano-fluid
σ_f Electrical conductivity of the base fluid
σ_n Electrical conductivity of the nanoparticles
σ_{nf} Electrical conductivity of the nanofluid
ψ Electrokinetic potential
ω Half channel height to Debye length ratio

From the above motivational investigations, it is remarkable to examine that the flow transmit phenomenon of nanofluid through the porous microchannel. The main purpose of the study is to incur a comprehensive analytical solution of the velocity distribution for the nanofluid flow in the porous microchannel with Navier slip boundary conditions, amalgamated with electroosmotic force and applied magnetic force. The constitutive governing equation is modified from the Brinkman momentum equation. The existence of Hall Effect and applied magnetic field on the skin friction coefficient and velocity profile has been explored. Further, the impact of porous medium and Navier slip parameter effect is also observed in this investigation.

Figure 1. Schematic sketch of the microchannel

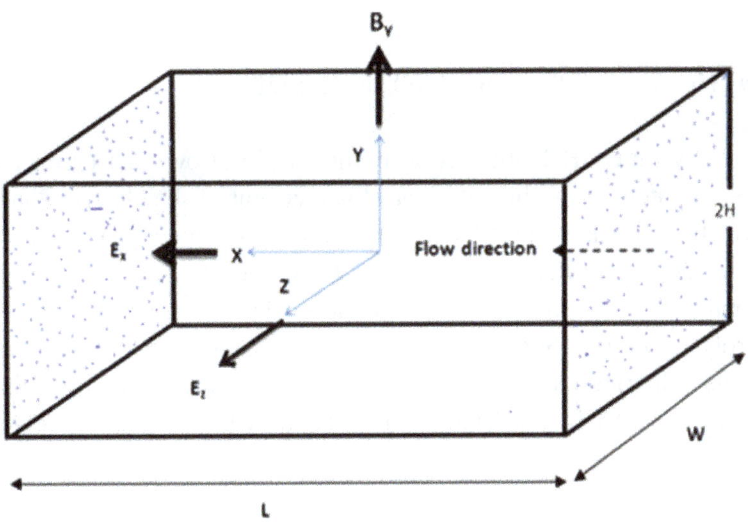

MATHEMATICAL MODEL AND PROBLEM FORMULATION

The hydrodynamic, pressure driven flow transport of a nanofluid through a rectangular porous microchannel is considered with prescribed dimension mentioned in Fig. 1. This study decorates the flow transport characteristics of electrically conducting fluid through a microchannel with porous medium, with a gap height of $2H$, width $W(2H \ll w)$ and length $L(2H \ll L)$, under constant pressure-gradient and combined electromagnetohydrodynamic effects. The electrical conductivity and density of the nanofluid are denoted by σ_{nf} and ρ_{nf}. In this study, the imposed pressure gradient and the applied electric field are acted along x-direction. The uniform magnetic field B_y is applied along the perpendicular direction of the axial direction and E_z is imposed as an external transverse electric field in z-direction. The microchannel is filled with a porous material with permeability K, in which the impermeable walls are at y=±H. Fully developed, steady flow is assumed in this investigation.

In this study, nanofluid is considered where blood is taken as a base fluid and Fe_3O_4 is taken as a nanoparticle. As per requirement of various bio-medical applications and lab-on-chip applications, this nanofluid is considered. The thermo-physical properties of the nanofluid are shown in the following Table 1.

Table 1. Thermo-physical properties of nanofluid (Abdalla et al., 2010; Sima et al., 2015)

Physical property	Base fluid (Blood)	Nano particles (Fe_3O_4)
$\rho(kg.m^{-3})$	1.08×10^{3}	5.2×10^{-3}
$\sigma(S.m^{-1})$	0.6	2.5×10^{4}
$\epsilon(F.m^{-1})$	2.03×10^{-7}	80

POTENTIAL DISTRIBUTION WITHIN EDL

In order to assess the EMHD influence on the microfluidic flow transport phenomenon, it may be enlightening to illustrate a brief antecedent of the electric double layer (EDL) effects, which is occurred near to the walls of the microchannel. Near to the walls of the microchannel, in physical contact with the electrolyte solution, develop surface charge due to ionization of the liquid or due to the absorption of ions of the electrolyte solution. Such kind of improvement of surface charge triggers a spontaneous redistribution of the ions, within the electrolyte solution, culminating in the formation of EDL. The EDL is constructed by two layers, one is stern layer and another is diffuse layer. The thickness of the EDL is typically of the order of few nm, depending on the concentration of the electrolyte solution.

The net charge density ρ_e inside the EDL of the microchannel flow is described from Poisson equation, (Malsliyah & Bhattacharjee, 2006) which is expressed as:

$$\nabla.(\nabla\psi) = -\frac{\rho_e}{\varepsilon_{nf}}, \tag{1}$$

where ε_{nf} is effective permittivity of the nanofluid, which is defined by the following mathematical expression: (Malsliyah & Bhattacharjee, 2006; Abdalla et al., 2010)

$$\frac{\varepsilon_{nf}}{\varepsilon_f} = 1 + \frac{2(\varepsilon_n - \varepsilon_f)\phi}{(\varepsilon_n + 2\varepsilon_f) - \phi(\varepsilon_n - \varepsilon_f)}, \tag{2}$$

where ε_n and ε_f are the permittivities of the nano particles and the base fluid respectively. Using the symmetry property of electrolytes (Malsliyah & Bhattacharjee, 2006), the electric charge density ρ_e is expressed by the following equation:

$$\rho_e = -2n_0 ez\sinh\left(\frac{ez\psi}{k_B T_{av}}\right), \tag{3}$$

where n_0 represents the ion density, k_B denotes the Boltzmann constant, z is the valance and the absolute temperature is T_{av}.

Since ψ is too much small, so $(ez\psi/k_B T_{av}) \ll 1$, the term $\sinh(ez\psi/k_B T_{av})$ can be estimated by $\left(\frac{ez\psi}{k_B T_{av}}\right)$. This principle is cognized as Dybye-Hückle linearization. After executing this Dybye-Hückle linearization, the linearized form of the Poisson equation is rewritten as:

$$\frac{\varepsilon_{nf}}{\varepsilon_f}\frac{d^2\psi}{dy^2} = \kappa^2\psi, \tag{4}$$

where $\kappa = ez\left(\frac{2n_0}{\varepsilon_f k_B T_{av}}\right)^{\frac{1}{2}}$ is Dybye-Hückle parameter and $\frac{1}{\kappa}$ be the thickness of the EDL and the appropriate boundary conditions are:

$$\psi=\zeta \text{ at } y=H,\ \frac{d\psi}{dy}=0 \text{ at } y=0. \tag{5}$$

At the channel wall, the zeta potential (ζ) is constant. To make dimensionless equation (4), $\psi^*=\psi/\zeta$, $y^*=y/H$, $x^*=x/H$ are employed as dimensionless variables and then the equation becomes:

$$\frac{\varepsilon_{nf}}{\varepsilon_f}\frac{d^2\psi^*}{dy^{*2}}=\omega^2\psi^*,\ \omega=\kappa H. \tag{6}$$

The appropriate boundary conditions for EDL potential are represented as:

$$\psi^*=1 \text{ at } y^*=1,\ \frac{d\psi^*}{dy^*}=0 \text{ at } y^*=0. \tag{7}$$

By solving equation (6) subjected to the boundary condition (7), the solution of non-dimensional EDL potential ψ^* is derived by

$$\psi^*=\frac{\cosh\left(\frac{\omega}{\varepsilon}y^*\right)}{\cosh\left(\frac{\omega}{\varepsilon}\right)},\ \text{where } \varepsilon^2=\frac{\varepsilon_{nf}}{\varepsilon_f}. \tag{8}$$

VELOCITY DISTRIBUTION

On the basis of the combined influence of electromagnetic force and constant pressure gradient, the governing equation of momentum for nanofluid flow in porous microchannel can be expressed as

$$-\nabla p + \mu_{nf}\nabla^2\vec{U} - \frac{\mu_{nf}}{K}\vec{U} + \vec{b} = 0\,, \tag{9}$$

where p be the pressure, K be the permeability constant of the porous material and μ_{nf} is the effective density of the nanfluid, which can be written as: (Xuan & Li, 2003; Yu & Chio, 2003)

$$\mu_{nf} = \frac{\mu_f}{(1-\phi)^{2.5}}\,, \tag{10}$$

The body force $\vec{b}$ per unit volume due to the combined EMHD effect is expressed by:

$$\vec{b} = \rho_e\vec{E} + \vec{F}\,, \tag{11}$$

$\vec{E}$ is the imposed electric field and $\vec{F}$ is Lorentz force which is generated due to the presence of magnetic field. Here $\vec{F} = \vec{J}\times\vec{B}$, Where $\vec{J}$ be the current density and $\vec{B}$ is the magnetic field. By considering the effect of Hall current, the current density $\vec{J}$ may be written as: (Sutton & Sherman, 1965)

$$\vec{J} + \frac{\sigma_{nf}}{en_e}\left(\vec{J}\times\vec{B}\right) = \sigma_{nf}\left(\vec{E} + \vec{u}\times\vec{B}\right), \tag{12}$$

where e is denoted by the charge of the electron, n_e be the number density of the free electron and σ_{nf} be the effective electric conductivity of nanofluid, which can be defined by the following mathematical expression (Xuan & Li, 2003; Yu & Chio, 2003):

$$\frac{\sigma_{nf}}{\sigma_f} = 1 + \frac{3(\sigma_n - \sigma_f)\phi}{(\sigma_n + 2\sigma_f) - (\sigma_n - \sigma_f)\phi}\,, \tag{13}$$

where σ_n and σ_f represent the electrical conductivities of the nanoparticle and base fluid respectively.

For the fully developed flow, having the electromagnetic field and the constant pressure gradient inside the EDL, the equation of momentum conservation (9) along x-direction is expressed as

$$\frac{\partial p}{\partial x}+\frac{\mu_{nf}}{K}u-\mu_{nf}\frac{d^2u}{dy^2}-\rho_e E_x+\frac{\sigma_{nf}B_y^2u}{1+m_{nf}^2}-\frac{\sigma_{nf}B_yE_z}{1+m_{nf}^2}=0, \tag{14}$$

E_x is the axial component and E_z is the lateral transverse component of applied electrical force and B_y is the magnetic field component along y-direction. The effective Hall parameter for nanofluid is represented by $m_{nf}=\sigma_{nf}B_y/(en_e)$. Due to the narrow confinement of the microchannel, only one momentum equation is considered in the x-direction. For physiological flow of blood, the no-slip boundary condition is not valid. Hence, the boundary conditions are derived as follows:

$$\frac{du}{dy}=0 \;\; at \;\; y=0, \tag{15}$$

$$u+\beta_s^*\frac{du}{dy}=0 \;\; at \;\; y=H. \tag{16}$$

The boundary condition is well known as Navier slip boundary conditions, where β_s^* is called slip length. If $\beta_s^*=0$, then the Navier slip boundary condition reduces to no-slip boundary condition.

Assuming that pressure gradient is fixed for a fully developed flow. By applying non-dimensional transformations, the dimensionless momentum equation (14) is reduced as follows:

$$\frac{\mu_{nf}}{\mu_f}\frac{d^2u^*}{dy^{*2}}-\left\{\frac{\sigma_{nf}}{\sigma_f}\left(\frac{Ha^2}{1+m_{nf}^2}\right)+\frac{\mu_{nf}}{\mu_f}\frac{1}{Da}\right\}u^*+\omega^2\psi^*+\frac{\sigma_{nf}}{\sigma_f}\left(\frac{1}{1+m_{nf}^2}\right)HaS=P, \tag{17}$$

where

$$u^* = \frac{u}{U_{HS}}, P = \frac{\left(-\frac{dp}{dx}\right)H^2}{\mu_f U_{HS}}, Da = \frac{K}{H^2}, \beta_s = \frac{\beta_s^*}{H},$$

$$Ha = B_y H\sqrt{\frac{\sigma_f}{\mu_f}}, U_{HS} = -\frac{\varepsilon_f \zeta E_x}{\mu}, S = \frac{E_z H}{U_{HS}}\sqrt{\frac{\sigma_f}{\mu_f}},$$

$m=(\sigma_f B_y)/(en_e)$ is called the Hall parameter. Ha, β_s and Da are Hartmann number, slip parameter and Darcy number respectively. The dimensionless boundary conditions are rewritten as:

$$\frac{du^*}{dy^*} = 0 \text{ at } y^*=0, \tag{18}$$

$$u^* + \beta_s \frac{du^*}{dy^*} = 0 \text{ at } y^*=1 \tag{19}$$

Equation (17) is solved subjected to the proper boundary conditions (18) and (19) to obtain solutions of dimensionless velocity, which is displayed as

$$u^* = \gamma\left(e^{\beta y^*} + e^{-\beta y^*}\right) + \frac{\varepsilon^2\omega^2\cosh\left(\omega y^*/\varepsilon\right)}{\left(\delta_a\varepsilon^2 - \omega^2\delta_\mu\right)\cosh\left(\omega/\varepsilon\right)} + \frac{\left(\delta_\sigma\delta_m HaS - P\right)}{\delta_a}, \tag{20}$$

where,

$$\gamma = \left\{ \frac{\omega^2\delta_\mu\left(P - \delta_\sigma\delta_m HaS\right) + \delta_a\varepsilon^2\left(\omega^2 - P + \delta_\sigma\delta_m HaS\right) - \delta_a\varepsilon\,\omega^3\beta_s\tanh\left(\frac{\omega}{\varepsilon}\right)}{\delta_a\left(\delta_a\varepsilon^2 - \omega^2\delta_\mu\right)\left\{\left(e^\beta + e^{-\beta}\right) - \beta_s\beta\left(e^\beta - e^{-\beta}\right)\right\}} \right\}$$

$$\delta_a = \left\{ \delta_\sigma \left(\frac{Ha^2}{1+m_{nf}^2} \right) + \delta_\mu \frac{1}{Da} \right\},\ \delta_\mu = \frac{\mu_{nf}}{\mu_f},\ \delta_\sigma = \frac{\sigma_{nf}}{\sigma_f},\ \delta_m = \left(\frac{m_{nf}^2}{1+m_{nf}^2} \right),\ \beta = \sqrt{\frac{\delta_a}{\delta_\mu}}.$$

SKIN FRICTION COEFFICIENT

In any viscous flow through a solid boundary walls, the flow velocity will be experienced the wall share stress. This wall share stress at the wall is called the skin friction. In our flow problem, the skin friction co-efficient of nanofluid flow on the microchannel surface is expressed as:

$$C_f = -\frac{\mu_{nf}}{\rho_{nf} U_{HS}^2} \left(\frac{du}{dy} \right)_{y=H} = -\left(\frac{\mu_{nf}}{\mu_f} \right) \left(\frac{\rho_f}{\rho_{nf}} \right) \left(\frac{1}{Re} \right) \left(\frac{du^*}{dy^*} \right)_{y=1}, \tag{21}$$

where *Re* is called the Reynolds number.

Figure 2. Comparison of the present solution of velocity distribution with (a) Result of Chakraborty et al. (Chakraborty et al., 2013) ω=4, S=50, Ha=0.5, P=1, β_s=0, (b) Result of Reza et al. (Reza et al., 2020a) ω=4, S=50, Ha=0.5, P=1, β_s=0.01

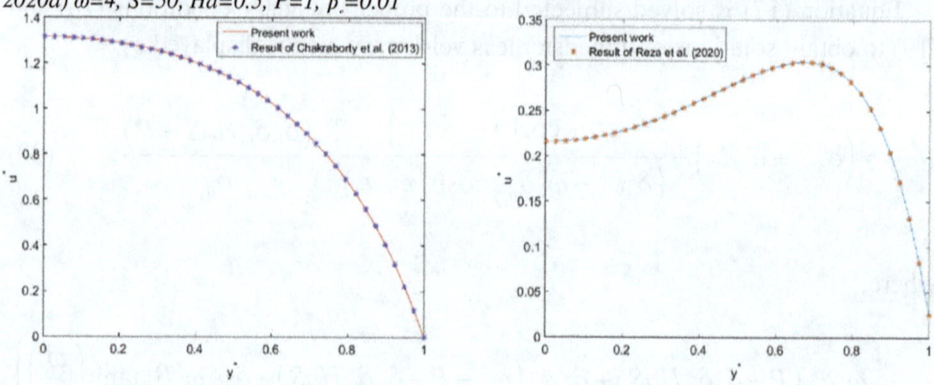

RESULTS AND DISCUSSIONS

In this article, the EMHD flow transport of nanofluid through the porous microchannel with interfacial slip effect has been examined. The equations (17) subject to the boundary conditions (18) and (19) are solved analytically. It is observed that EMHD flow characteristics represented by non-dimensional flow velocity (u^*) and skin friction coefficient (C_f) depend on the dimensionless physical parameters the Hall parameter (m), porosity parameter (Darcy Number) (Da), applied magnetic field (Hartman Number) (Ha), transverse electric field (S) and naoparticle volume friction parameter (ϕ). To examine and analyze the impact of Hall parameter on the electromagnetically driven flow, the following typical parametric values are used (Chakraborty et al., 2013; Reza et al., 2020a; Buren et al., 2014): $O(B_y)$~1–50T; $O(E_x)$~0–$10^4 V/m$; $O(E_z)$~0–$10^4 V/m$; $O(p)$~10Pa/m.

Figure 3. (a) Impact of Hartmann number on velocity distribution, (b) Impact of Darcy number on velocity distribution.

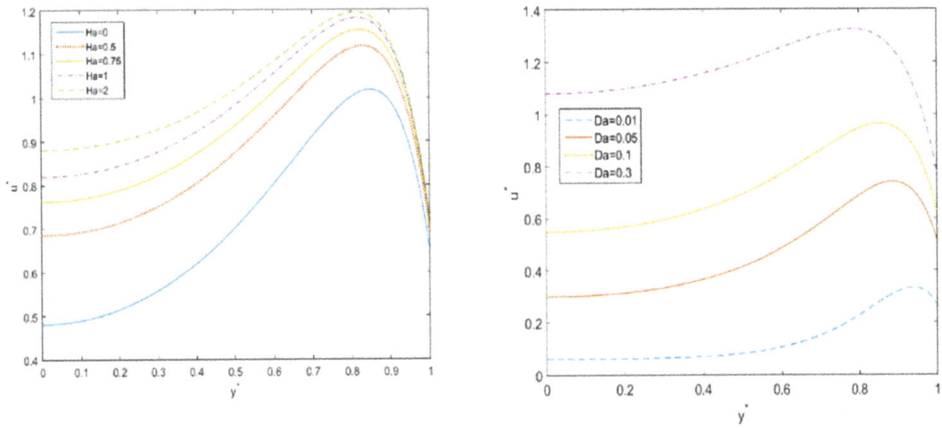

To validate the present computed results, we compare the velocity profile with the results of Chakraborty et al. (Chakraborty et al., 2013) and Reza et al. (Reza et al., 2020a). We have set m=0, β_s=0, ϕ=0, and Da→∞ and other parameters are indicated in the figure caption of Figure-2(a) so that our work is coincide with the work of Chakraborty et al. (Chakraborty et al., 2013). They analyzed the EMHD flow transport phenomenon with no-slip boundary condition. It can be observed from Fig. 2(a) that the present results for the velocity distribution show an excellent agreement with the analytical solutions by Chakraborty et al. (Chakraborty et al., 2013), when the slip parameter, Hall parameter and nano-particles are ignored without taking the porous medium. Similarly, by setting m=0 and ϕ=0 the results will compared with Reza et al. (Reza et al., 2020a). Further, from Figure 2(b) depicted that the results of velocity distribution is also agreed with the results of Reza et al. (Reza et al., 2020a), when the Hall parameter and nano-particles are ignored.

Figure 4. (a) Impact of Hall parameter on velocity distribution, (b) Impact of nanoparticle volume friction on velocity distribution.

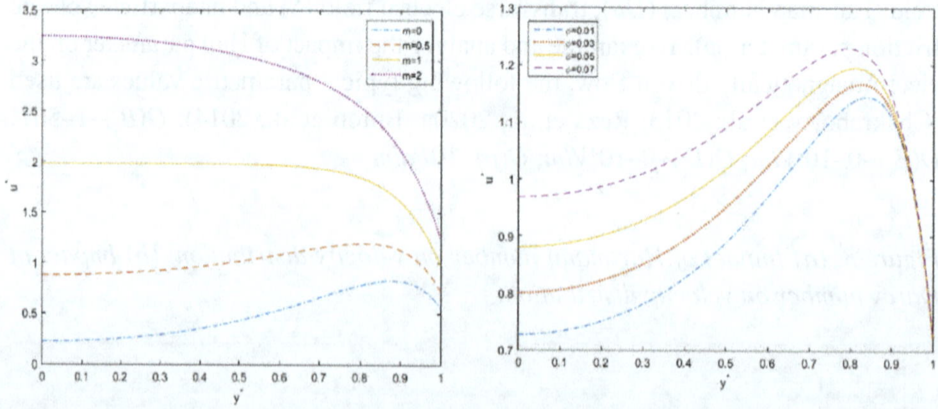

Figure 5. Impact of slip parameter on (a) velocity distribution, (b) skin friction.

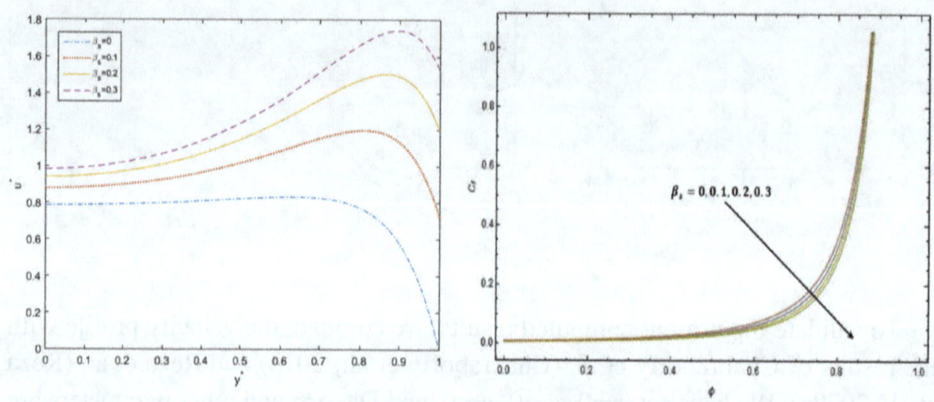

This investigation deals the analytical solutions of the velocity distribution of nanofluid where the base fluid is blood and Al_2O_3 is used as nanoparticle under the influence of combined EMHD effects through a porous microchannel. In this present segment, we discussed the effects of various dimensionless parameters on nanofluid flow transport through the porous microchannel. The interactive influences of different non-dimensional parameters on the flow transport phenomenon are examined. The impacts of Hall current, electromagnetic effect, the influence of Navier slip parameter and the effect of porous medium are elaborated on nanofluid flow performance. The graphical interpretation of velocity profiles and skin friction coefficients are displayed from Figure 3-6 with the dimensionless parameter values $\omega=10$, $S=10$, $Da=0.2$, $P=0.5$, $\phi=0.05$, $m=0.5$.

The non-dimensional velocity profiles for several values of Ha (Hartmann number) is shown in Figure 3. In the appearance of the lateral electric field (i.e. S=10), it gives an increasing velocity profile with an increase of Ha which is shown in the Figure 3(a). It can be noted that the nanofluidic flow does not occur fully developed with the variation of applied magnetic field. It can be observed that from equation (8) the electromagnetic force is a linear combination of the transverse electric field (E_z) and the square of the magnetic field (B_y). Thus, in this case, it consists two adding forces acting on the fluid flow motion, the component of pressure gradient and the component of electromagnetic force. Hence, the vicinity of the lateral electric field (S) i.e. because of the flow adding force the fluid velocity is enhanced. Figure 3(b) depicts the velocity profiles with the variations of different values of Darcy number, which indicates the strength of the porous medium permeability. Increase in porous medium permeability signifies a rise in the magnitude of the fluid velocity. It is noteworthy that due to the increment of the permeability, more fluid is passed through the porous medium and therefore the velocity is increased.

Figure 4(a) shows that the increment of the Hall parameter is enhanced the fluid velocity. The enhancement of Hall parameter reduces the effective electrical conductivity of ferromagnetic nanoparticles, which are mixed with blood (base fluid). Therefore, the damping force effect of the applied magnetic field reduces and as a result, fluid velocity is enhanced. It is very interesting to know that for large value of Hall parameter, the velocity profile becomes fully developed and it is noteworthy that the velocity is also maximum in the middle layers of the flow. Nevertheless, in the presence of transverse electric field the velocity profile has an enhancement with the increase values of the nanoparticle volume friction parameter, which is shown in Figure 4(b). In this analysis, it can be observed that the fluid velocity has an increasing trend with the increment of slip velocity parameter; corresponding figure has been presented in Figure 5(a). However, the interfacial velocity slip does not have much impact on velocity in the middle layers of the flow transport but there is a significant effect is visible near to the wall of the microchannel.

The variation of skin friction (C_f) with nanoparticle volume friction (ϕ) for different values of slip parameter (β_s) is displayed in Figure 5(b). The skin friction co-efficient increases with the increases of volume friction parameter but after a certain value of (ϕ), it increases rapidly. The enhancement of slip parameter decreases the skin friction. The diversity in skin friction for a comprehensive range of Darcy number with the variation of other dimensionless parameters are depicted in Figure 6. It is noticeable that skin friction is enhanced rapidly with the enhancement of Darcy number (Da). Due to the presence of applied magnetic field, the skin friction is increased with increases of Darcy number, which is shown in Figure 6(a). Figure

6(b) delineates the impression of Hall parameter on the skin friction and it can be noticeable that when the Hall parameter increases then the skin friction coefficient increases rapidly. Increment of volume friction parameter enhanced the skin friction with Darcy number, which is represented in Figure 6(c). The variation of skin friction (C_f) with Darcy number (Da) for different values of slip parameter (β_s) is displayed in Figure 6(d). The enhancement of slip parameter reduces the skin friction and it becomes maximum for no-slip boundary condition.

Figure 6. Impact of Darcy number on skin friction with (a) Variation of hartmann number, (b) variation of Hall parameter, (c) variation of nanoparticle volume friction, (d) variation of slip parameter.

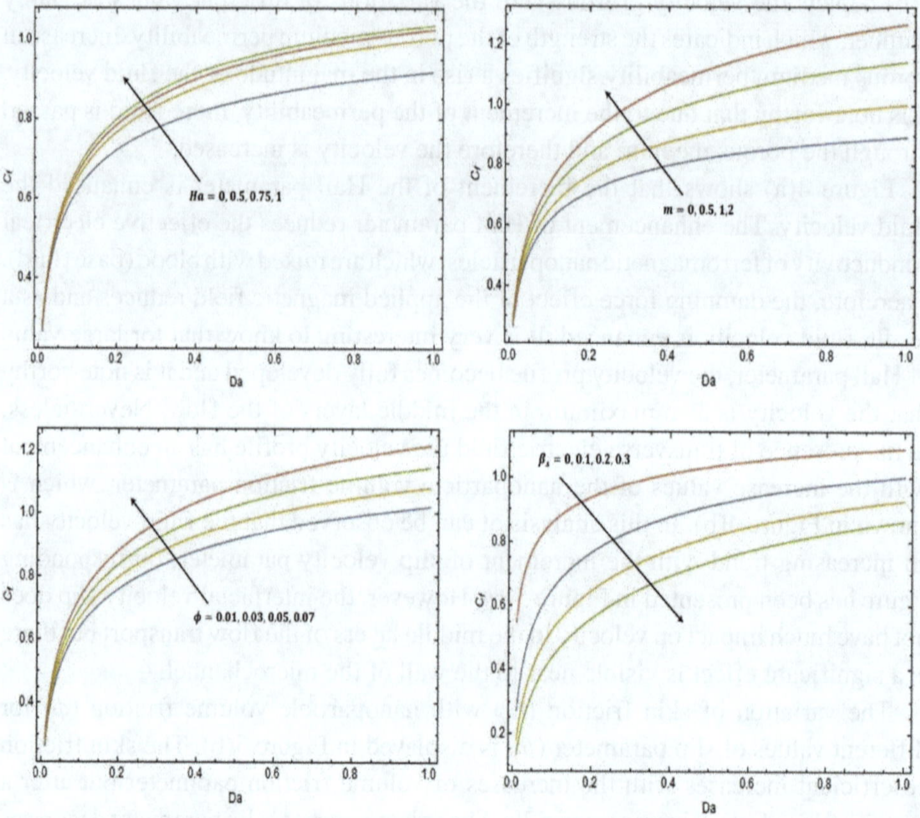

CONCLUSION

In this theoretical study, velocity distribution and skin friction co-efficient is investigated analytically for the nanofluid flow through a porous microchannel with EMHD effects. The computational work deals with the following observations:

- The flow is not a fully developed flow, but it is fully developed for no slip boundary condition and large value of Hall parameter.
- The flow is impressed by the Navier slip boundary conditions and the velocity becomes maximum near to the boundary.
- As a consequence of Hall current effect, the velocity and the skin friction both are enhanced.
- The flow through porous microchannel augments with the enhancement of Darcy number.
- Nanofluidic flow velocity is improved with the increase in volume fraction of nanoparticle.

The presented results can be considered as the approximate estimation of blood flow through capillaries. The study deals with the important applications in bio-medical areas, drug delivery, clinical applications and cancer treatment during chemotherapy.

ACKNOWLEDGMENT

This work was supported by SERB, Govt of India. (Grant File No. EMR/2016/006383). The authors would like to acknowledge this support.

REFERENCES

Abdalla, S., Al-Ameer, S. S., & Al-Magaishi, S. H. (2010). Electrical properties with relaxation through human blood. *Biomicrofluidics*, *4*, 034101.

Azari, M., Sadeghi, A., & Chakraborty, S. (2020). Electroosmotic flow and heat transfer in a heterogeneous circular microchannel. *Applied Mathematical Modelling*. doi:10.1016/j.apm.2020.06.022

Buren, M., Jian, Y., & Chang, L. (2014). Electromagnetohydrodynamic flow through a microparallel channel with corrugated walls. *Journal of Physics. D, Applied Physics*, *47*, 425–501.

Chakraborty, R., Dey, R., & Chakraborty, S. (2013). Thermal characteristics of electromagnetohydrodynamic flows in narrow channels with viscous dissipation and Joule heating under constant wall heat flux. *International Journal of Heat and Mass Transfer*, *67*, 1151–1162.

Chakraborty, S., & Paul, D. (2006). Microchannel flow control through a combined electromagnetohydrodynamic transport. *Journal of Physics. D, Applied Physics*, *39*, 5364–5371.

Das, S. K., Dao, T. P., & Perumal, T. (Eds.). (2021a). *Nature-Inspired Computing for Smart Application Design*. Springer Singapore.

Das, S. K., Das, S. P., Dey, N., & Hassanien, A. E. (Eds.). (2021c). *Machine learning algorithms for industrial applications*. Springer.

Das, S. K., Sadasmanta, S., Dey, N., & Kumar, R. (Eds.). (2020). *Design frameworks for wireless networks*. Springer Singapore.

Das, S. K., Samanta, S., Dey, N., Patel, B. S., & Hassanien, A. E. (Eds.). (2021b). *Architectural Wireless Networks Solutions and Security Issues*. Springer Singapore Pte. Limited.

Das, S. K., Tripathi, S., & Burnwal, A. P. (2015). Intelligent energy competency multipath routing in wanet. In *Information systems design and intelligent applications* (pp. 535–543). Springer.

De, D., Mukherjee, A., Das, S. K., & Dey, N. (Eds.). (2020). *Nature inspired computing for wireless sensor networks*. Springer.

Jang, J., & Lee, S. S. (2000). Theoretical and experimental study of MHD (magnetohydrodynamic) micropump. *Sensors and Actuators. A, Physical*, *80*, 84–89.

Laster, D. J., & Santiago, J. G. (2004). A review of micropumps. *Journal of Micromechanics and Microengineering*, *14*, R35–R64.

Liu, M., & Yang, J. (2009). Electrokinetic effect of the endothelial glycocalyx layer on two-phase blood flow in small blood vessels. *Microvascular Research*, *78*(1), 14–19. doi:10.1016/j.mvr.2009.04.002

Liu, Y., & Jian, Y. (2019). Electroviscous effect on electromagnetohydrodynamic flows of maxwell fluids in parallel plate microchannels. *Applied Mathematics and Mechanics*, *40*, 1457–1470.

Liu, Y. Jian, Y., & Tan, W. (2018). Entropy generation of electromagnetohydrodynamic (EMHD) flow in a curved rectangular microchannel. *International Journal of Heat and Mass Transfer, 127*(A), 901-913.

Masliyah, J. H., & Bhattacharjee, S. (2006). Electrokinetic and Colloid Transport Phenomena. Wiley–Interscience. https://doi.org/10.1002/cjce.5450840612.

Rana, A., Reza, M., & Shit, G. C. (2022). Electromagnetohydrodynamic thermo-fluidic transport in a porous microchannel with wall roughness. *Colloids and Surfaces. A, Physicochemical and Engineering Aspects, 650*, 129336.

Rashid, M., Shahzadi, I., & Nadeem, S. (2018). Corrugated walls analysis in microchannels through porous medium under Electromagnetohydrodynamic (EMHD) effects. *Results in Physics, 9*, 171–182.

Reza, M., & Rana, A. (2020). *Analysis of Exact Solutions of Electromagnetohydrodynamic Flow and Heat Transfer of Non-Newtonian Casson Fluid in Microchannel with Viscous Dissipation and Joule Heating*. Advances in Fluid Mechanics and Solid Mechanics. doi:10.1007/978-981-15-0772-4_11

Reza, M., Rana, A., & Patra, R. (2020a). Thermal Characteristics of a casson fluid in microchannel through porous medium with an interfacial slip with electromagnetohydrodynamic effect. *ISHMT Digital Library, 2019*, 959–964. doi:10.1615/IHMTC-2019.1610

Reza, M., Rana, A., & Patra, R. (2020b). Thermal performance for electromagnetohydrodynamic flow of non-Newtonian Casson fluid through porous microtube. *IACSIT International Journal of Engineering and Technology, 9*(3), 804–811.

Reza, M., Rana, A., & Shit, G. C. (2021). Thermo-fluidic transport of electromagnetohydrodynamic flow in a corrugated porous medium microchannel. *The European Physical Journal Plus, 136*(5), 496.

Sarkar, S. Ganguly, S., & Chakraborty, S. (2017). Influence of combined electromagnetohydrodynamics on microchannel flow with electrokinetic effect and interfacial slip. *Microfluid. Nanofluid, 21*(3), 1-16.

Sarkar, S., Ganguly, S., & Dutta, P. (2017). Electrokinetically induced thermofluidic transport of power-law fluids under the influence of superimposed magnetic field. *Chemical Engineering Science, 171*, 391–403.

Sima, W., Shi, J., Yang, Q., Huang, S., & Cao, X. (2015). Effects of conductivity and permittivity of nanoparticle on transformer oil insulation performance: Experiment and theory. *IEEE Transactions on Dielectrics and Electrical Insulation*, *22*(1), 380–390.

Stone, H. A., Stroock, A. D., & Ajdari, A. (2004). Engineering flows in small devices microfluidics toward a lab-on-a-chip. *Annual Review of Fluid Mechanics*, *36*(1), 381–411. doi:10.1146/annurev.fluid.36.050802.122124

Sutton, G. W., & Sherman, A. (1965). *Engineering Magnetohydrodynamics*. Mc Graw-Hill.

Vargas, C., Arcos, J., Bautista, O., & Mendez, F. (2017). Hydrodynamic dispersion in a combined magnetohydrodynamic-electroosmotic driven flow through a microchannel with slowly varying wall zeta potentials. *Physics of Fluids*, *29*, 092002.

Wang, L., Jian, Y., Liu, Q., Li, F., & Chang, L. (2016). Electromagnetohydrodynamic flow and heat transfer of third grade fluids between two micro-parallel plates. *Colloids and Surfaces. A, Physicochemical and Engineering Aspects*, *494*, 87–97.

West, J., Karamata, B., Lillis, B., Gleeson, J. P., & Alderman, J. K. (2002). Application of magnetohydrodynamic actuation to continuous flow chemistry. *Lab on a Chip*, *2*, 224–230.

Weston, M. C., Gerner, M. D., & Fritsch, I. (2010). Magnetic Fields for Fluid Motion. *Analytical Chemistry*, *82*, 3411–3418.

Xuan, Y., & Li, Q. (2003). Investigation on convective heat transfer and flow features of nanofluids. *ASME Journal of Heat Transfer*, *125*, 151–155.

Yu, W., & Chio, S. U. S. (2003). The role of interfacial layers in the enhanced thermal conductivity of nanofluids: A renovated Maxwell model. *Journal of Nanoparticle Research*, *5*, 167–171.

Zhao, G., Jian, Y., & Li, F. (2016b). Streaming potential and heat transfer of nanofluids in microchannels in the presence of magnetic field. *Journal of Magnetism and Magnetic Materials*, *407*, 75–82.

Zhao, G. Y., & Jian, F. L. (2016a). Heat transfer of nanofluids in microtubes under the effects of streaming potential. *Applied Thermal Engineering*, *100*, 1299–1307.

Chapter 6
Enhanced Belief Function-Based Decision Blending for Detecting Fault in Wireless Sensor Networks

Bhabani Sankar Gouda
National Institute of Science and Technology, Berhampur, India

Sudhakar Das
National Institute of Science and Technology, Berhampur, India

Ruchika Padhi
National Institute of Science and Technology, Berhampur, India

Debendra Muduli
C. V. Raman Global University, Bhubaneswar, India

ABSTRACT

In wireless sensor networks (WSN), various software and hardware issues can lead to various fault types. The issue can be found using many forms of fault detection. The diverse obstacles determine the distinct fault kinds and need to find out effective fault detection and problem-solving are required. This chapter discusses four main types of faults: gain fault, offset fault, stuck-at fault. In this work, the authors use the notion of decision blending to categorize the blending outcomes and to assess the accuracy in order to save energy and make better use of the available bandwidth for data transmission. Three performances are assessed by the decision blending function: detection accuracy (DA), sensitivity, and rate of error. Different methods, such as k-nearest neighbor (KNN), enhanced extreme learning machine (EELM), enhanced support vector machine (ESVM), and enhanced recurrent extreme learning machine, are used in the belief function approach (ERELM). Here, the authors applied decision blending approaches in WSNs to emulate these techniques for improving belief function.

DOI: 10.4018/978-1-6684-7343-6.ch006

INTRODUCTION

Wireless Sensor Networks (WSNs) collect different types of data from a different environment. WSN has a different topology consisting of any number of sensor nodes deployed as different patterns like random, static, dynamic, dense or sparse. WSNs are implemented in different fields according to the rapid development field. WSN is used in data science fields, where a large amount of data is processed. Machine learning algorithms implemented for categories collected large data from sensor nodes. Ensemble learning algorithms are used to aggregate the predicted data of all sensor nodes. This aggregation concept is known as decision fusion, minimizing the wrong model selection by implementing multiclass classification problems (El Hindi et al., 2018). Sensor nodes in WSN are communicated with others directly or indirectly. The head node merges the collected data and decisions are made accordingly. The head sensor node must be at the center position of the network otherwise at decentralized case data accuracy is a problem. Local sensor nodes cannot deal with data accuracy problems. Therefore, fusion rules are implemented to fuse the reliable data. To analyze the local sensor data detection and the false alarm to confirm the decision fusion rule proposed for the data accuracy (Sriranga et al., 2018). In (Samet, Lefevre, Yahia, 2014) belief function fusion approaches an estimation classifier used to define the forest image on a classification method to categorize problems of images and overcome methods. Here to find distance between targets distance estimation model classifier used. The collected data from the sensor node is sent to the center head node in binary form. For accurate data all the data compare with threshold value set by the system and reach at the final decision implementing optimal decision. Optical fusion rules are used to decode and forward relaying for different channels. There are different techniques based on intuitionist algorithms i.e., data based IFS and weight based TOPSIS intuitionist decision algorithm on fuzzy. Fusion rule used to find the amount of required energy and working condition based on statistical channel information and on-off keying (Zhang et al., 2017). Here the Gaussian mixture is used for multi-input-output decision fusion based on energy detection over different classes of fading distribution methods. Multiple classifiers and techniques are used to evaluate the problems related to energy dissipation. Deep learning algorithms and classifiers are used here. Initially the decision tree, Fourier based, was used as a basic classifier to perform on data loss, aggregation error, calibration fault in WSN (Das et al., 2021; De et al., 2020). There are different types of faults in WSN which can be detected using machine learning techniques. Types of faults defined according to sensing data.

(a). **Offset fault:** Calibration's error and actual data added in expected data at sensing unity called offset fault.

(b). **Gain fault:** Sensed data rate varies in specific time known as gain fault.
(c). **Stuck-at fault:** Null valued data found at sensing and that not changed with time known as stuck at fault.
(d). **Out of bounds:** Data value will be different and that out of boundary from normal value.

According to different types of faults, the machine learning algorithm applied to improve accuracy in the fusion of belief functions. Complex Basic Belief Assignment (BBA) approach used to upload the BBA structure to the fusion center using each sensor node. Here used a different method to get a better result in fault detection i.e., Enhanced Support Vector Machine (ESVM), Enhanced K-Nearest Neighbor (EKNN), Enhanced Extreme Learning Machine (EELM) and Enhanced Recurrent Extreme Learning Machine (ERELM) (Das et al., 2021).

RELATED WORKS

In this section some literature review is discussed based on two techniques: first is fault detection and another is machine learning based fault detection system in WSN.

Fault Detection in WSN

Fault in WSN is many types according to data sensed from the network like data thrashing, aggregation blunder and calibration liability. Battery used in nodes also causes error which is a more important factor. This failure leads to a malfunctioning of sensors in the whole network. Fault in WSN occurs due to data simultaneously or separately. It can happen continuously over a period of time or within instantly. The control on fault, deals with fault are more difficult to deal with simultaneous and instant occurrence of fault. Some random faults occur in the network. At this condition faults or failures need to be discovered. The detection of faults in WSN can be a demanding issue because of different characteristics of nodes after deployment in the network (Rossi et al., 2014). Different Fault detection methods can be proposed as centralized approach, distributed and hybrid network based on statics approach, neighbors, self-detection, and machine learning approaches. Machine learning seems to be the most efficient way for detecting different faults. As data mining technique categorization is the most sufficient and that used technique for decision making assistance, which is implemented in automatic diagnosis. For decision making different algorithms are categorized for data. Data learning algorithms allowed learning automatically and recognizing complex structure for different intelligent decisions. There are three classes of learning techniques:

(a). **Supervised Learning:** The class of data is predetermined and examples are known. The classification is based on labeled data.
(b). **Unsupervised Learning:** It processed non labeled data. It classifies a set of objects without having advanced expertise. It is also called clustering.
(c). **Semi-supervised Learning:** Both labeled and unlabeled data learned.

Machine Learning Approaches Implemented for Fault Detection

There are many different researches in the field of machine learning with different algorithms in WSN (Zidi et al., 2017). Convolution Neural Network (CNN) classifiers used as data organization method. The data classifier is tested on data accuracy. CNN classifier performance is tested by 10-fold cross validation method. There are two classifiers compared for better performance i.e., SVM classifier and CNN classifier. WSN has lots of applications like environment monitoring, home automation, underwater acoustics, ship hull monitoring and biomedical diagnosis (Zhang & Zhang, 2015). Artificial intelligence is the base of machine learning techniques used for computational task not for conventional task (Jiao et al., 2018). Different ranges of dataset tested for accuracy of linear regression, Naïve Bayes (NB) and Stochastic Gradient Descent (SGD) (Gupta et al., 2017). Das et al. (2020) proposed a framework application for handling several operations of the network. It helps to manage several designed models of the network. Each of the models is based on some factors of application. It indicates solution prospective point of view operation in the area of wireless network. It helps to guide several operation and model of the network that helps to incorporate the system in smart application. ECG signal processing and SVM were combined to classify the data. Principal Component Analysis (PCA), Artificial Neural Network (ANN), a knowledge based system; KNN and SVM are used with different parameters such as ECG and HRV for classification (Venkatesan et al., 2018; De et al., 2017). Advanced Extreme Learning Machine (AELM) technique proposed for heterogeneous ensemble approach. AELM is based on the regularized ELM, L2 norm optimized ELM (ELML2) and kernel-ELM (Abuassba et al., 2017). The decision boundary is identified using WMV. Multilayer Perceptron (MLP) used to predict wireless sensor outputs (Omairi & Ismail, 2016). In Structured MLP, the first layer consisted of only one node that received inputs from all components. All layer's input and output were connected. MLP contains special learning algorithms which produce different things as weights for different situations. In this paper two models were constructed using MLP. The models used for as in emergent a monitor framework for streaming data analysis in deep learning. In WSN data are not accurate for many reasons like incomplete and outliers. According to these reasons, if data get accurate than result also be accurate implementing deep learning. In (Pan & Li, 2010) the proposed method maintains accuracy and makes queries to get accurate results. If some absent data is found then it satisfies the time constraints. S. K. Das (2021) designed a review article for handling several challenges of smart application. The work also mentioned several techniques and issues modeling for factor analysis based on smart application system. It helps to deals several issues and application that are used in smart application design. It also helps to model several parameters that are directly or indirectly influence in the application. In (Jan et al., 2017) four type faults are named as erratic, drift, hard-over, spike and stuck faults. In (Al-Jarrah et al., 2018; Niu et al., 2004) decision fusion is implemented in distributed co-operation with a large sensor network. Deep learning and belief functions are used for classification of surgical site infections, forest images and real time data and optimized using SVM, NB and neural networks (Samet et al., 2014; Strauman et al., 2018; Lee et al., 2017; Swain & Khilar, 2017; Swain & Khilar, 2017; Yuan et al., 2018). In this paper, aim is to improve the decision fusion approach by apply four different machine learning classifiers i.e., SVM, KNN, RELM, ELM with the enhanced versions i.e., ESVM, EKNN, ERELM, EELM. De et al. (2020) designed a survey model for wireless sensor network for application management. Each of the work is based on several algorithms and challenges system. But it helps to resolve several issues based of factors of information. Several solutions are mentioned in this proposed to resolve the challenges based on framework analysis of wireless sensor network. Binh and Dey (2018) designed soft computing review system based on wireless sensor network and its application. The work is fully based on several techniques of soft computing technique that helps to model its related application. Each of the work helps to model and resolve the issues that are raised during the operation of wireless sensor network. Hence, there are several works have been proposed for the purpose of fault diagnosis and its modeling. Das and Giacalone (2022) proposed a book for the purpose of fuzzy logic and its related usage in the areas of engineering and management. It contains several related works that helps to deals with several types of issues and problems that helps in the real-life applications.

SYSTEM MODEL

At first the sensor senses the target and the sensor classifies the target as different classes and the output is called local decision (u). The output proceeds according to Belief function theory. In output of targets has many uncertainty data found and the probability of support to the function discussed below. Belief function theory involve three steps i.e., BBA construction, BBA combination, decision making.

Figure 1. System model of belief function

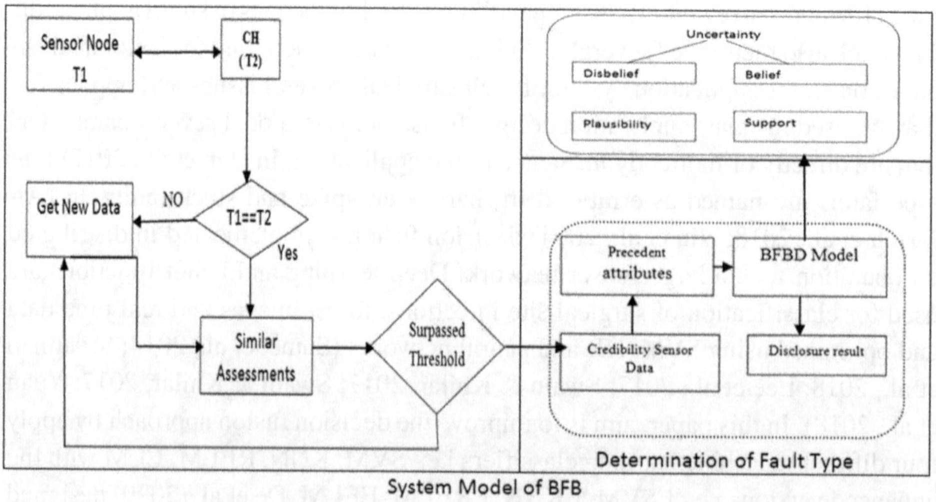

Belief Function Theory

The other name is Dempster-Shafer evidence theory which gives multisource information and decentralized targets for solutions. This approach fuses different results from sources to get a final decision. It reduces the burden of dealing with uncertainty. It includes two slots: mass construction and BBA combination.

BBA Construction

For mass construction a frame of judgment, it includes the non-empty finite set with mutually exclusive. We assumed Ω = S1, ..., Sn is the frame of judgment, and its power set is 2^{Ω}. The mass function is 2^{Ω} [0, 1], which satisfies the following equation (Rossi et al., 2014).

$$\sum_{S\subset 2^{\Omega}} m(S) = 1 \; m(\phi) = 0 \tag{1}$$

where S is a subset of 2^{Ω} and m(S) is a basic belief task that shows the rank of the subset S. There are two metrics that define the reliability of the hypothesis S, which are given below (Rossi et al., 2014).

$$B(S) = \sum_{S_i\subset S} m(Si) \tag{2}$$

$$P(S) = \sum_{S_i \cap S \neq \phi} m(Si) \tag{3}$$

B(S) indicates belief. For hypothesis S of the evidence, the support degree is B(S); plausibility P(S) is the degree that is not in conflict with hypothesis S for the evidence. It is constructed based on the space interval between the target and each sample of the class. BBA construction obtained the probability vector from the confusion matrix. In local decision u_i, probability vector is $P_i(u_i| w_k)$ is the condition probability of the class label w_k. If probability is more than Belief is larger. This type of decision is called a hard decision and the decision reliability is evaluated according to assumption. Let there be two targets with less distance than both are from the same class. According to the KNN classifier, the distance between the object and reliability is calculated at each label. If the relative distance $d_{i,j}$ is large then w_j is not the class label of the target class. If distance $d_{i,j}$ is small, there is more possibility that w_j is the class label of the target. The final decision is calculated according to the maximum BBA's.

BBA Combination

It is the rule of belief function for decision fusion for combine different numbers of BBA. Hypothesis S is calculated for BBA using the following equation(Rossi et al., 2014).

$$m(S) = \frac{\sum \bigcap_{j=1}^{M} Sj = S \prod_{i=1}^{M} m(S_i)}{1-k} \tag{4}$$

where k is the conflict degree of final BBA. If k=1, then conflict is high among the combined BBAs and fusion results are unreliable class labels are high probability for final decision as (Rossi et al., 2014).

$$Bt\,P(S) = \sum_{Si \subset S} \frac{m(S)}{|Si|} \tag{5}$$

Fusion Rule

There are two types of fusion rules.

(a) **NB fusion:** Here decisions are mutually independent. In binary fusion, the fusion method is considered as optimal fusion rule. Fusion decision is labeled with maximum fusion statistic, as by (Rossi et al., 2014).

$$l_d = \arg\max_{l \le k \le c}\left\{\prod_{i=1}^{n} P_{i,k}\right\} \tag{6}$$

(b) **WMV:** WMV denotes decision $u_{i,k}$ on w_k label of sensor s_i. The target is $w_{k,}$ $u_{i,k}$ =1 and $u_{i,k}$ = 0. $u_{i,k}$ is the weighted majority voting with coefficient b_i and decision is made by (Rossi et al., 2014).

$$l_d = \arg\max_{l \le k \le c}\left\{\sum_{i=1}^{n} b_i u_{i,k}\right\} \tag{7}$$

Weight Wi can be calculated as:

$$W_i \propto \log\log\left(\frac{p_i}{1-p_i}\right) \tag{8}$$

Here sensor classification accuracy is pi. A sensor with larger accuracy assigned a larger weight. It is better and gives better results than voting rules.

PROBLEM FORMULATION

KNN, ELM, SVM, and RELM are the four different classifier types that have been improved for data classification and error detection.

KNN Classifier

It is employed for neighbor searches (Venkatesan et al., 2018). To measure distance, it employs the Manhattan function of Euclid. To locate the closest neighbor, the Brute force algorithm (BF) is utilized. Three neighbors made up the Euclidean distance function. Utilizing the labeled representation of unidentified cases, the distance function weight is assessed. Distance calculation method for locating KNN instances. KNN examines which examples fall into which classes. The closest neighbors belong to the same grade. Final class is the KNN set with the highest weight.

ELM Classifier

It is an extension of the single-hidden layer feed forward (SLFM) algorithm for classification (Gupta et al., 2017). When employing non-differentiable activation functions, direct solution calculation is done via ELM. The classical, gradient-based learning techniques that use activation functions have limitations such as over-fitting and the local optimum learning rate. Using a single hidden layer instead of over-fitting produces good results. When computing bigger amounts of datasets, it uses more computer power and performs less efficiently. For these datasets, 20 hidden layers are used.

SVM Classifier

A supervised machine learning algorithm is called SVM. The kernel technique transforms data and determines a single final ideal boundary value for potential outputs. the feature map in large dimensions with estimated input based on features. SVM divides data into labeled categories and produces hyper planes for categorization. The accuracy of the belief function technique is examined by SVM. The dataset's characteristics were normalized between [-1,1]. The weight associated with each feature is determined by it. The training set weights were reweighted if the dataset wasn't organized and a sigmoid kernel function was employed. The fundamental rules of weight evaluation were as follows: if a feature was absent, its weight was zero, and a feature that was more significant appeared more frequently.

RELM Classifier

It consists of only one hidden layer (SHLRN). Context neurons once connected input to output in a reverse direction. Neurons use input to store delayed values from output neurons. The weights in ELM were updated using it. Weights are chosen at random. This algorithm uses biases and weights that are optimal. Delay is introduced to the hidden layer H-matrix as input on the output. Moore-Penrose generalized inverse was used to evaluate the weight of the hidden layer.

FEATURE COMPARISON

The enhanced versions of the four classifiers KNN, ELM, SVM, and RELM are known as enhanced KNN, enhanced ELM, enhanced SVM, and enhanced RELM.

Enhanced KNN (EKNN)

Clusters made for nearest neighbor points. For calculating nearest points in a cluster three types of datasets are used i.e., training, evaluation and testing. To find cluster area and points inside the cluster EKNN follow certain steps. K-mean algorithm implemented for calculating cluster head between random numbers of samples. Implementing voting majority class label of cluster identified. Euclidean distance function used for calculating distance between points. By evaluating the distance and cluster head, the cluster is analyzed and determined the nearest cluster. These methods repeated several times for getting a good cluster head.

Algorithm 1: Pseudocode of EKNN

Result: Classification and DA
Input: Original dataset
i= 1;
while Maximum Iteration
do
Classifying the training set into k no. of clusters; Identifying the cluster centers using majority voting; Estimating the quality of cluster centers;
increment i by 1;
end
designing the resultant classifier

Enhanced ELM (EELM)

In enhanced ELM to discover freshly created nodes as hidden layers by fusing the top hidden nodes into the layer underneath them. Bias values are used to evaluate the accuracy of classifier training. L stands for the hidden neuron count, while n stands for the training data. to improve the settings for the ELM optimization technique. The dataset is split into five subsets, each of which includes one test set and four training sets. The bias dimension is L. input weight value L × n. The dimension of the solution is L X (n+1). Hidden layer neurons are initialized using random values. Evaluation of fitness value as:

$$f = 1 - \frac{\sum_{i-1}^{n} misclassificationrate_i}{n} \tag{9}$$

The fitness layer calculated time to time for hidden neurons.

Algorithm 2: Pseudocode of EELM

Result: Classification and DA
Input: Original dataset;
event target is identified by n sensors;
i = 1;
while n Iteration do Xi is received by Si;
increment i by 1;
end

Enhanced SVM (ESVM)

The genotype data parameter g and the data error penalty coefficient C were defined by the upgraded SVM. In the beginning, N genotype data were generated, sorted, and used as a starting point by GA. To assess each optimization criterion, fitness was calculated. Here, the three techniques of selection, crossover, and variation are used. The selection process examined and highlighted fitness value. The groups with the highest fitness levels were advanced. Features in the group were matched in the crossover technique. The exchange of the pair values is based on probability. In the variation technique used to evaluate the combination, solution, and variation probability.

Algorithm 3: Pseudocode of ESVM.

Result: Classification and DA
Input: Original dataset; prepossessing of Data;
Initialize all input parameters;
Use of a hypothesis space in the high-dimensional feature space kernel function;
Training of the model Calculation of the Fitness value;
Termination of the Satisfactory condition;
Optimal parameter (C, g); Establishment of the Model; Performance evaluation;

Enhanced RELM (ERELM)

It was examined using the performance of the belief function-based fusion. RELM initialized with learned weights of RELM. Three layers implement the ERELM algorithm.

(a). **Input layer:** ERELM begins operating from the input layer because it is the first layer. The input layer does not take input value from the preceding layer, nor does it have input weighting. The input layer's number of neurons is equal to the dataset's number of features.

(b). **Hidden layer:** It is the model's second layer. The hidden layer receives a value from the input layer. The number of hidden layers determines the number of iterations.

(c). **Output layer:** Implemented as the input of the output layer is a hidden layer. In the Sigmoid function, the output is converted into a probability score for each class.

Algorithm 4: Pseudo code of ERELM

Classification and DA are the results.
Original dataset as an input;
use a differential evaluation method to optimize the weights;
the input biases bi and weights wi as received from RELM;
calculating the output matrix for the hidden layer H, where H = hij(i = 1,..., N) and j = (1,..., K) and hij= g(wj xi + bj);
using the Moore-Penrose-implemented inverse of the H matrix to calculate the output weight matrix;
giving the input and hidden layers the modified weights as context neurons;

All the classifiers used to find the fault in WSN. Offset, gain, stuck-at, and out of bounds faults are the different sorts of faults. Different rate of dataset implemented. Here fault calculation used an equation $\alpha+\beta x+\eta$, where α as offset, β as gain, x as non-faulty at time t and η is noise.

Models for offset fault included:

$$x' = \alpha + x + \eta \tag{10}$$

The gain fault model was:

$$x' = \beta x + \eta \tag{11}$$

Modeling of Stuck-at fault included:

$$x' = \alpha \tag{12}$$

Models for out-of-bounds fault included:

$$x'\langle\theta_1\, or\, x'\rangle\theta_1 \tag{13}$$

For each fault, a dataset was created in order to assess how well the EKNN, EELM, ESVM, and ERELM performed. Calculated as: Detection Accuracy (DA) of Fault:

$$DA = \frac{\text{number of faulty node detected}}{\text{Total number of faulty node}} \tag{14}$$

RESULT AND ANALYSIS

Based on classification accuracy, we contrasted KNN, SVM, ELM, and RELM with the energy-enhanced classifiers, i.e., ERELM, ESVM, EELM, and EKNN. In the suggested method, a single sensor functions as a believer. The accuracy of KNN-based fusion is shown in Figure 2. The EKNN results are displayed in Figure 3 along with comparisons to other fusion classifiers. It has been demonstrated that the EKNN classifier outperforms the KNN in terms of accuracy. Due to their weight and likelihood of results, NB and WMV do not accurately represent accuracy, so EKNN is employed to improve the performance of the suggested Belief function (Our). EKNN performance is inferior to KNN performance.

Figure 2. Implementing KNN for sensor number versus classification accuracy

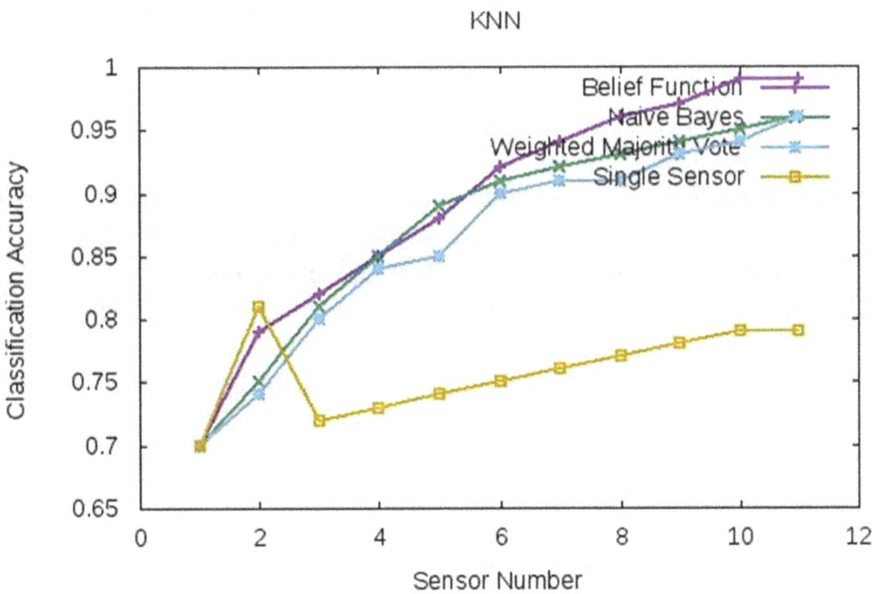

Figure 3. Implementation of Enhanced KNN algorithm.

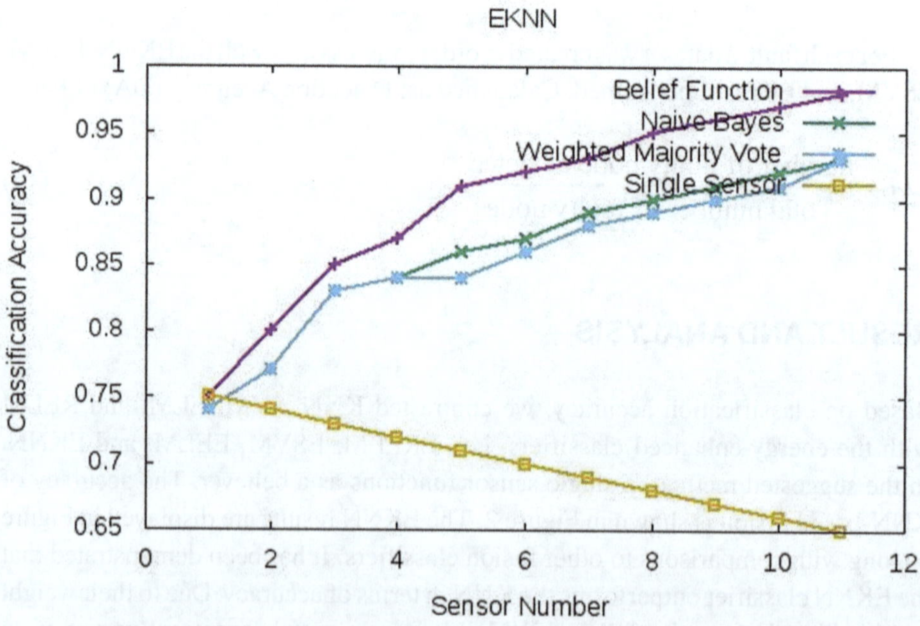

We take into account distance information when choosing the KNN; KNN carry the same weights whether they are too close or too far, based on the vote. If the KNN value is picked, even the chance of a tie in the vote results from the KNN's poor performance. We determined the classes of classifiers during the testing of samples based on the results of each sample test for each KNN that was integrated. In this EKNN procedure, the distance between the sample data is used to calculate the weight of the evidence. Based on the K training samples that correspond to the set of KNNs, we estimated the Belief function (our) for each KNN. In this situation, the K closest samples in our suggested techniques rose. Due to the excellent accuracy of EKNN, if an EKNN tie occurred, the decision was made using distance information.

The results of ELM and EELM are displayed in Figures 4 and 5. EELM's Belief function, including the NB, WMV, and current Belief function.

The new strategy increases the accuracy of our belief function. 24 hidden neurons are employed in this ELM method to demonstrate how the classifier has been over fit. Check the corresponding labels of the learning process in the event that the EELM has a significant number of concealed nodes. In order to avoid the over-fitting issue and validate the labels and results displayed with the utmost precision, the EELM method ignored the pertinent concealed nodes.

Figure 4. Implementing ELM for sensor number versus classification accuracy.

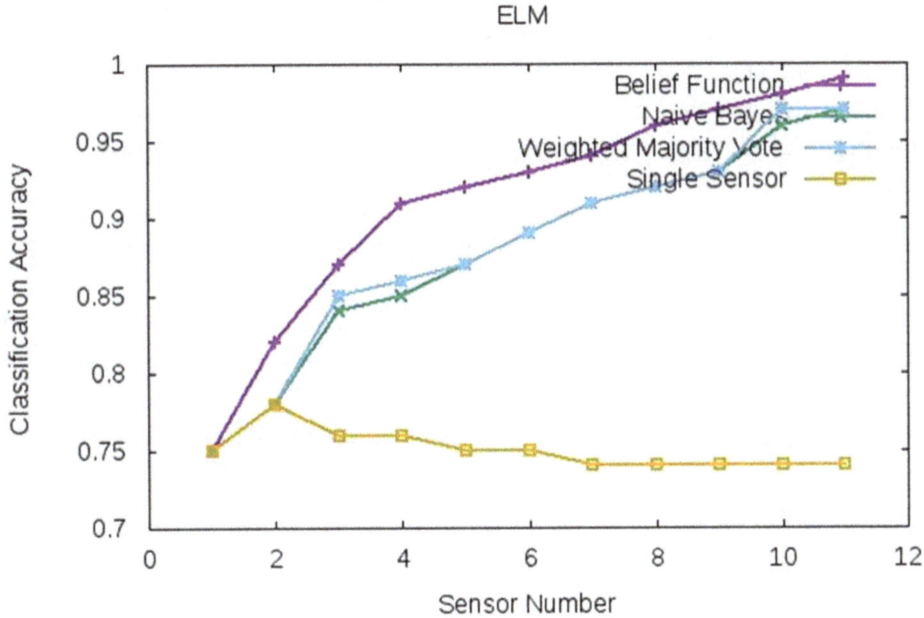

Figure 5. Implementation of EELM.

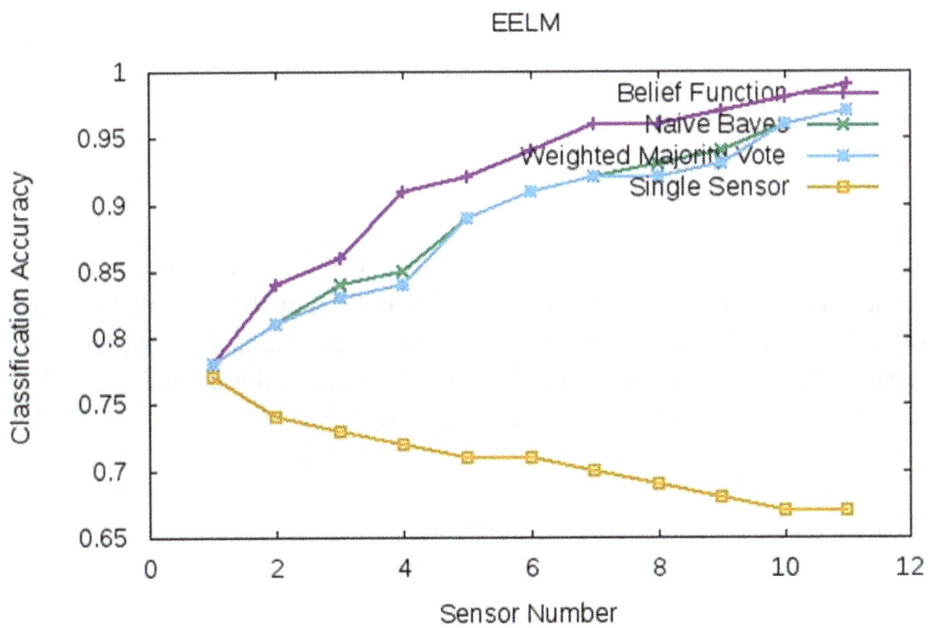

Figure 6 shows the results of the SVM classifier and Belief function (our) outperformed the other two techniques. SVM presents the improved efficiency of the SVM classifier. In figure 7 ESVM gives better results than the SVM and improved the public presentation of the proposed Belief function. In the case of SVM, the data is divided into two classes and gives the smallest intervals; it is shown less classification and improved the data accuracy due to the smallest interval.

Figure 6. Implementation of SVM for sensor number versus classification accuracy.

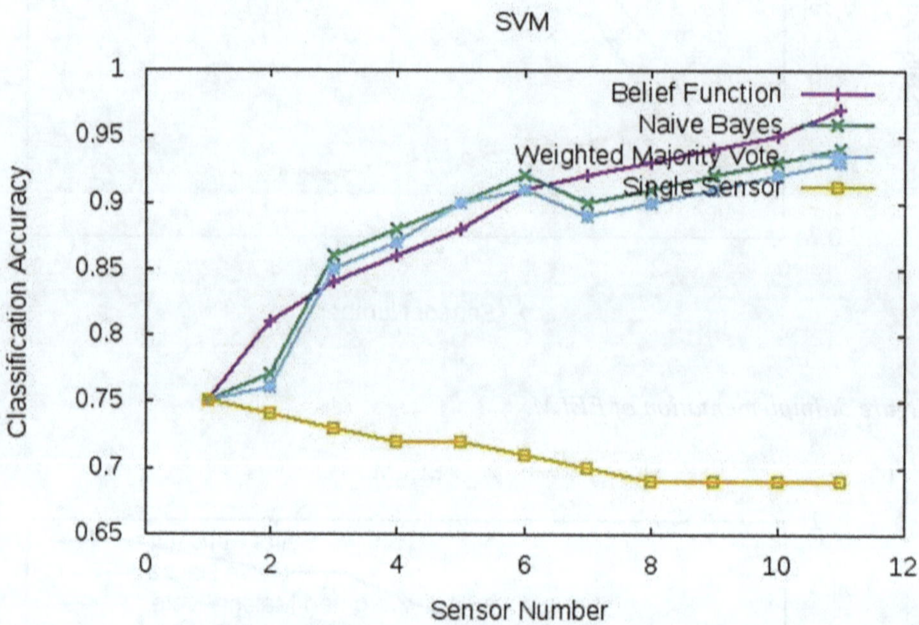

Figure 8 results show that the RELM method and Figure 9 showed that the Belief function with others and represents the WMV and NB approach. Belief function beats others. Here RELM uses fewer neurons and connects through the back from the output to the input based on the specified task. In Figure 9, it is plotted by ERELM where it shows the justified classification and specific fusion results based on that RELM conquered the results of all the proposed techniques.

Figures 10 and 11 display the ESVM's detection precision for all induced faults. The detection accuracy increases at a fault probability of 0.3, decreases over time, and ESVM performs better than ERELM as the fault probability rises.

Figure 7. Implementation of ESVM.

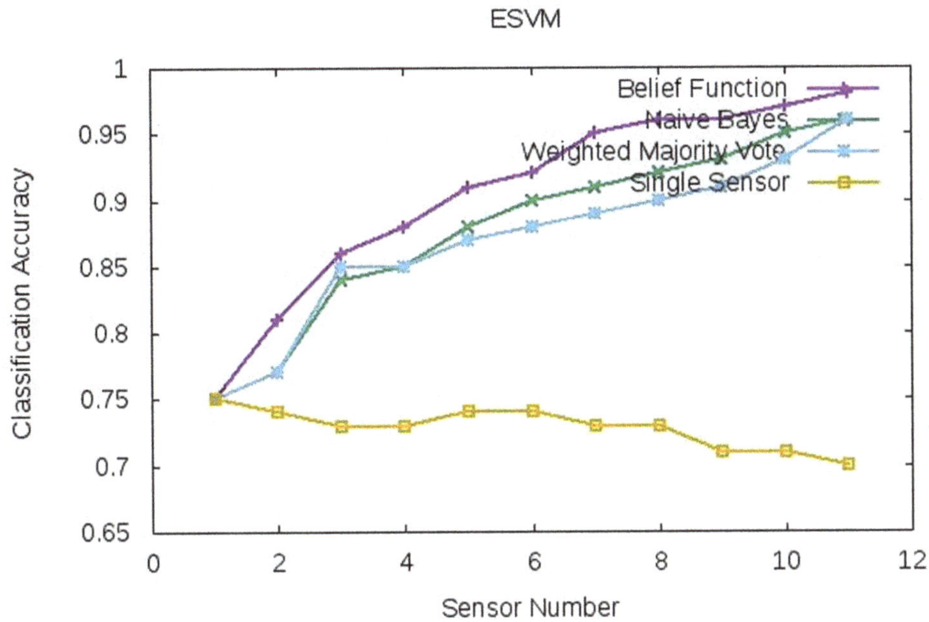

Figure 8. Implementing RELM.

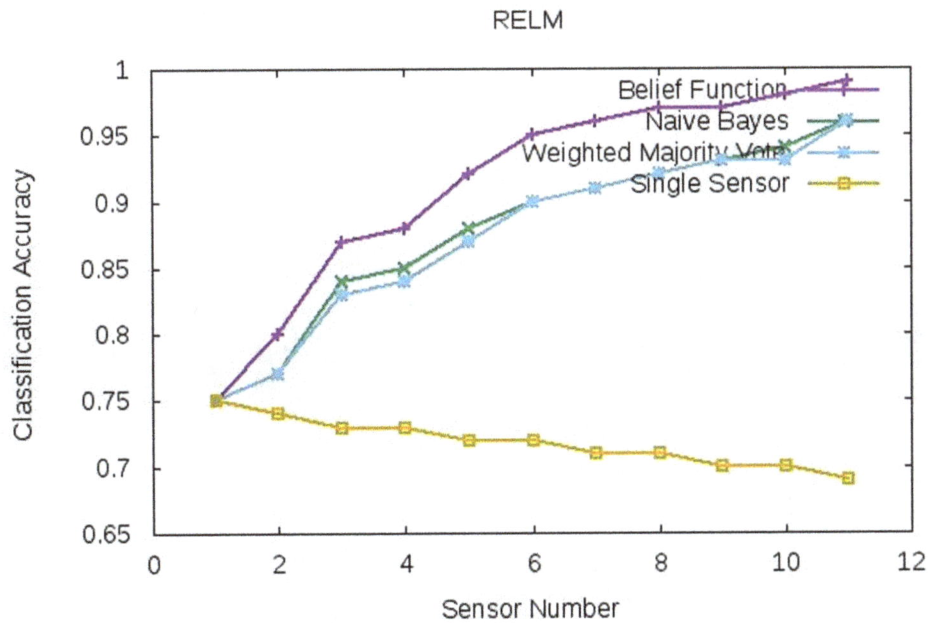

Figure 9: Implementing ERELM.

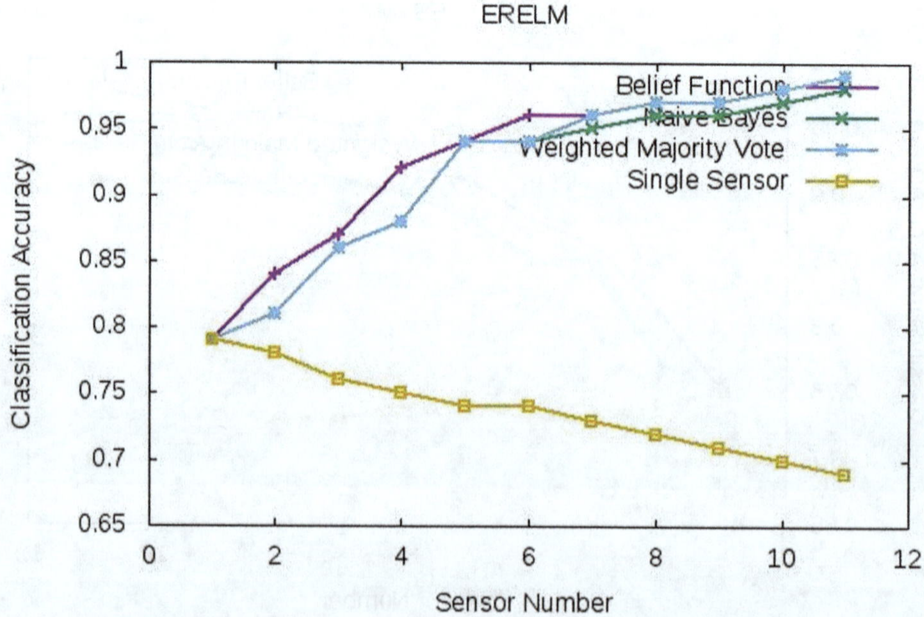

Figure 10. Fault probability versus Detection accuracy in ERELM.

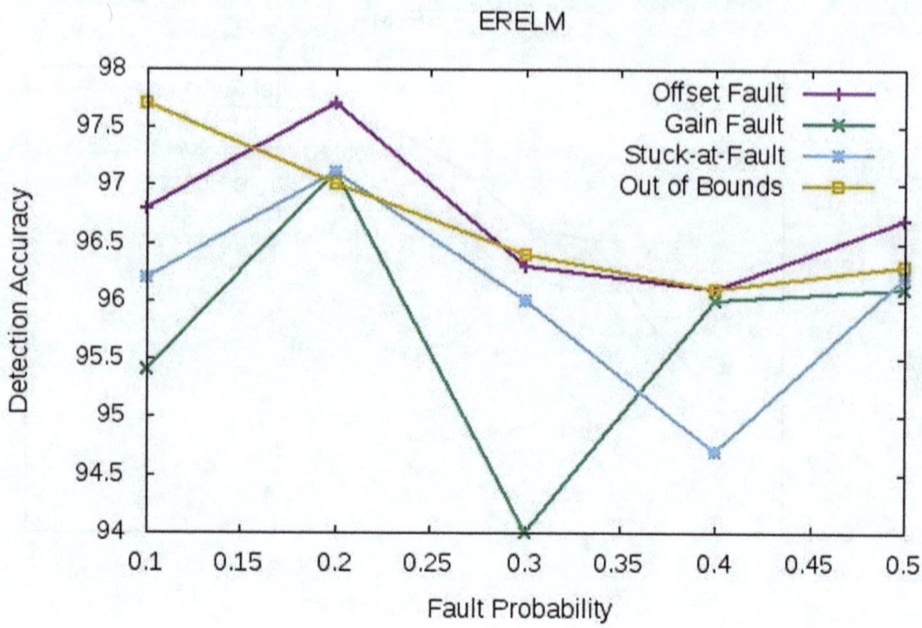

Figure 11. Fault probability verses Detection accuracy in ESVM.

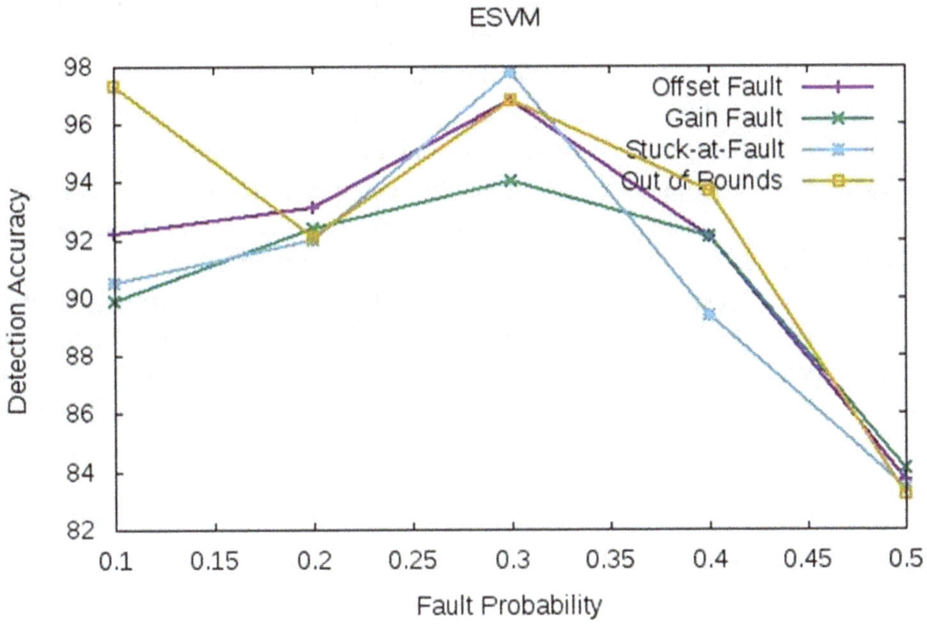

Figure 12. Fault probability versus Detection accuracy in EELM.

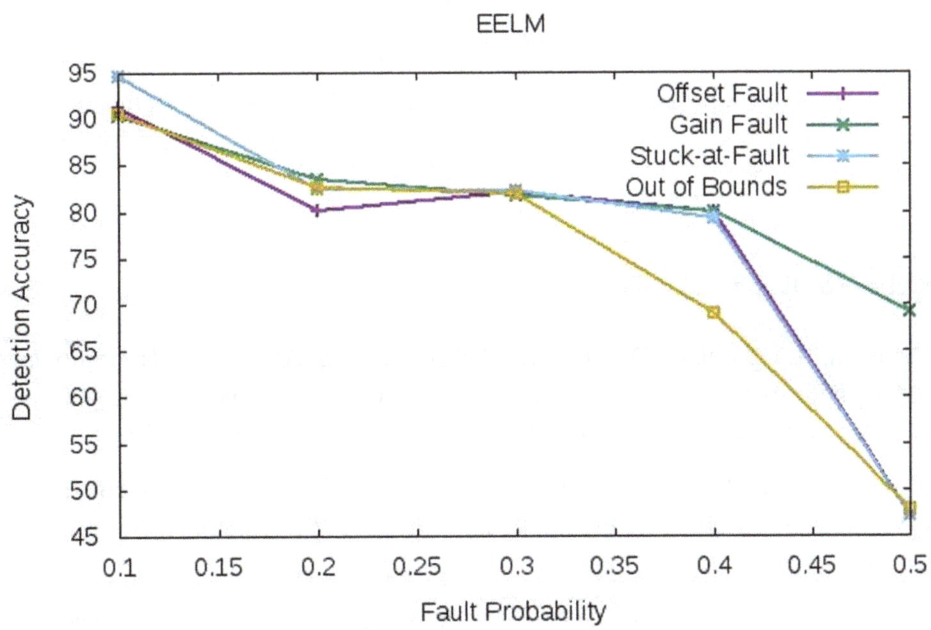

Figure 12 demonstrates that the data accuracy DA of EELM is higher. However, as the likelihood of a fault increased, DA gradually dropped. The neurons are diminished as a result of the lack of a trading sample of data. The EELM performs worse than the EREXLM and ESVM in this instance.

It has been noted that the WSN nodes' battery life is insufficient to handle the various computations. The fusion Centre node and all other network nodes are shown in this section. Only Belief functions (our) are sent to the data fusion Centre in this case; nodes won't transfer data directly to the fusion Centre. All of the computations done in the fusion Centre are based on the laws and regulations, which BBA has confirmed. The intricate calculations won't need to be done at the fusion Centre (Rules). To do operations in tandem, in the BBA of the fusion Centre. The battery power of the fusion is preserved in the suggested manner. Only the connection and data transfer will not be compromised in WSN, it detects and shows. We presume that each node has enough processing capacity to run the classifier for a particular task.

CONCLUSION

We improve the decision-making method of the belief function. The easiest way to find the flaws in our suggested method is to use upgraded versions of the EKNN, ESVM, EELM, and ERELM classifiers. A belief function fusion technique will be used to implement these procedures in comparison to the originals in the preceding approaches for defect detection. Faults featured a variety of datasets, including offset, stuck-at, gain, and out of bounds faults. ERELM outperforms the competition in terms of performance. Future multi-class fusions such as multi-symbol signal modulation, remote sensing, and picture fusion will come from the decision reliability utilized to enhance fusion.

REFERENCES

Abuassba, A. O., Zhang, D., Luo, X., Shaheryar, A., & Ali, H. (2017). Improving classification performance through an advanced ensemble based heterogeneous extreme learning machines. *Computational Intelligence and Neuroscience*.

Al-Jarrah, M. A., Al-Dweik, A., Kalil, M., & Ikki, S. S. (2018). Decision fusion in distributed cooperative wireless sensor networks. *IEEE Transactions on Vehicular Technology*, *68*(1), 797–811.

Binh, H. T. T., & Dey, N. (Eds.). (2018). *Soft computing in wireless sensor networks*. CRC Press.

Das, S. K. (2021). Smart Design and Its Applications: Challenges and Techniques. In *Nature-Inspired Computing for Smart Application Design* (pp. 1–6). Springer.

Das, S. K., Das, S. P., Dey, N., & Hassanien, A. E. (Eds.). (2021). *Machine learning algorithms for industrial applications*. Springer.

Das, S. K., & Giacalone, M. (Eds.). (2022). *Fuzzy Optimization Techniques in the Areas of Science and Management*. CRC Press.

Das, S. K., Samanta, S., Dey, N., & Kumar, R. (Eds.). (2020). *Design frameworks for wireless networks*. Springer Singapore.

Das, S. K., Samanta, S., Dey, N., Patel, B. S., & Hassanien, A. E. (Eds.). (2021). *Architectural Wireless Networks Solutions and Security Issues*. Springer Singapore Pte. Limited.

De, D., Mukherjee, A., Das, S. K., & Dey, N. (Eds.). (2020). *Nature inspired computing for wireless sensor networks*. Springer.

De, D., Mukherjee, A., Das, S. K., & Dey, N. (2020). Wireless sensor network: applications, challenges, and algorithms. In *Nature inspired computing for wireless sensor networks* (pp. 1–18). Springer.

De Cock, M., Dowsley, R., Horst, C., Katti, R., Nascimento, A. C., Poon, W. S., & Truex, S. (2017). Efficient and private scoring of decision trees, support vector machines and logistic regression models based on pre-computation. *IEEE Transactions on Dependable and Secure Computing*, *16*(2), 217–230.

El Hindi, K., AlSalman, H., Qasem, S., & Al Ahmadi, S. (2018). Building an ensemble of fine-tuned naive Bayesian classifiers for text classification. *Entropy (Basel, Switzerland)*, *20*(11), 857. doi:10.3390/e20110857 PMID:33266581

Gupta, S., Mittal, M., & Padha, A. (2017, December). Predictive analytics of sensor data based on supervised machine learning algorithms. In *2017 International Conference on Next Generation Computing and Information Systems (ICNGCIS)* (pp. 171-176). IEEE.

Jan, S. U., Lee, Y. D., Shin, J., & Koo, I. (2017). Sensor fault classification based on support vector machine and statistical time-domain features. *IEEE Access: Practical Innovations, Open Solutions*, *5*, 8682–8690.

Jiao, L., Wu, H., Bie, R., Umek, A., & Kos, A. (2018). Multi-sensor golf swing classification using deep CNN. *Procedia Computer Science*, *129*, 59–65.

Lee, K. S., Lee, S. R., Kim, Y., & Lee, C. G. (2017). Deep learning–based real-time query processing for wireless sensor network. *International Journal of Distributed Sensor Networks*, *13*(5).

Niu, R., Moore, M., & Klamer, D. (2004). *Decision fusion in a wireless sensor network with a large number of sensors*. Academic Press.

Omairi, A., & Ismail, Z. H. (2016). Modeling battery state of charge in wireless sensor networks based on structured multi-layer perceptron. *Journal of Advanced Research in Applied Sciences and Engineering Technology*, *5*(2), 36–45.

Pan, L., & Li, J. (2010). K-nearest neighbor based missing data estimation algorithm in wireless sensor networks. *Wireless Sensor Network*, *2*(02), 115.

Rossi, P. S., Ciuonzo, D., Kansanen, K., & Ekman, T. (2014). On energy detection for MIMO decision fusion in wireless sensor networks over NLOS fading. *IEEE Communications Letters*, *19*(2), 303–306.

Samet, A., Lefevre, E., & Yahia, S. B. (2014, November). Belief function classification with conflict management: Application on forest image. In *2014 Tenth International Conference on Signal-Image Technology and Internet-Based Systems* (pp. 14-20). IEEE. 10.1109/SITIS.2014.34

Sriranga, N., Nagananda, K. G., Blum, R. S., Saucan, A., & Varshney, P. K. (2018, July). Energy-efficient decision fusion for distributed detection in wireless sensor networks. In *2018 21st International conference on information fusion (FUSION)* (pp. 1541-1547). IEEE. 10.23919/ICIF.2018.8454976

Strauman, A. S., Bianchi, F. M., Mikalsen, K. Ø., Kampffmeyer, M., Soguero-Ruiz, C., & Jenssen, R. (2018, March). Classification of postoperative surgical site infections from blood measurements with missing data using recurrent neural networks. In *2018 IEEE EMBS International Conference on Biomedical & Health Informatics (BHI)* (pp. 307-310). IEEE.

Swain, R. R., & Khilar, P. M. (2017, November). Soft fault diagnosis in wireless sensor networks using PSO based classification. In *TENCON 2017-2017 IEEE Region 10 Conference* (pp. 2456–2461). IEEE.

Swain, R. R., & Khilar, P. M. (2017). Composite fault diagnosis in wireless sensor networks using neural networks. *Wireless Personal Communications*, *95*(3), 2507–2548.

Venkatesan, C., Karthigaikumar, P., Paul, A., Satheeskumaran, S., & Kumar, R. (2018). ECG signal preprocessing and SVM classifier-based abnormality detection in remote healthcare applications. *IEEE Access: Practical Innovations, Open Solutions*, *6*, 9767–9773.

Yuan, Y., Li, S., Zhang, X., & Sun, J. (2018, July). A comparative analysis of svm, naive bayes and gbdt for data faults detection in wsns. In *2018 IEEE International Conference on Software Quality, Reliability and Security Companion (QRS-C)* (pp. 394-399). IEEE.

Zhang, W., & Zhang, Z. (2015). Belief function based decision fusion for decentralized target classification in wireless sensor networks. *Sensors (Basel)*, *15*(8), 20524–20540.

Zhang, Z., Hao, Z., Zeadally, S., Zhang, J., Han, B., & Chao, H. C. (2017). Multiple attributes decision fusion for wireless sensor networks based on intuitionistic fuzzy set. *IEEE Access: Practical Innovations, Open Solutions*, *5*, 12798–12809. doi:10.1109/ACCESS.2017.2722483

Zidi, S., Moulahi, T., & Alaya, B. (2017). Fault detection in wireless sensor networks through SVM classifier. *IEEE Sensors Journal*, *18*(1), 340–347.

Chapter 7
An Optimal Data Communication System in Wireless Sensor Network Using Artificial Neural Network

Ashish Kumar Dass
National Institute of Science and Technology, Berhampur, India

Santosh Kumar Das
https://orcid.org/0000-0001-6852-1361
Sarala Birla University, Ranchi, India

ABSTRACT

Wireless sensor networks (WSN) can be referred to as a group of spatially dispersed and dedicated sensors that are self-capable of collecting environmental information, processing it as per requirement, and forwarding it to a base station for different application. In this chapter, the authors focus on secure data transformation in wireless sensor node. The intrusion detection system using sensor node is complicated as sensors lifetime depends upon power consumption and for which the network's lifetime also get affected. Security as well as network lifetime can be enhanced by relocating the computational load of the operation from the sensor nodes to the base station. At base station by encrypting and compressing the data using cryptography and artificial neural network (ANN), the authors increase the security and increase the throughput of the network. On the server side it is decompressed and then decrypted.

DOI: 10.4018/978-1-6684-7343-6.ch007

INTRODUCTION

Wireless Sensor Networks (WSNs) consist of a huge range of tiny sensor/detector nodes that send the collected data using the wireless channels. Most sensor networks consist of many low powers, less cost nodes deployed to keep an eye on the environment. This sensor network is a mix of different variety of sensors and actuators with variable computational power. WSNs have many wide range applications ranging from military surveillance to under water monitoring, critical health supervising, urban traffic controlling etc. (Yick et al., 2008; Akyildiz et al., 2002). Most of the applications of WSNs require se- cure transmission of data from node to server and vice versa. As sensor nodes have limited storage capacity and energy for processing the data, security in WSNs becomes a vital concern. But it is common problem that Sensor nodes can face different unwanted attacks during deployment or functioning in any hostile environments (Chen et al., 2009; Patel & Aggarwal, 2003).

For securing that data or information from various types of attacks normally cryptography is used which is an art of hiding the data from intruder. It is a technique in which the security to the data is provided by encrypting the whole data which cannot be read by the attacker .Now a day's attacker uses very complicated and modern technologies. To deal with those attacks it requires higher computational cryptographic algorithm but the sensor nodes have limited computational power to execute those processes (Kahate, 2014).

From the above context, a need of a new technology which is more se- cure than the existing technique is realized. In this Chapter, an efficient algorithm has proposed in which a hybrid technology of cryptography and Artificial Neural Network (ANN) are used for offering additional security with more secure exchange of information with less computational power.

Wireless Sensor Network (WSN) can be refer as a group of heterogeneous sensors with less energy, storage capacity and computation power. It has a wide range of applications in our day-to-day life (El-Bendary, 2015; Abdollahzadeh & Navimipour, 2016).

As wireless network is more prone to outer threats and attacks so security is an important concern of these types of networks. One of the ways to tackle this kind of situation is the use of cryptography. Cryptography encrypts the data before transmitting. Due to physical size and energy level of sensor nodes it has restricted processing power, storage capacity and communication bandwidth. These things have to be considered particularly while dealing with security issues (El-Bendary, 2015; Yang, 2014; Kodali & Chougule, 2013). The security model which are: confidentiality, integrity, and availability. Each of these goals is in brief explained below.

(a). **Confidentiality:** Confidentiality is to stop illegal access of the information in data communication

(b). **Integrity:** Integrity makes sure that data being sent to the server in a correct and complete way such that the rightness and completeness of the data is sustained

(c). **Availability:** The timely availability of data to authorized users (El-Bendary, 2015; Xiao et al., 2009).

LITERATURE REVIEW

In the last few decades, several works have been proposed for the purpose of wireless network and its related learning and optimization system. Some of the works are discussed as Anand et al. (2021) proposed a method for controlled framework for mobile ad-hoc network. It is based on multicast routing system. It is a reliable routing method for ad-hoc network. It overcomes the limitation of unicast routing system. The proposed method efficiently maintains the communication between sender and receiver based on multicast routing method. It also helps to make transparency of the system. It helps to increase execution speed of the model and decrease the center lifetime of the network. Das et al. (2021) proposed a list of application in the area of wireless network and its related application. Main purpose of this book to enhance the activity and application of wireless network. Each of the work in this book indicates architectural solution based on particular issues. It helps to design some related security issues based on different factors and its information system. It helps to manage several information efficiently. Patsariya and Rajavat (2021) proposed a method for route selection in mobile ad-hoc network. It is based on node capability management system. It uses randomness behavior of the network based on traffic management. It uses several parameters of network such as residual energy, buffer system, mobility, and traffic information for handling routing of the network. It control random behavior of the network based on entry level management in the network. Finally, it helps to validate and compare with existing routing protocols. Das et al. (2021) proposed a list of algorithms based on efficiency of machine learning methodologies. It helps to manage several machines learning information to deals with some problems. It helps to manage several operations and its application to solve real-life problem. Several machine learning algorithms are mentioned in this book that handle several areas and give new direction to the readers. Eksert et al. (2021) proposed a clustering method for ad-hoc network based on fusion of intra and inter communication system. It also helps in link management and scheduling purpose of the network. The proposed method is based on CUPS architecture system and modeling. It is based on plane separation and user control management. It is based on flat topology management system. It helps to control several challenges based on free medium system. The proposed method uses non-linear optimization system based on instant communication system. De et al. (2020) proposed a method that contains several nature inspired application in the area of wireless sensor network. Several issues and problems are mentioned in this article to handle several operation of the network. The work is based on architecture system that handles several computation applications. Each of the application is inspired by nature with the context of algorithm. Thiagarajan et al. (2021) proposed a multipath routing technique based on on-demand system for mobile ad-hoc network. It is based on distance vector system based on routing protocol. It helps to increase the quality of service of the network based on three parameters such as bandwidth, end-to-end delay and hop count. The proposed method is based on evolutionary optimization technique. It is helps to detect intrusion based on grammatical evolution and helps to established multipath based on network connectivity. Das et al. (2020) proposed a framework application for handling several operations of the network. It helps to manage several designed model of the network. Each of the models is based on some factors of application. It indicates solution prospective point of view operation in the area of wireless network. It helps to guide several operation and model of the network that helps to incorporate the system in smart application. Kumar et al. (2021) designed a method for congestion control and management for mobile ad-hoc network. It is based on random waypoint method and analysis and management. It is an adaptive routing which is used to maintain load balance of the system. There are several parameters are used in this work such as topology of network, mobility, nature of network. Finally, it helps to optimize packet loss, energy consumption, connection fails, and time information. It also helps to manage failure link of the system and overhead information management. De et al. (2020) designed a survey model for wireless sensor network for application management. Each of the work is based on several algorithms and challenges system. But it helps to resolve several issues based of factors of information. Several solutions are mentioned in this proposed to resolve the challenges based on framework analysis of wireless sensor network. Dragonas et al. (2020) proposed a method for 5G network for fairness management and analysis. It is based on time independent networking system. It is based on MAC policy management that depends on several types of constrained of ad-hoc network. This work is also applicable for IoT and TDMA policy management. Finally, it helps in re-transmission management and delay management for emerging analysis. Qurashi et al. (2020) designed a method for ad-hoc network and management. It is designed for resilient system for intrusion detection system. It is based on IPv6 management for enabling services of the system. It is used for redundancy management for performance analysis. It helps in collaborating information management. It also helps in network topology management and average degree management of the nodes. It is based on distributed agent deploy information management. It helps to optimize energy consumption and overhead management of the network. S. K. Das (2021) designed a review article for handling several challenges of smart application. The work also mentioned several techniques and issues modeling for factor analysis based on smart application system. It helps to deals several issues and application that are used in smart application design. It also helps to model several parameters that are directly or indirectly influence in the application. Robinson et al. (2019) proposed a method for fault tolerant system and management. The nature of the work is based on on-demand system that is used for multi-path analysis based on distance vector system. It is a routing algorithm that is used for disjoint analysis for managing different types fault. It helps in overhead management and route discovery handling system. It helps to reduce delay of the system based on overhead of the network management. Bhattacharya and Sinha (2017) designed an efficient method for load balancing system and management for mobile ad-hoc network. It is based on least common system for multiple based routing systems. It is used to design and analysis multiple path between source and destination node of the network. It is used for route packet design based on data packet analysis. It is used for route discovery information management for multiple route design system. So, it easily helps to manage route discovery information system. Lou and Zhuang (2016) designed a method for efficient routing system for coordinated system of sleep scheduling management in ad-hoc network. It is based on sleep scheduling management system of the network. It is based on energy efficiency system and management. It helps in performance design based on topology analysis based on time division multiple access based on medium access control system. It is based on sleep scheduling system based on network condition management. It helps to design several network condition management based on traffic density and management. Binh and Dey (2018) designed soft computing review system based on wireless sensor network and its application. The work is fully based on several techniques of soft computing technique that helps to model its related application. Each of the work helps to model and resolve the issues that are raised during the operation of wireless sensor network. Sarkar and Datta (2016) proposed an energy efficient method based on multipath system for mobile ad-hoc network. It is based on self-organized way and management. The proposed method is based on stochastic management system. The system is design based on autonomous system and information designed system. It helps to communicate management based on energy efficiency. The proposed method is based on Markov chain system based on forwarding information system. This is based on data packet management based on wireless channel management. Das and Giacalone (2022) proposed a book for the purpose of fuzzy logic and its related usage in the areas of engineering and management. It contains several related works that helps to deals with several types of issues and problems that helps in the real-life applications.

THREAT MODEL IN WIRELESS SENSOR NETWORKS

Wireless sensor network is always deployed in unsuitable or unfavorable conditions (2016). There may be a chance that the attacker might steal the data directly from the physical node (2013). So WSNs nodes lack confidentiality. The attacks to the WSN, based on their access levels is active and passive attacks and based on places is internal and external attacks.

Passive vs Active Attacks

In passive attacks, the invader only reads messages by means of a radio sender/receiver unit and harm is caused in the network. These invaders just gather important information. Contrary to this is the active attack, the invader send false information into the network and disturb the transmission causing information load.

Internal versus external attacks (attacks based on location). In the internal attack, the attacker operates as a sensor node and disrupt network by communicating with all the other networks thus inserting false data. The external attacks have no role to play in a network yet it can cause disruption in the network.

Some Attacks Based on WSN

(a). **Physical attacks:** The attacker might physically damage the sensor nodes causing permanent damage to the sensor network.

(b). **Denial-of-Service (DoS) Attacks:** This attack stops one or more node to respond to the request.

(c). **Sybil attacks:** This attack put the security, reliability of data at risk.

(d). **Sinkhole/Black hole attacks:** In black hole attack a node act as a black hole and deletes all the data that passes through that node. The sinkhole is also similar to black hole and it also includes the attacker.

(e). **Wormhole attacks:** In this attack invader sends false message with a delay to a node creating more delay to the server end. In this way invader creates misconception thus increasing traffic.

(f). **Selective forwarding attack**: This attack is somewhat similar to black hole attack, the invader deletes some particular packet which makes hard to detect the attack.

DESIGN OF THE PROPOSED SYSTEM

We used neural network as encryption and decryption algorithm. We used a multilayer back propagation neural network, and it contain one in- put layer, one output layer and as many as hidden layer (more than one). Input layer and output layer consist of 14 node we used sigmoid function in the entire layer (hidden, input, output), whereas rectified linear unit (ReLU) activation used in hidden layer only.

For integrity of the system we are using digital signature using elliptical curve, as key of elliptical curve is small and secure, the public and private key of the sensor node are given during deployment. Hence, by this we achieve confidentiality, integrity and authentication.

Neural network is trained using the datasets given in the table, after the network is trained the weights of different layers in the neural network is deploy in the wireless sensor nodes. For encryption, pre-processing is done, i.e. the plain text is divided into blocks of size two, which is further represented in the 14 bit binary representation and it is given as input to the network. The output of neural network is pseudo random cipher for the decryption, digital signature is verified then the pseudo random cipher is passed to neural and plain text is obtained.

Working of the Whole System

Sender

(a). The plain text is divided into block of size 2. Example:" *hello! subham*" it is divide into block, "he", " l#", " lo", "!# "...."am". here '#' is delimiter. [step 1 and step2 are pre-processing step]

(b). The each block of letter has ASCII code and represent in binary form. For the above example 'he' is converted into '1001000 1001101'

(c). Those are passing as input and based on the training it is converted into pseudo random cipher.

(d). Which is digital sign by the sender using the EEC, the private and public key are given before depletion of the sensor node

(e). Finally, the encrypted digital sign message is pass in the open channel.

Receiver

(a). The encrypted digital sign message is received from the open channel.

(b). Which is digital sign by the sender using the EEC, it is verify in the receiver side. If the signature is match then it is correct and the user is authorized.

(c). The message receive from above step is then block into of size 14 and give as input to the neural

(d). Based on the training it gives some output. Output is the message after pre-processing. The output is then converted into readable format.

(e). Finally, the message is received from the open channel.

CRYPTOGRAPHY METHODOLOGIES

Cryptography could be a system of exploitation progressed scientific standards in putting away and sending information in a quite certain sort so exclusively those whom it's alleged will sweep and technique it. Encryption could be a key idea in cryptography-it's a strategy whereby a message is encoded partner exceedingly in a very} group that can't be check or comprehended by a snooper. The system is past and was introductory utilized by Caesar to encipher his messages exploitation Caesar figure. A plain content from a client are regularly scrambled to a figure content, at that point send through a correspondence channel and no spy can meddle with the plain content. When it achieves the collector complete, the figure content is decoded to the first plain content.

Symmetrical Encryption

This is the main very coding that includes just 1 mystery key to figure and interpret information. Symmetrical encryption is an old and best- known procedure. It utilizes a mystery key that may either be assortment, a word or a string of arbitrary letters. It is an emulsified with the plain content of a message to adjust the substance in an unmistakable way. The sender and furthermore the beneficiary should perceive the key that is acclimated encipher and translate every one of the messages.

Asymmetrical Encryption

Awry coding is also called open key cryptography, which is a generally new technique, contrasted with symmetric encryption. Unbalanced coding utilizes 2 keys to encipher an obvious content. Mystery keys square measure changed over the web or an outsized system. It guarantees that malignant people don't abuse the keys. It is important to see that anybody with a mystery key will interpret the message and this can be the reason hilter kilter coding utilizes 2 associated keys to boosting security

ARTIFICIAL NEURAL NETWORKS

Neural networks are modeled loosely after the human brain, which are designed to recognize pattern. It can also be defined as the way human brain processes the information.

Working Principles of Brain

Human Brain consists of millions of neurons or nerve cells through which information is processed and received from our sense organs and transmitted. The way human brain make decisions based on the feature of the object, it gives weight (or priority). Decisions with higher weights are considered.

Let's consider an example of 500 denominations. Is original or duplicate? Based on the feature such as watermark, color, size, etc. it makes decision. If it is not correct we learn from it. Based on this neural network or artificial neural network is designed, where each neuron interprets the data sensory through a kind of machine perception.

There are many ways to classify the types of artificial neural network but a useful distinction is based on training, it can be supervised or unsupervised training. In supervised training, a set of input and output data are given, in other words, here the data set are labeled, in contract to unsupervised learning where no input and output data set are given, i.e. data set are not labeled or classified. Artificial neural network consists of input layer, output layer, and hidden layer weight. Descriptions are given as:

(a). **Input layer:** Sensory or pre-processed data are given as input and its output act as input to other layer.
(b). **Hidden layer:** It get the input from other layer, it processes and act as output to other layer.
(c). **Output layer:** It gets the input form the hidden layer or input layer and processes the input.
(d). **Weight:** Each connection has some weight. W_{ab} weight from node a to node b.

Working Principles of Artificial Neuron Network

A basic computational unit is called neuron or node. It receive the in- formation form the other node or external source. Each input has some weight W_{ab}, which can be modified. Sum of input to the neuron is calculated as:

$$\text{in} = \sum_b W_{ab} \times b$$

where W_{ab}: weight from node a to node b Output of the neuron is calculated as Y=f (in).

If sum of input is passed as an argument to the activation function and its output is greater than threshold, then it activates the neuron. Sigmoid function and rectifier linear unit (ReLU) are used as the activation function.

Sigmoid function: It is the logistic function. Its value ranges in between 0 to 1 for all the real value.

CRYPTOGRAPHY USING EEC

Key Generation

Key generation is a very important half wherever we've to get each public key and personal key. The sender is going to be encrypting the message with receiver's public key and therefore the receiver can rewrite its personal key.

Now, we've to pick out variety'd' at intervals vary of 'n'. Using the subsequent equation we are able to generate the general public key.

$$Q = d * P$$

d = the random range that we've elect at intervals the vary of (1 to n-1).
P is the point on the curve.
'Q' is that the public key and'd' is that the personal key.

Encryption

Let 'm' be the message that we tend to square measure causing. We have to represent the 'm' message on the curve.

Consider 'm' has the purpose 'M' on the curve 'E'. Randomly select 'k' from [1 – (n-1)].

Two cipher texts are going to be generated let or not it's C1 and C2.

$$C1 = k*P$$

$$C2 = M + k*Q$$

C1 and C2 will be sending.

Decryption

We have to urge back the message 'm' that was send to America,

M = C2 – d * C1

M is that the original message that we've send.

RESULT DISCUSSION

It is based on the two neural networks being able to synchronize by mutual learning. Here, two networks learn from the dataset simultaneously, one network learns and sends pseudo to other network and other network from the pseudo, it finds the plain text. Synchronization of two neural networks is a complex dynamic algorithm but if third party interferes in between, then it is unable to decrypt the message as it is random.

Two sensor nodes p, q wants to communicate. As the medium is open, any interference can modify message that is no confidentiality is there by using the present cryptographic algorithm, those require heavy computational power and moreover senor node has limited power. So using the neural network, form encryption and decryption we can increase the lifetime of the sensor node more over with low energy consumption we are getting confidentiality.

CONCLUSION

We have tested the behavior of the system proposed above successfully; we establish a secure communication system. We encrypt the message using the neural network and decrypt the cipher compares the result, we found it is error free and computation of system is less compared to the system implement using RSA. It is efficient compare to present algorithm in terms of power consumption for computation.

Table 1.

Plain text	encrypt	Cipher text	Plain text	encrypt	Cipher text
AF	JP	10010111010000	HY	VU	10101101010101
AG	AS	10000011010011	HZ	VX	10101101011000
AH	GV	10001111010110	IN	YN	10110011001110
AI	AY	10000011011001	IO	YQ	10110011010001
AY	HU	10101101010101	IQ	YW	10110011010111
AZ	AX	10000011011000	IR	YZ	10110011011010
BA	DA	10001001000001	IS	YC	10110011000011
B0	D0	10001000110000	IT	YF	10110011000110
BC	DG	10001001000111	JL	BH	10000101001000
BD	DJ	10001001001010	JM	BK	10000101001011
CG	GS	10001111010011	JN	BN	10000101001110
CH	GV	10001111010110	JO	BQ	10000101010001
CI	GY	10001111011001	JP	BT	10000101010100
CJ	GB	10001111000010	JQ	BW	10000101010111
CK	GE	10001111000101	JR	BZ	10000101011010
CL	GH	10001111001000	JS	BC	10000101000011
CM	GK	10001111001011	PK	TE	10101001000101
DF	JP	10010101010000	PL	TH	10101001001000
DG	JS	10010101010011	PM	TK	10101001001011
DH	JV	10010101010110	PN	TN	10101001001110
DI	JY	10010101011001	PO	TQ	10101001010001
EJ	MB	10011011000010	P0	T0	10101000110000
EK	ME	10011011000101	RE	ZM	10110101001101
EL	MH	10011011001000	RF	ZP	10110101010000
FG	PS	10100001010011	RG	ZS	10110101010011
FH	PV	10100001010110	RH	ZV	10110101010110
FI	PY	10100001011001	RI	ZY	10110101011001
GO	SQ	10100111010001	US	IC	10010011000011
GP	ST	10100111010100	UT	IF	10010011000110
GQ	SW	10100111010111	U0	I0	10010010110000
GR	SZ	10100111011010	WZ	OX	10011111011000
HG	VS	10101101010011	XA	RA	10100101000001
H0	V0	10101100110000	XB	RD	10100101000100
HI	VY	10101101011001	YG	US	10101011010011
HV	VL	10101101001100	YH	UV	10101011010110
HW	VO	10101101001111	ZU	XI	10110001001001
HX	VR	10101101010010	ZX	XR	10110001010010

REFERENCES

Abdollahzadeh, S., & Navimipour, N. J. (2016). Deployment strategies in the wireless sensor network: A comprehensive review. *Computer Communications*, *91*, 1–16. doi:10.1016/j.comcom.2016.06.003

Akyildiz, I. F., Su, W., Sankarasubramaniam, Y., & Cayirci, E. (2002). A survey on sensor networks. *Communications Magazine, IEEE*, *40*(8), 102–114. doi:10.1109/MCOM.2002.1024422

Al Qurashi, M., Angelopoulos, C. M., & Katos, V. (2020). An architecture for resilient intrusion detection in ad-hoc networks. *Journal of Information Security and Applications*, *53*, 102530.

Anand, M., Balaji, N., Bharathiraja, N., & Antonidoss, A. (2021). A controlled framework for reliable multicast routing protocol in mobile ad hoc network. *Materials Today: Proceedings*.

Bhattacharya, A., & Sinha, K. (2017). An efficient protocol for load-balanced multipath routing in mobile ad hoc networks. *Ad Hoc Networks*, *63*, 104–114.

Binh, H. T. T., & Dey, N. (Eds.). (2018). *Soft computing in wireless sensor networks*. CRC Press.

Chen, X., Makki, K., Yen, K., & Pissinou, N. (2009). Yen, Pissinou N. Sensor Network Security: A Survey. *IEEE Communications Surveys and Tutorials*, *11*(2), 52–73. doi:10.1109/SURV.2009.090205

Das, S. K. (2021). Smart Design and Its Applications: Challenges and Techniques. In *Nature-Inspired Computing for Smart Application Design* (pp. 1–6). Springer.

Das, S. K., Das, S. P., Dey, N., & Hassanien, A. E. (Eds.). (2021). *Machine learning algorithms for industrial applications*. Springer.

Das, S. K., & Giacalone, M. (Eds.). (2022). *Fuzzy Optimization Techniques in the Areas of Science and Management*. CRC Press.

Das, S. K., Samanta, S., Dey, N., & Kumar, R. (Eds.). (2020). *Design frameworks for wireless networks*. Springer Singapore.

Das, S. K., Samanta, S., Dey, N., Patel, B. S., & Hassanien, A. E. (Eds.). (2021). *Architectural Wireless Networks Solutions and Security Issues*. Springer Singapore Pte. Limited.

De, D., Mukherjee, A., Das, S. K., & Dey, N. (Eds.). (2020). *Nature inspired computing for wireless sensor networks*. Springer.

De, D., Mukherjee, A., Das, S. K., & Dey, N. (2020). Wireless sensor network: applications, challenges, and algorithms. In *Nature inspired computing for wireless sensor networks* (pp. 1–18). Springer.

Dragonas, V., Tsoumanis, G., Koufoudakis, G., Papamichail, A., Oikonomou, K., & Stavrakakis, I. (2020). A Fairness-Aware topology independent TDMA MAC policy in time constrained wireless ad hoc networks. *Computer Networks, 171*, 107157.

Eksert, M. L., Yücel, H., & Onur, E. (2021). Intra-and inter-cluster link scheduling in CUPS-based ad hoc networks. *Computer Networks, 185*, 107659.

El-Bendary, M. A. (2015). *Developing security tools of WSN and WBAN networks applications*. Springer Japan.

Kahate, A. (2014). *Cryptography Techniques. In Cryptography and Network Security* (3rd ed.). McGraw.

Kodali, R. K., & Chougule, S. (2013). Hybrid key management technique for WSNs. In *International Conference on Heterogeneous Networkingfor Quality, Reliability, Security and Robustness*. Springer. doi: 10.1007/978-3-642-37949-9_74

Kumar, J., Singh, A., & Bhadauria, H. S. (2021). Congestion control load balancing adaptive routing protocols for random waypoint model in mobile ad-hoc networks. *Journal of Ambient Intelligence and Humanized Computing, 12*(5), 5479–5487.

Lou, C., & Zhuang, W. (2016). Energy-efficient routing over coordinated sleep scheduling in wireless ad hoc networks. *Peer-to-Peer Networking and Applications, 9*(2), 384–396.

Patel, M. M., & Aggarwal, A. (2013). Security attacks in wireless sensor net- works: A survey. *Intelligent Systems and Signal Processing (ISSP), International Conference on 2013*, 329-333.

Patsariya, M., & Rajavat, A. (2021). Node capability-based route selection on mobile ad hoc network. *Materials Today: Proceedings*.

Rathore, H. (2016). Case study: A review of security challenges, attacks and trust and reputation models in wireless sensor networks. In *Mapping Biological Systems to Network Systems* (pp. 117–175). Springer.

Robinson, Y. H., Julie, E. G., Saravanan, K., Kumar, R., & Son, L. H. (2019). FD-AOMDV: Fault-tolerant disjoint ad-hoc on-demand multipath distance vector routing algorithm in mobile ad-hoc networks. *Journal of Ambient Intelligence and Humanized Computing, 10*(11), 4455–4472.

Sahingoz, O. K. (2013). Large scale wireless sensor networks with multi- level dynamic key management scheme. *Journal of Systems Architecture*, *59*(9), 801–807. doi:10.1016/j.sysarc.2013.05.022

Sarkar, S., & Datta, R. (2016). A secure and energy-efficient stochastic multipath routing for self-organized mobile ad hoc networks. *Ad Hoc Networks*, *37*, 209–227.

Thiagarajan, R., Babu, M. R., & Moorthi, M. (2021). Quality of Service based Ad hoc On-demand Multipath Distance Vector Routing protocol in mobile ad hoc network. *Journal of Ambient Intelligence and Humanized Computing*, *12*(5), 4957–4965.

Xiao, Y., Chen, H., Yang, S., Lin, Y. B., & Du, D. Z. (2009). Wireless network security. *EURASIP Journal on Wireless Communications and Networking*, *2009*(1), 1–3.

Yang, K. (2014). *Wireless sensor networks*. Academic Press.

Yick, J., Mukherjee, B., & Ghosal, D. (2008). Wireless sensor network survey. *Computer Networks*, *52*(12), 2292–2330. doi:10.1016/j.comnet.2008.04.002

Chapter 8
An Intelligent Agent-Based Localization System in Underwater Wireless Sensor Networks

Ashish Kumar Dass
National Institute of Science and Technology, Berhampur, India

Sudhakar Das
National Institute of Science and Technology, Berhampur, India

Sudhir Ranjan Pattanaik
National Institute of Science and Technology, Berhampur, India

ABSTRACT

Underwater wireless sensor networks (UWSN) comprise of numerous relatively low-cost, self-organizing, and static as well as mobile sensor nodes. The wireless sensor node is used to sense natural phenomenon in terms of humidity, temperature, etc. BS is used to collect required information with proper location information from the sensor nodes and process it for future references. The problem of localization in underwater sensor nodes has led to proposal of many techniques over the past few decades. In this chapter, an intelligent scan model method has been proposed for localization system under UWSN. The localization system is used to identify the actual location of sensor node for identifying sense data properly trough Data MULE (mobile ubiquitous LAN extensions), which is an unmanned floating vehicle that has a sufficient amount of storage area and rechargeable energy to store the data of the sensors in the field. Finally, the proposed method is compared with existing method and outperforms the results efficiently.

DOI: 10.4018/978-1-6684-7343-6.ch008

INTRODUCTION

In modern era, the applications of wireless network increases rapidly due to its several flexibility and usage. It has several variations in terms of applications like wireless ad-hoc network or WANET, mobile ad-hoc network or MANET, vehicular ad-hoc network or VANET, the combination of several ad-hoc networks or HANET and wireless sensor network or WSN. The current paper is based on UWSN which stands for Underwater Wireless Sensor Network. It is a subset of WSN. UWSN is a collection of several autonomous nodes named as sensor nodes, vehicle nodes, and Base Station (BS) nodes (Yang et al., 2019). UWSN is a collection of different nodes based on several variations based on their working principle and location in the application area. Many sensor nodes interconnect each other and send the collected information to the surface sink or the buoy station. BS collection basic information from all cluster and process it for future references through satellite or onshore sink node. The combination of all nodes spreads or deployed in particular acoustic region to achieve a specific goal. The purpose of the nodes is to monitor underwater environmental condition and sense different homogeneous or heterogeneous elements and collect this data and send it to the BS that is situated outside of the water. In underwater, there are three types nodes are available such as static nodes, dynamic nodes, and semi-static nodes (Jain & Hussain, 2020; Akyildiz et al., 2005).

The UWSN rapidly used in several applications such as ocean sampling, distributed tactical surveillance, various military application, assisted navigation, disaster management, undersea exploration, seismic diagnosis and imagination, etc. Although, there are several usages of the UWSN, but without position information of each node in the network applications are illogical.

Many algorithms in WSN use GPS connectivity for the localization. But in UWSN GPS is very cost effective and in some cases GPS cannot be applicable in the harsh under water environment. So there requires availability of one smart node which moves in the sensing area by creating a proper path such that it can communicate to all the un-localized nodes. This node is called mobile ubiquitous LAN extensions (MULE). various authors have proposed different path planning models such as namely, random way point, Scan, D-Scan, Hilbert, and Circles having some issues such as location accuracy, coverage etc. (Mishra et al., 2018).

Hence, there is need of some intelligent and efficient technique that reduces the above mentioned issues and performs any operation efficiently. The current paper is based on localization system using an intelligent agent. The intelligent agent called data MULE play the role of optimization and solve different location accuracy, energy conservation and extending life time of the network of the network. In last decade, several heuristic, meta-heuristic, nature-inspired algorithms are proposed by the different authors to solve different issues of the WSN as well as UWSN.

The proposed method is the extension of the works (Mishra et al., 2018; Mishra et al., 2018) which is based on "Scan" model of the WSN named as "Improved Scan Model". Details of the proposed work are described later section.

The roadmap the paper as follows. Unit 2 described some information about existing works. Unit 3 illustrated the main proposed method. Unit 4 described simulation analysis. And Unit 5 concludes the paper.

LITERATURE REVIEW

In several years, various works are proposed in the context of wireless networks based on several issues and their constraints like Das et al. (2021) proposed a list of application in the area of wireless network and its related application. Main purpose of this article is to enhance the activity and application of wireless network. Each of the work in this book indicates architectural solution based on particular issues. It helps to design some related security issues based on different factors and its information system. It helps to manage several information efficiently. Yang et al. (2018) designed an intelligent system for transportation system in wireless network. This is based on an existing transportation system based on process structured system. Finally, it helps to enhance network capabilities and services of the network. It also helps to user function and usages in the network and network metrics properly to maintain the network. Loganathan and Subbiah (2020) designed an energy-based communication system for device-to-device communication in the network. It is based on multicriteria decision making system where multiple criteria are involved for integrating the network metrics efficiently. Finally, it helps to enhance the network lifetime and helps in communication system. P. Chandrakar (2019) designed an authentication system for the users in wireless network. This is basically based on healthcare system and used for medical purpose. This proposal is used for sensing patient body information and send to the doctor for treatment and diagnosis purpose. It also helps in user authentication, privacy and data security purpose, so that efficient result is comes from the diagnosis system. Das et al. (2021) proposed a list of algorithms based on efficiency of machine learning methodologies. It helps to manage several machines learning information to deals with some problems. It helps to manage several operations and its application to solve real-life problem. Several machine learning algorithms are mentioned in this book that handle several areas and give new direction to the readers. Jat et al. (2018) designed an intelligent technique for QoS in WLAN. This proposal is based on video delivery system. This is based on multimedia application for video data processing and analyzing. The data is analyze here based on real time data that generated by the internet. It also helps in video data transmission, storage, evaluating, and broadcasting. In WSN, data is gathered from multiple homogeneous or heterogeneous sources because, real life data is connected with different IoT, IoV or cloud environment. So, it is difficult to keep the natures of the data in same structure. Information retrieval (Rasheed & Banka, 2018) is very important part in modern re-search areas which indicate collect information that are stored in unstructured form based on multiple local languages and process it in particular pattern after observing. De et al. (2020) proposed a method that contains several nature inspired application in the area of wireless sensor network. Several issues and problems are mentioned in this article to handle several operation of the network. The work is based on architecture system that handles several computation applications. Each of the application is inspired by nature with the context of algorithm. Hao et al. (2019) designed an evaluation system for big data analysis. This data is based on IoV where it means internet of vehicle. This proposal is based on K-means algorithm that is used here as a clustering. In this work, different behavior of the driving are involve for controlling vehicle. Finally, it helps in reducing fuel consumption and helps in transportation globally. Some ad-hoc network and WSN based works also proposed like Gawas and Govekar (2019) proposed ACO based routing method for VANET. This is a nature inspired based method that helps to find an optimal route between source and destination nodes. Basically, this proposal is based on IoV that stands for Internet of Vehicle; this is one part of VANET. This proposal is based on traditional routing protocol which named as AODV routing. ACO algorithm helps to optimize network parameter and produce the optimal path. Das et al. (2020) proposed a framework application for handling several operations of the network. It helps to manage several designed model of the network. Each of the models is based on some factors of application. It indicates solution prospective point of view operation in the area of wireless network. It helps to guide several operations and model of the network that helps to incorporate the system in smart application. Sarkar et al. (2018) designed an ant-based routing protocol for MANET that is based on traditional routing protocol AODV. These ants are working based on the principle of AOC algorithm. Two basic parameters are used in this routing method like hop-count and residual energy. Simulation results outperform some existing routing mechanisms such as AODV, DSR, and EADSR by reducing congestion and enhancing quality of services as well as transmission rate. Abbas and Fan (2018) proposed a clustering method for VANET which purpose is for low latency. This is based on ACO technique. In this method, traditional AOMDV routing method is used for handling latency and security. The main purpose is to find cluster-head and cluster nodes for data communications. Finally, it enhances some network metrics such as energy consumption, QoS, throughput, and delay of the network. De et al. (2020) designed a survey model for wireless sensor network for application management. Each of the work is based on several algorithms and challenges system. But it helps to resolve several issues based of factors of information. Several solutions are mentioned in this proposed to resolve the challenges based on framework analysis of wireless sensor network. Mandhare et al. (2016) designed a meta-heuristic-based routing protocol for MANET. The purpose of this routing is to enhance QoS of the network. The proposed method reduces the issue of non-deterministic NP hard issue. The key technique is used in this method that is Cuckoo search method. The method used in the AODV routing technique with the help RREQ and RREP packet for finding the shortest path. Finally, it compares with some nature-inspired techniques such as PSO, ACO and simple AODV and enhance the matrices scalability and mobility, and reduce congestion of the network. Some localization based works also proposed by authors (Binh et al., 2020; Binh et al., 2018) for maximizing workable area of the network. S. K. Das (2021) designed a review article for handling several challenges of smart application. The work also mentioned several techniques and issues modeling for factor analysis based on smart application system. It helps to deals several issues and application that are used in smart application design. It also helps to model several parameters that are directly or indirectly influence in the application. Wagh and Gomathi (2018) proposed a method for route designing in VANET. In this method, a novel nature inspired technique is used such as lion optimization technique. This novel optimization technique mixed with fuzzy logic for reducing the uncertainty of the network. Fuzzification method of the fuzzy logic is used for reducing several costs of the network such as awareness, collision, travel, congestion and increases the reliability as well as QoS of the network. Tripathi and Das (2018) designed a robustness method for routing in ad-hoc network. Here, network is based on transparent system and heterogeneous in nature. Basic key elements of this network are fusion of linear programing and game theory. It consists of two objective functions for two players where both players are competitive to each other. Finally, it helps to reduce conflicting nature and pave the complexity of the network and find optimal route. Yang et al. (2018) designed an intelligent system for transportation system in wireless network. This is based on an existing transportation system based on process structured system. Finally, it helps to enhance network capabilities and services of the network. It also helps to user function and usages in the network and network metrics properly to maintain the network. Loganathan and Subbiah (2020) designed an energy based communication system for device to device communication in the network. It is based on multi-criteria decision making system where multiple criteria are involved for integrating the network metrics efficiently. Finally, it helps to enhance the network lifetime and helps in communication system. Binh and Dey (2018) designed soft computing review system based on wireless sensor network and its application. The work is fully based on several techniques of soft computing technique that helps to model its related application. Each of the work helps to model and resolve the issues that are raised during the operation of wireless sensor network.

P. Chandrakar (2019) designed an authentication system for the users in wireless network. This is basically based on healthcare system and used for medical purpose. This proposal is used for sensing patient body information and send to the doctor for treatment and diagnosis purpose. It also helps in user authentication, privacy and data security purpose, so that efficient result is comes from the diagnosis system. Jat et al. (2018) designed an intelligent technique for QoS in WLAN. This proposal is based on video delivery system. This is based on multimedia application for video data processing and analyzing. The data is analyze here based on real time data that generated by the internet. It also helps in video data transmission, storage, evaluating, and broadcasting. In WSN, data is gathered from multiple homogeneous or heterogeneous sources because; real life data is connected with different IoT, IoV or cloud environment. So, it is difficult to keep the natures of the data in same structure. Information retrieval (Rasheed & Banka, 2018) is very important part in modern research areas which indicates collect in-formation that are stored in unstructured form based on multiple local languages and process it in particular pattern after observing. Hao et al. (2019) designed an evaluation system for big data analysis. This data is based on IoV where it means internet of vehicle. This proposal is based on K-means algorithm that is used here as a clustering. In this work, different behavior of the driving are involve for controlling vehicle. Finally, it helps in reducing fuel consumption and helps in transportation globally. Tripathi and Das (2018) proposed five input parameters based intelligence routing using multiple criteria of ad-hoc network. This is based on soft set method which mixed by extended fuzzy set i.e., intuitionist fuzzy set and two techniques of the multi-criteria decision system. Each input parameter is mapped into the soft set in terms of three elements such true membership value, false membership value and between both which known as hesitation member-ship value. Finally, it helps to resolve the uncertainty of the network efficient and derive optimal route of the network. Sun et al. (2018) proposed an optimization technique for handling attack in WSN. This is based on PSO optimization along with multiple objectives of the network. It contains to minimize the energy consumption of the nodes and maximize the load balancing of the network. The natures of both objectives are non-linear formulation and nature of the PSO optimization is binary. Finally, it helps to dynamically maintain all objectives and their constraints and maintain the convergence of the system and finally helps to enhance the network lifetime. Cao et al. (2018) designed a PSO based distributed optimization technique for WSN. It is based on the deployment system of the network. In this system, nodes are heterogeneous types and divided into two types as relay nodes and simple nodes. This system helps to maximize coverage of the network and maximize the network lifetime with the help of relay nodes. In this paper, nodes are deploying in the 3D environment system and finally it helps to reduce time of the passing message and cost of the network. Phoemphon et al. (2018) designed a hybrid method for optimization in WSN using fusion of PSO, fuzzy logic and machine learning. This fusion helps in finding global positioning system of the network. The main key issue of this system is residual energy of the sensor node. And the fusion helps to reduce it comparatively of the network. This fusion also helps to analyze several factors of the network such as variation of the topology, coverage sensing, and density of the nodes. Das and Giacalone (2022) proposed a book for the purpose of fuzzy logic and its related usage in the areas of engineering and management. It contains several related works that helps to deals with several types of issues and problems that helps in the real-life applications.

PROPOSED METHOD

In this section, the proposed method is discussed briefly. It divided into some steps that contain different inherent components that help to under-stand the proposed method properly.

Application of Data MULE

Due to restrict information measure, severely non-negligible node quality and impaired channel, localization for Underwater Wireless sensing element Networks (UWSNs) is incredibly difficult to help positioning most of existing localization schemes assumes that the network incorporates a plurality of auxiliary anchor nodes. As Geo Positioning System (GPS) and different existing direct positioning technologies don't work expeditiously in underwater atmosphere, sometimes take with information of node with special instrumentation as anchor node. Because of the high price and poor flexibility it has not well appropriate for UWSNs. In order to beat this disadvantage, it's a high requirement for proposing an associate Anchor-Free Localization algorithmic program which will be self-sustain and is intended for effectively than pre-existing models.

A self-sufficient mobile node with rechargeable energy and trained with movements dynamically interacts with static agents to transfer, download, or transport knowledge to a distinct physical location. It have a tendency to take into account a system comprising associate degree Autonomous Underwater Vehicle (AUV) and lots of static underwater sensing element Nodes (USN) networked along optically and acoustically. The AUV can also called data MULE (Mobile Ubiquitous LAN Network) as it self- sustains and self-located node with highly equipped software and hardware. It can locate the static nodes using vision and hover above the static nodes for data up-load.

According to static or dynamic and according to requirement of algorithm. AUVs move in an exceedingly pre-determined or dynamically controlled movement fashion. So as to realize insight into the results of AUVs' quality, we tend to associate analyze the network performance with a varied AUVs' movement pat-terns and thus the range of AUVs that assures that having an optimum AUVs' movement pattern is incredibly necessary to understand a peak network performance.

The mobile anchor node moves mistreatment completely different models like the equidistant spiral, Random, Zigzag model and broadcast the in-formation packets every time amount, containing its own current coordinates, and conjointly the serial form of the anchor-nodes. The Underwater detector node receives the coordinate's data from the mobile anchor-node or different detector nodes. The coordinate data is unbroken and forwarded if the position of the node has not been calculated. Otherwise, the node only forwards the information received.

When the undetected sensor node receives beacon signals from an AUV, it initiates negotiation with the AUV to work out the channel for knowledge transmission and the modulation theme. Then, exploitation the negotiated short-range/high-rate channel from the AUV the sink node receives the aggregated knowledge. Knowledge MULE node collects knowledge exploitation equipped sensors and sporadically send the info to the sink. For path choice we are able to conjointly apply different routing metrics like ETT (Expected Transmission Time) and ETX (Expected Transmission Count).

Distance Calculation Between Dumb Node and Anchor Node

Anchor node or vessel node broadcasts the message in a predefined for-mat. That message format includes the location (x, y, z coordinate), time of message creation, and message itself. If any unsettled node is within the range of that anchor node, then that node receives that message and store in the memory. Now using the below mentioned distance formula dumb node can calculate their distance from that of beacon node shown in Eq. 1 (Mishra et al., 2018).

$$D = \sqrt{(dx - x)^2 - (dy - y)^2} \tag{1}$$

where, D is the distance calculated between anchor node and dumb node; x is the X-coordinate of anchor node; dx is the X-coordinate of dumb node; y is the Y -coordinate of anchor node; dy is the Y-coordinate of dumb node. Equation of the circle i cantered at x and y is shown in Eq. (2) (Mishra et al., 2018).

$$(x - x_i)^2 + (y - y_i)^2 = R_i^2 \tag{2}$$

Figure 1. Principle of spiral model

Algorithm 1: Principle of spiral model.	
Step 1:	The vessel node containing data MULE determines its position with the help of the GPS system.
Step 2:	Then, it will broadcast a message which will be received by all unknown nodes which are in the range of the vessel node.
Step 3:	Then, the data MULE has moved to maintain a spiral and update x co- ordinate and y-co-ordinate.
Step 4:	It will change its direction in reverse and it travels in the same way up to its previous coordinates according to the path pattern shown in Fig. 2.
Step 5:	During the travel of data MULE all of the dumb nodes store maximum three messages. And with the help of three positional coordinates of data MULE, the dumb node calculates its position.

Where, x_i and y_i is the location of the anchor node also called as the beacon node, and Ri is the distance from anchor location to node location. The proposed model is validated scan model (Mishra et al., 2018; Mishra et al., 2018) and spiral model. Description of the spiral model shown as pseudo code in Algorithm 1.

But it has some limitations as in spiral model the data MULE moves by making angular velocity. So, halt point of the data MULE makes very difficult and measurement of the equal distance of halt is cumbersome. So, implementing spiral model makes very difficult.

The proposed method is named as Improved Scan Model (ISCAN) which is the modified version of the scan model. Now, considering the main objective of data collection from the nodes of sensor network, an efficient distributed localization model has proposed. In this model, it has attempted to improve the percentage of nodes to be localized along with the energy consumption to carry out localization. Further, it has attempted to resolve the linearity problem by receiving another message either from beacon node or any neighboring node upon request. The innovative model focuses to improve the percentage of k- coverage by placing the beacon nodes in a random sized quadrilateral position instead of rectangular or squared manner. In this model, the data MULE node follows random co-axial path instead of moving its direction on perpendicular axis when it reaches nearer to the boundary. Detail of the proposed method shown in Algorithm 2.

RESULT AND DISCUSSION

This section is illustrated simulation results and analysis based on some existing algorithms. In this simulation, nodes are taken in variation range like 50, 100, 150, 200, and 250 and transmission range varies as 100, 150, and 200. The performance metrics that analyze are residual energy, number of nodes settled, and localization error. Finally, it outperforms the existing methods.

Figure 2. The proposed ISCAN method

Algorithm 2: The proposed ISCAN method
Step 1: The vessel node containing data MULE determines its position with the help of the GPS system. **Step 2:** Then, it will broadcast a message which will be received by all unknown nodes which are in the range of the vessel node. **Step 3:** Then, the value of the x-coordinate of the data MULE node increases by addition of range in x-coordinates value. Simultaneously the data MULE makes small angel by changing slightly the y-coordinate. **Step 4:** Further, the data MULE moves in horizontal direction until it touches the boundary. **Step 5:** Then afterwards it changes the direction in upwards by changing its y- coordinate. Here also the small changes of x-coordinate occur as shown in the Fig. 3. **Step 6:** It will change its direction in reverse and it travels in the same way up to its previous coordinates according to the path pattern shown in Fig. 3. **Step 7:** During the travel of data MULE all of the dumb nodes store maximum three messages. And with the help of that coordinates of MULE, the dumb node calculates its position.

Figure 3. Beacon node movement during localization (Improved SCAN)
(Mishra et al., 2018).

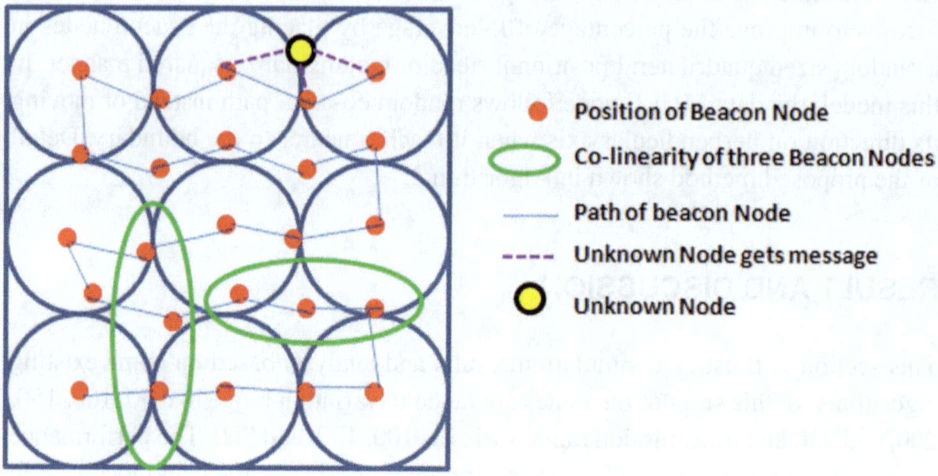

Figure 4. Estimated versus real position

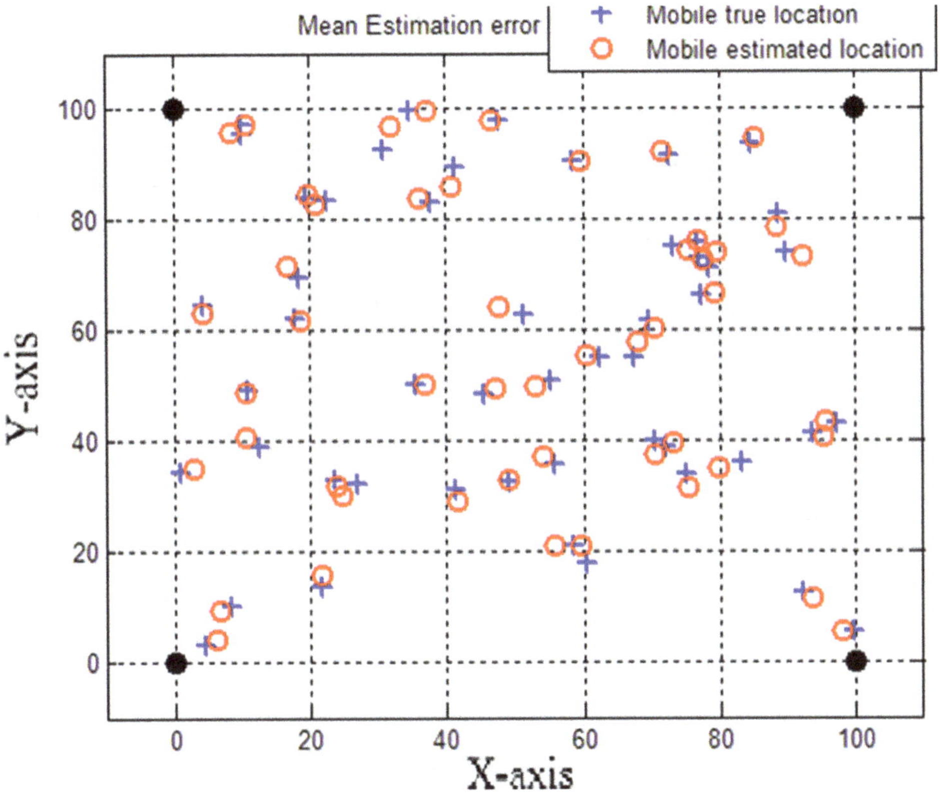

The Localization Error (LE) is the Mean Square Error (MSE) between the estimated location and the true location of the un-known node, as per the Eq. 3 shown below.

$$LE = \frac{1}{K}\sum_{i=1}^{k}\sqrt{\left(X_{est_i} - X_{real_i}\right)^2 - \left(Y_{est_i} - Y_{real_i}\right)^2} \tag{3}$$

where K is the number unknown nodes; $\left(X_{est_i}, Y_{est_i}\right)$ is the estimated coordinates of unknown nodes; $\left(X_{real_i}, Y_{real_i}\right)$ is the real coordinates of unknown nodes.

Figure 4 illustrates the estimated location of topology of 50 sensor nodes with the actual location by SCAN algorithm where range of data MULE is 100m. Most of the estimated positions are nearly equal to the actual positions of the sensor nodes. The red circle shows the estimated location. Similarly, localization error has been calculated for topology contains 50,100,150,200,250 nodes for SCAN, SPIRAL, and ISCAN algorithm for variable Data MULE range of 100m, 150m and 200m. The experimented result has shown in the Table 1.

Table 1. Comparison of Algorithms for different Data MULE

	Range =100m			Range =150m			Range =200m		
Travelling Models	SCAN	SPIRAL	ISCAN	SCAN	SPIRAL	ISCAN	SCAN	SPIRAL	ISCAN
50	1.12	1.25	1.05	1.35	1.42	1.31	1.38	1.52	1.34
100	1.19	1.33	1.12	1.43	1.5	1.39	1.45	1.59	1.41
150	1.22	1.36	1.16	1.46	1.54	1.42	1.47	1.62	1.45
200	1.24	1.37	1.18	1.47	1.57	1.43	1.47	1.63	1.47
250	1.24	1.38	1.19	1.47	1.58	1.44	1.47	1.63	1.47

From the Table 1, the graphs have generated for comparison of above three mentioned algorithms for different data MULE range as shown in Figures 5 to 7.

Figure 5. Localization Errors (range 100 meter)

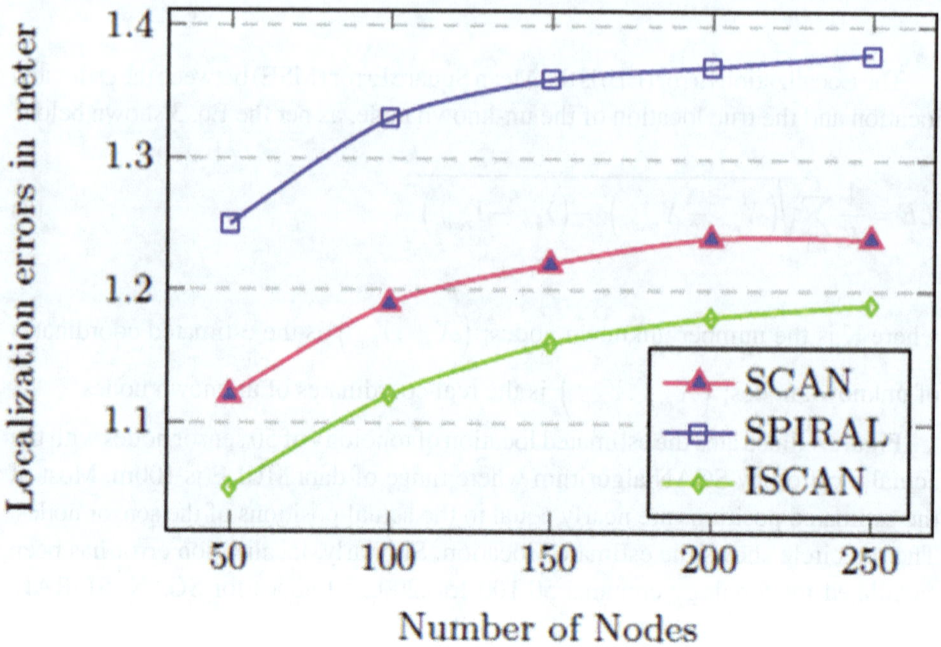

Figure 6. Localization Errors (range 150 meter)

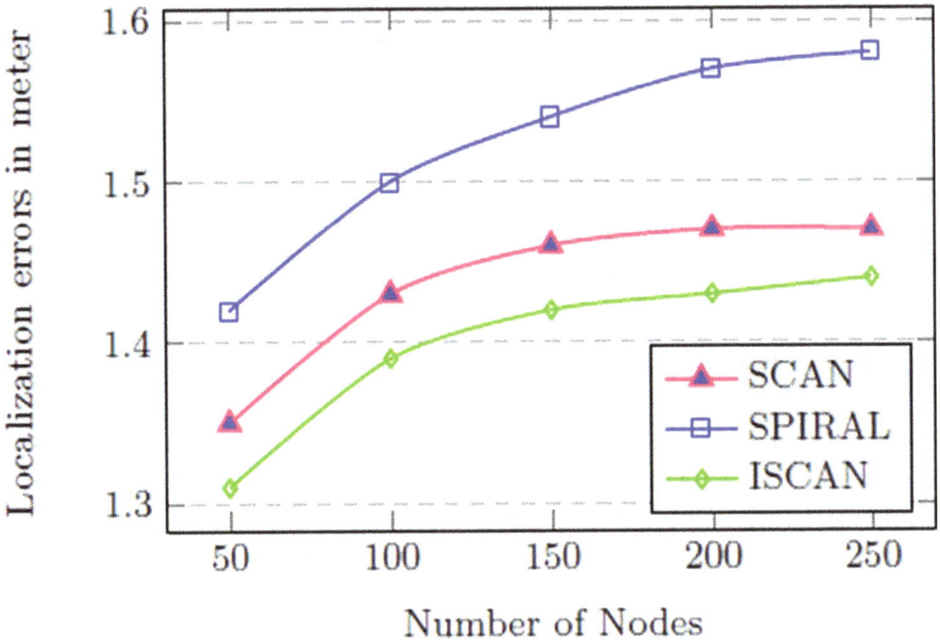

Figure 7. Localization Errors (range 200 meter)

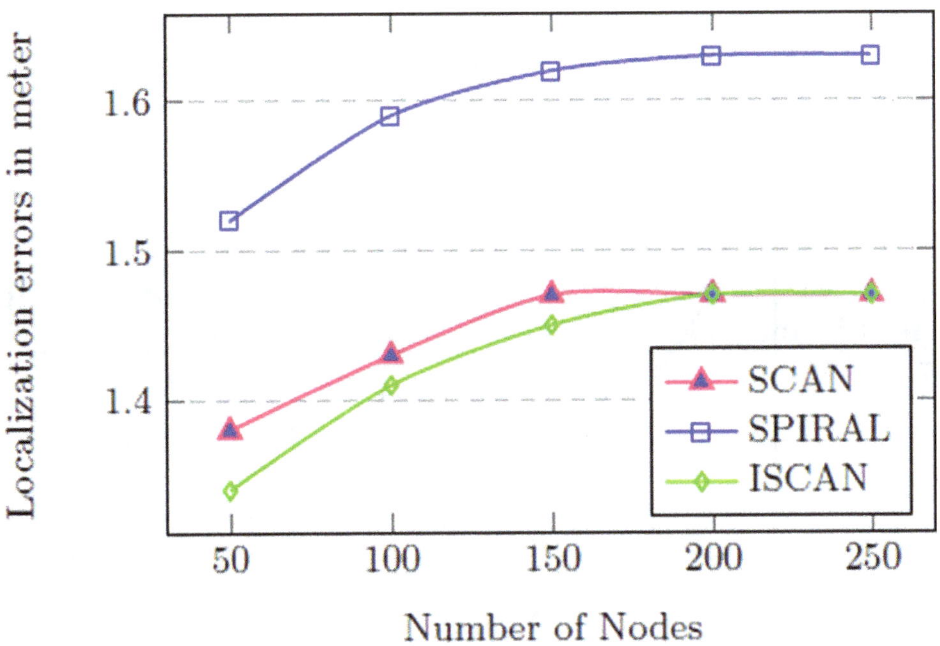

Table 2. Changed Data MULE Range

	Range =100m			Range =150m			Range =200m		
Travelling Models	SCAN	SPIRAL	ISCAN	SCAN	SPIRAL	ISCAN	SCAN	SPIRAL	ISCAN
50	36	35	37	32	29	34	31	29	33
100	68	66	68	61	59	63	60	58	61
150	105	102	105	93	88	95	91	89	92
200	142	138	147	128	125	131	121	120	121
250	183	178	187	148	142	149	144	143	144

Number of Nodes Settled

By using different algorithm and taking the variable size of unknown node such that 50,100,150,200,250 total settled nodes has calculated. Here also we have changed the data MULE range like 100m, 150m, and 200m. The result shown in the Table 2.

According to the above table the percentage of settled nodes are shown and compared for the all three discussed algorithms and shown in Figures 8 to 10.

Figure 8. Percentage of Nodes settled (range 100 meter)

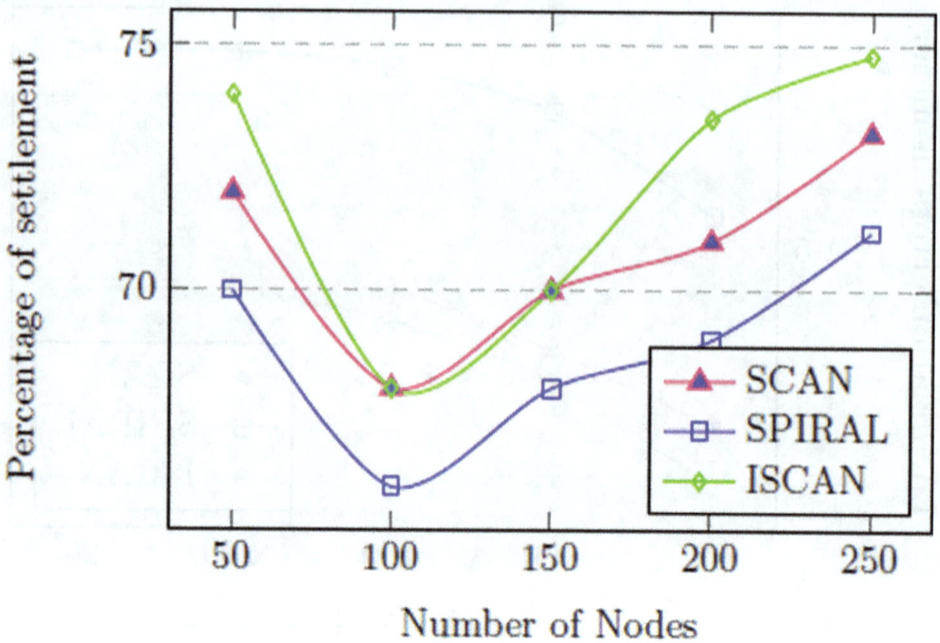

Figure 9. Percentage of Nodes settled (range 150 meter)

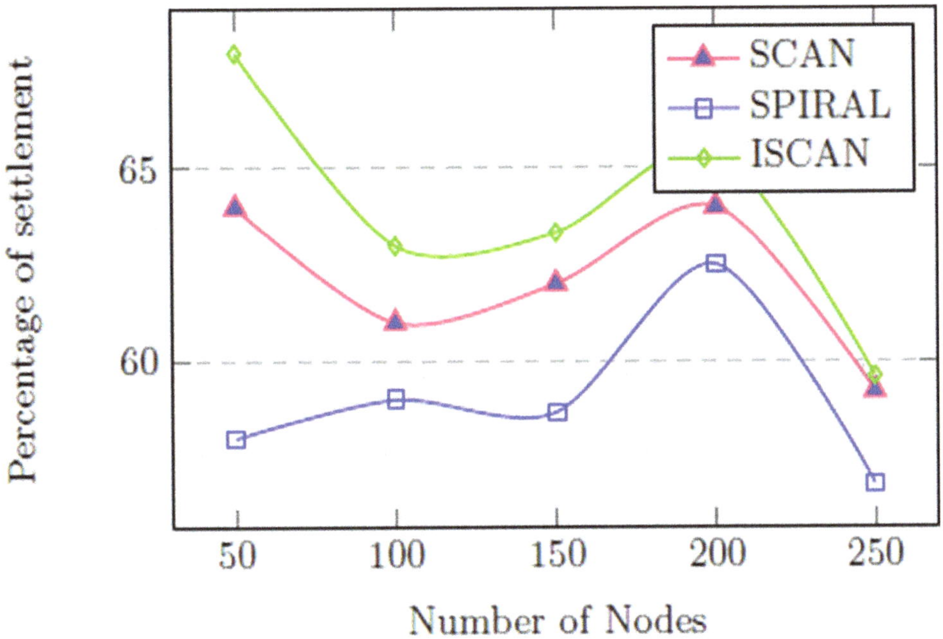

Figure 10. Percentage of Nodes settled (range 200 meter)

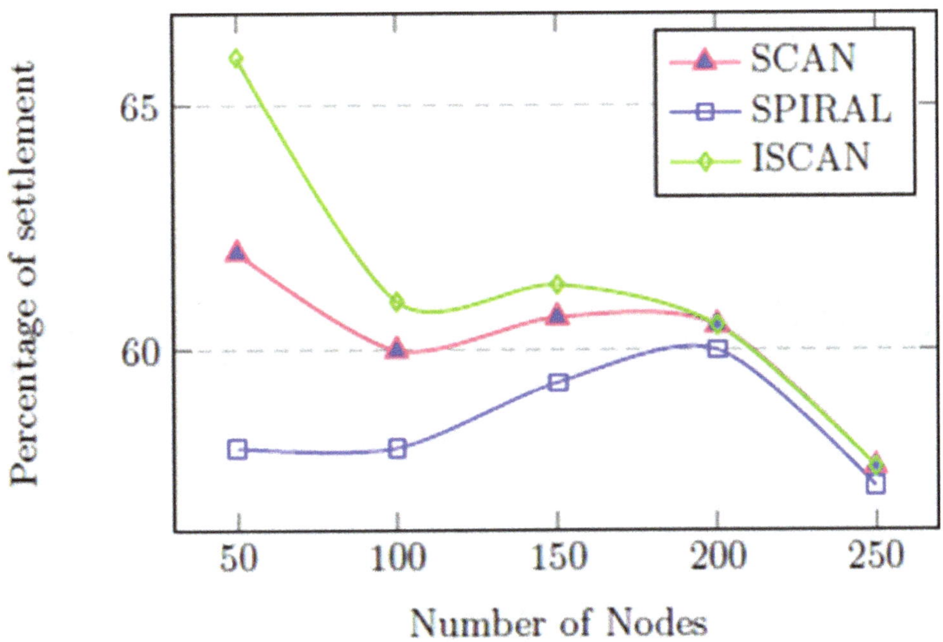

Table 3: Different types of models in different situation

	Range =100m			Range =150m			Range =200m		
Travelling Models	**SCAN**	**SPIRAL**	**ISCAN**	**SCAN**	**SPIRAL**	**ISCAN**	**SCAN**	**SPIRAL**	**ISCAN**
50	0.74293	0.74191	0.74157	0.74877	0.74677	0.74477	0.75342	0.75241	0.76218
100	0.75054	0.75049	0.74910	0.75930	0.74490	0.75320	0.75939	0.75839	0.76228
150	0.74677	0.74662	0.74677	0.76027	0.74852	0.75688	0.76039	0.76032	0.76519
200	0.74490	0.74430	0.74351	0.75932	0.75832	0.75557	0.76017	0.76018	0.76377
250	0.74496	0.74396	0.74240	0.75848	0.75658	0.75473	0.7655	0.76128	0.76839

Energy Remaining After Nodes Settled

The residual energy can be defined as the remained energy after all the nodes deployed. By applying the above mentioned algorithm and the collected information the Table 3 has created. It shows different types of model in different situation and figures shown in Figures 11 to 13.

Figure 11. Residual Energy (range 100 meter)

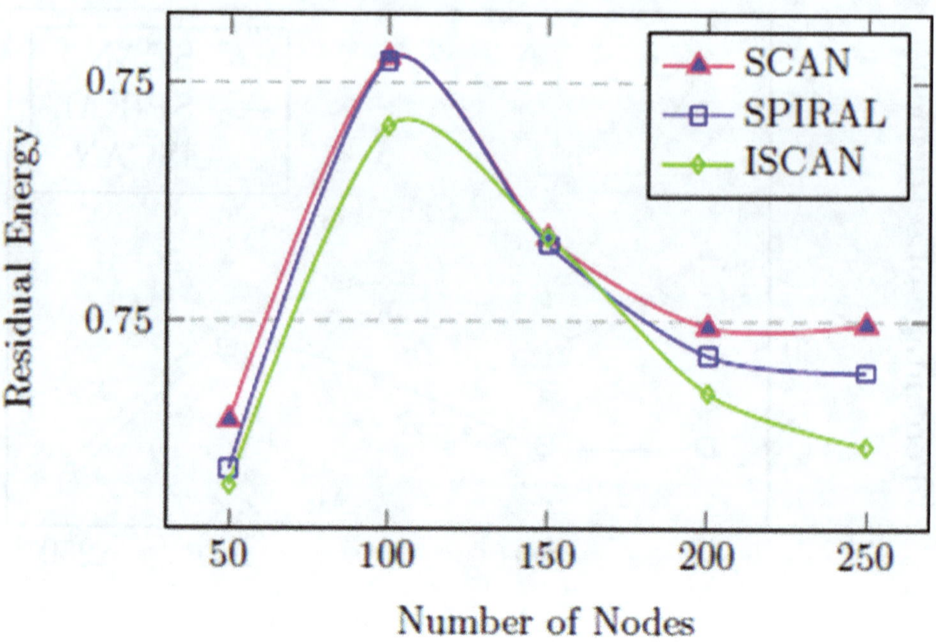

Figure 12. Residual Energy (range 150 meter)

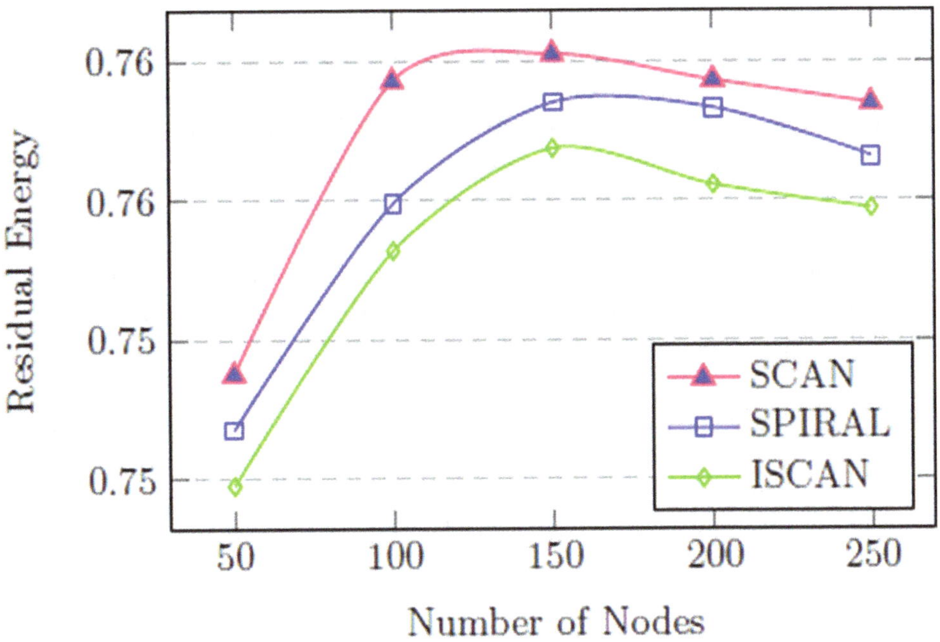

Figure 13. Residual Energy (range 200 meter)

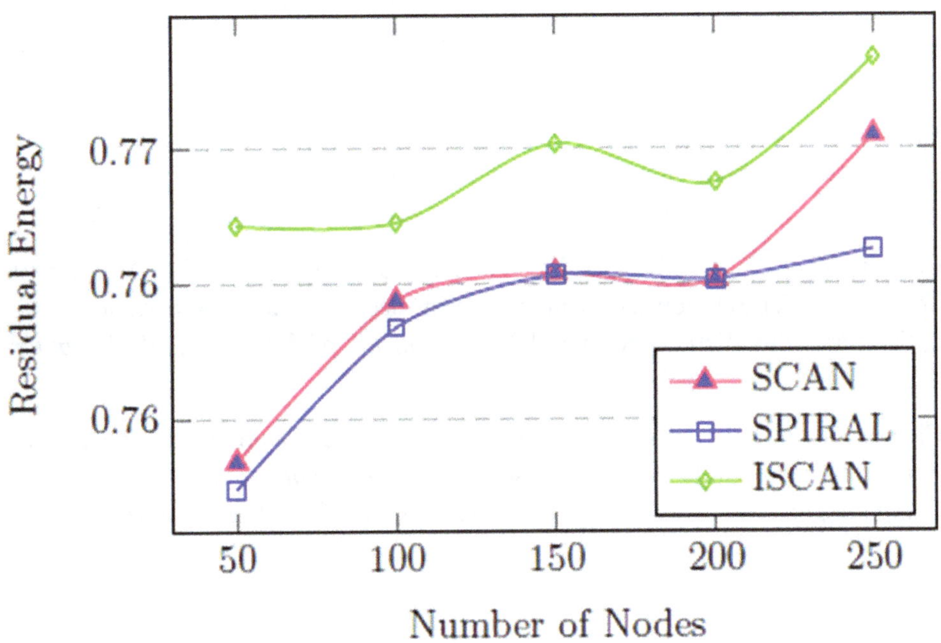

CONCLUSION

The proposed research work is based on an efficient localization method using data MULE. The algorithm has validated in term of matrices such as localization errors, number of nodes settled, and residual energy deployed randomly for different sensor nodes with different transmission variations. The different models were compared and contrasted with their advantages and disadvantages and the ISCAN model compared with existing algorithms had outperformed in accordance to the results. For the future scope of the research can be extended with other parameters to find the location accuracy in presence of faulty and dynamic node implementation.

REFERENCES

Abbas, F., & Fan, P. (2018). Clustering-based reliable low-latency routing scheme using ACO method for vehicular networks. *Vehicular Communications*, *12*, 66–74.

Akyildiz, I. F., Pompili, D., & Melodia, T. (2005). Underwater acoustic sensor networks: Research challenges. *Ad Hoc Networks*, *3*(3), 257–279. doi:10.1016/j.adhoc.2005.01.004

Binh, H. T. T., & Dey, N. (Eds.). (2018). *Soft computing in wireless sensor networks*. CRC Press.

Binh, H. T. T., Hanh, N. T., Nghia, N. D., & Dey, N. (2020). Metaheuristics for maximization of obstacles constrained area coverage in heterogeneous wireless sensor networks. *Applied Soft Computing*, *86*, 105939.

Binh, H. T. T., Hanh, N. T., Van Quan, L., & Dey, N. (2018). Improved cuckoo search and chaotic flower pollination optimization algorithm for maximizing area coverage in wireless sensor networks. *Neural Computing & Applications*, *30*(7), 2305–2317.

Cao, B., Zhao, J., Lv, Z., Liu, X., Kang, X., & Yang, S. (2018). Deployment optimization for 3D industrial wireless sensor networks based on particle swarm optimizers with distributed parallelism. *Journal of Network and Computer Applications*, *103*, 225–238.

Chandrakar, P. (2019). A secure remote user authentication protocol for healthcare monitoring using wireless medical sensor networks. *International Journal of Ambient Computing and Intelligence*, *10*(1), 96–116. doi:10.4018/IJACI.2019010106

Chandrakar, P. (2019). A secure remote user authentication protocol for healthcare monitoring using wireless medical sensor networks. *International Journal of Ambient Computing and Intelligence*, *10*(1), 96–116.

Choudhury, S., Majumder, A., & Sarkar, D. (2018). Enhanced-Ant-AODV for optimal route selection in mobile ad-hoc network. *Journal of King Saud University-Computer and Information Sciences*.

Das, S. K. (2021). Smart Design and Its Applications: Challenges and Techniques. In *Nature-Inspired Computing for Smart Application Design* (pp. 1–6). Springer.

Das, S. K., Das, S. P., Dey, N., & Hassanien, A. E. (Eds.). (2021). *Machine learning algorithms for industrial applications*. Springer. doi:10.1007/978-3-030-50641-4

Das, S. K., & Giacalone, M. (Eds.). (2022). *Fuzzy Optimization Techniques in the Areas of Science and Management*. CRC Press.

Das, S. K., Samanta, S., Dey, N., & Kumar, R. (Eds.). (2020). *Design frameworks for wireless networks*. Springer Singapore.

Das, S. K., Samanta, S., Dey, N., Patel, B. S., & Hassanien, A. E. (Eds.). (2021). *Architectural Wireless Networks Solutions and Security Issues*. Springer Singapore Pte. Limited.

Das, S. K., & Tripathi, S. (2018). Adaptive and intelligent energy efficient routing for transparent heterogeneous ad-hoc network by fusion of game theory and linear programming. *Applied Intelligence*, *48*(7), 1825–1845.

Das, S. K., & Tripathi, S. (2018). Intelligent energy-aware efficient routing for MANET. *Wireless Networks*, *24*(4), 1139–1159.

De, D., Mukherjee, A., Das, S. K., & Dey, N. (Eds.). (2020). *Nature inspired computing for wireless sensor networks*. Springer.

De, D., Mukherjee, A., Das, S. K., & Dey, N. (2020). Wireless sensor network: applications, challenges, and algorithms. In *Nature inspired computing for wireless sensor networks* (pp. 1–18). Springer.

Gawas, M. A., & Govekar, S. S. (2019). A novel selective cross layer based routing scheme using ACO method for vehicular networks. *Journal of Network and Computer Applications*, *143*, 34–46. doi:10.1016/j.jnca.2019.05.010

Hao, R., Yang, H., & Zhou, Z. (2019). Driving Behavior Evaluation Model Base on Big Data From Internet of Vehicles. *International Journal of Ambient Computing and Intelligence*, *10*(4), 78–95. doi:10.4018/IJACI.2019100105

Hao, R., Yang, H., & Zhou, Z. (2019). Driving Behavior Evaluation Model Base on Big Data from Internet of Vehicles. *International Journal of Ambient Computing and Intelligence, 10*(4), 78–95.

Jain, U., & Hussain, M. (2020). Underwater Wireless Sensor Networks. In *Handbook of Computer Networks and Cyber Security* (pp. 227–245). Springer. doi:10.1007/978-3-030-22277-2_9

Jat, D. S., Bishnoi, L. C., & Nambahu, S. (2018). An Intelligent Wireless QoS Technology for Big Data Video Delivery in WLAN. *International Journal of Ambient Computing and Intelligence, 9*(4), 1–14. doi:10.4018/IJACI.2018100101

Jat, D. S., Bishnoi, L. C., & Nambahu, S. (2018). An Intelligent Wireless QoS Technology for Big Data Video Delivery in WLAN. *International Journal of Ambient Computing and Intelligence, 9*(4), 1–14.

Jayakumar Loganathan, J., & Subbiah, J. (2020, February). Energy aware Dynamic Mode Decision for Cellular D2D Communications by using Integrated Multi-Criteria Decision Making model. *International Journal of Ambient Computing and Intelligence, 11*(3), 7.

Loganathan, J., & Subbiah, J. (2020). Energy aware Dynamic Mode Decision for Cellular D2D Communications by using Integrated Multi-Criteria Decision Making model. *International Journal of Ambient Computing and Intelligence, 11*(3). doi:10.4018/978-1-5225-5643-5.ch028

Mandhare, V. V., Thool, V. R., & Manthalkar, R. R. (2016). QoS Routing enhancement using metaheuristic approach in mobile ad-hoc network. *Computer Networks, 110*, 180–191.

Mishra, T. K., Dass, A. K., & Panda, S. K. (2018, December). Enhanced Path Planning Model for Anchor-Free Distributed Localization in Wireless Sensor Networks. In *2018 Fifth International Conference on Parallel, Distributed and Grid Computing (PDGC)* (pp. 430-435). IEEE. 10.1109/PDGC.2018.8745802

Mishra, T. K., Kumar, S., & Dass, A. K. (2018, December). An Enhanced Path Planning Model for Anchor-Free Localization in Wireless Sensor Networks. In *2018 International Conference on Information Technology (ICIT)* (pp. 204-209). IEEE.

Phoemphon, S., So-In, C., & Niyato, D. T. (2018). A hybrid model using fuzzy logic and an extreme learning machine with vector particle swarm optimization for wireless sensor network localization. *Applied Soft Computing, 65*, 101–120.

Rasheed, I., & Banka, H. (2018, March). Query Expansion in Information Retrieval for Urdu Language. In *2018 Fourth International Conference on Information Retrieval and Knowledge Management (CAMP)* (pp. 1-6). IEEE.

Rasheed, I., & Banka, H. (2018, March). Query Expansion in Information Retrieval for Urdu Language. In *2018 Fourth International Conference on Information Retrieval and Knowledge Management (CAMP)* (pp. 1-6). IEEE.

Sun, Z., Liu, Y., & Tao, L. (2018). Attack localization task allocation in wireless sensor networks based on multi-objective binary particle swarm optimization. *Journal of Network and Computer Applications*, *112*, 29–40.

Wagh, M. B., & Gomathi, N. (2018). Route discovery for vehicular ad hoc networks using modified lion algorithm. *Alexandria Engineering Journal*, *57*(4), 3075–3087.

Yang, G., Dai, L., Si, G., Wang, S., & Wang, S. (2019). Challenges and security issues in underwater wireless sensor networks. *Procedia Computer Science*, *147*, 210–216. doi:10.1016/j.procs.2019.01.225

Yang, W., Wang, X., Song, X., Yang, Y., & Patnaik, S. (2018). Design of intelligent transportation system supported by new generation wireless communication technology. In Intelligent Systems: Concepts, Methodologies, Tools, and Applications (pp. 715-732). IGI Global.

Yang, W., Wang, X., Song, X., Yang, Y., & Patnaik, S. (2018). Design of intelligent transportation system supported by new generation wireless communication technology. In Intelligent Systems: Concepts, Methodologies, Tools, and Applications (pp. 715-732). IGI Global.

Chapter 9
Multiple Nontrivial Solutions to a Nonlocal Critical Exponent Problem

Amita Soni
National Institute of Technology, Rourkela, India

D. Choudhuri
National Institute of Technology, Rourkela, India

ABSTRACT

The authors use the manifold technique to demonstrate that the issue combining a nonlinear integro-differential operator and a term with critical exponent has three nontrivial weak solutions. The operator considered in the problem covers the case of the fractional p-Laplacian operator and is possibly degenerate. The outcomes include a sign-changing solution and a positive and a negative solution. Also, presence of critical exponent term increases the complexity of the problem as compactness property fails in this case.

INTRODUCTION

Compactness plays an important role in variational methods and functional analysis. In last few years, problems involving critical exponent is gaining a lot of attention because of the difficulty in dealing with the loss of compactness. Concentration-compactness principle is one of the important and fundamental tool to overcome this loss introduced by (Lions, 1985). An application of this principle can be seen in the paper of (Azorero & Peral, 1991) who to our knowledge were the first to work in this

DOI: 10.4018/978-1-6684-7343-6.ch009

direction. Here in [4], the authors have shown the existence of solutions for different cases of λ involved in the equation along with multiplicity of solutions of noncritical problem with non-symmetric functional. This principle can be further modified in (Bonder et. al., 2018) where authors have obtained concentration-compactness lemma for fractional p-Laplacian. The researcher is then got interested about multiplicity of solutions of such kind of problems. One of the methods to get the multiple solutions is based on the concepts involving manifolds. It was (Struwe, 1991) who worked on such problems defined on manifolds to guarantee the existence of three nontrivial weak solutions for a Pseudo-Laplace-operator without critical exponent term. Following the work of (Struwe, 1991), there are a few articles devoted to the existence of multiplicity of solutions by combining the concept of concentration-compactness lemma and manifolds. An important such contribution can be found in (Nápoli et. al, 2009). Here (Nápoli et. al, 2009) the authors have proved the existence of three nontrivial solutions of problem involving p-Laplacian and critical exponent on manifolds. In (Bonder & Silva, 2009) and (Silva, 2011) the authors have extended this result for p(x) Laplacian operator. The authors in (Cantizano & Silva, 2019) obtained the same results for the problem involving a fractional p-Laplacian with a critical exponent term. The given problem is a decision-making problem as for the problem under study it is being decided whether solutions will exist in the manifolds considered or not.

Motivated by the above papers, in this paper we are going to extend our result in a more general setting, i.e. for a general nonlocal integro-differential operator. It is a generalization owing to the fact that it covers the case of fractional p-Laplacian. This type of operator appeared in many applications and models such as in quasi-geostrophic dynamics, nonlocal diffusion and modified porous medium equations, phase transition models, dislocation problems (refer Caffarelli & Silvestre, 2007; Holden & Karlsen, 2012 and references therein). For more information on this operator the readers may refer to (Kuusi et. al., 2015). Note that we will not impose the condition that f is odd. The problem we will address is as follows.

$$\begin{aligned} -\mathcal{L}_{\Phi}u &= |u|^{p_s^*-2} u + \lambda f(x,u) \, in \Omega, \\ u &= 0 \, in \mathbb{R}^N \setminus \Omega, \end{aligned} \tag{1}$$

where, we have the following assumptions on the term f (x, u).

$(f_1)\, f:\Omega\times\mathbb{R}\to\mathbb{R}$ is a measurable function with respect to the first variable and continuously differentiable with respect to second variable for almost every x $\in \Omega$. Moreover f (x, 0) = 0 for every x $\in \Omega$.

(f_2) For any $u \in L^q(\Omega)$ and $p < q < p_s^*$, there exist constants

$$c_1\left(0,\frac{1}{p_s^*-1}\right), c_2\left(p,p_s^*\right), 0<c_3<c_4$$

such that

$$c_3\left\|u\right\|_{L^q}^q(\Omega)\leq c_2\int_{\Omega}^{\cdot}F(x,u)\,dx\leq\int_{\Omega}^{\cdot}f(x,u)\,u\,dx\leq c_1\int_{\Omega}^{\cdot}f_u(x,u)\,u^2\,dx\leq c_4\left\|u\right\|_{L_q}^q(\Omega).$$

The main result of this article is given is the form of the following theorem.

Theorem 1.1. With the assumptions (f_1) and (f_2) there exist $\lambda'>0$ depending on n, p, p_s^* and the constant c_3 in (f_2) such that for every $\Lambda\in\left[1,\left(\frac{c_2}{p}\right)^{1/4}\right)$, where c_2 is also from (f_2) and $\lambda>\lambda'$, there exist three distinct nontrivial weak solutions of the problem (1). Moreover, these solutions are positive, negative and with a non-constant sign.

FUNCTIONAL ANALYTIC SETUP AND MAIN TOOLS

Linear Elliptic operator [19]: A linear operator L of order k on $\Omega\subset\mathbb{R}^N$ defined as $Lu=\sum_{|\alpha|\leq k}a_\alpha(x)\partial^\alpha u$ is elliptic if for each x ∈ Ω and every nonzero

$$\xi=(\xi_1,\xi_2,\cdots,\xi_N)\in\mathbb{R}^N,$$

$$\sum_{|\alpha|=k}a_\alpha(x)\xi^\alpha\neq 0,$$

where $\xi^\alpha=\xi_1^{\alpha_1}\cdots\xi_N^{\alpha_N}$.

Nonlinear Elliptic operator [19]: A nonlinear differential operator

$$L(u)=F(x,u,\left(D^\alpha u\right)_{|\alpha|\leq k})$$

is elliptic if its first-order Taylor expansion with respect to u and its derivative about any point is a linear elliptic operator with respect to u and its derivatives.

Palais-Smale condition [18]: Let $I : X \to \mathbb{R}$ be a C^1 functional with X being a Branch space. Then it is said to satisfy Palais-Smale condition if the following holds: whenever (u_n) is a sequence in X such that $I(u_n)$ is bounded $I'(u_n) \to 0$ strongly in X* then (u_n) has a subsequence which converges strongly.

We now discuss a few definitions and the functional analytic setting which will be used in this paper. A nonlocal operator $-\mathcal{L}_\Phi$ is defined as:

$$\langle -\mathcal{L}_\Phi u, \varphi \rangle = \int \int_{\mathbb{R}^{2N}} \Phi(u(x) - u(y))(\varphi(x) - \varphi(y)) K(x, y)\, dxdy, \tag{2}$$

where $\varphi \in C_c^\infty(\mathbb{R}^N)$ i.e., a smooth function with compact support. The function Φ satisfies $\Phi(0)=0$ and the following monotonicity property and is a real valued continuous function across the range $\mathbb{R}$,

$$\Lambda^{-1} |t|^p \leq \Phi(t)t \leq \Lambda |t|^p, \forall t \in R. \tag{3}$$

We assume that Φ is convex to get the following conditions.

$$a^{-1} |t|^p \leq \Phi'(t)t^2 \leq a |t|^p \tag{4}$$

For some a>0. The kernel $K : \mathbb{R}^N \times \mathbb{R}^N \to R$ is a measureable function satisfying the following ellipticity property

$$\frac{1}{\Lambda |x-y|^{N+sp}} \leq K(x, y) \leq \frac{\Lambda}{|x-y|^{N+sp}}, \forall x, y \in \mathbb{R}^N, x \neq y, \tag{5}$$

Where

$$\Lambda \geq 1, s \in (0,1), p > 2 - \frac{s}{N} \, with\, q \in (p, p_s^*), where\, p_s^* = \frac{Np}{N-sp} \, and\, p' = \frac{p}{p-1}.$$

The nonlocal operator $-\mathcal{L}_\Phi$ which is based on the assumptions (3) and (5) is an elliptic operator. Take note that in (2), we recover the fractional p-Laplacian when the specific case $\Phi(t)=|t|^{p-2}t$ with $K(x, y) = |x-y|^{-(N+sp)}$ is taken, for every $\varphi \in C_c^\infty(\mathbb{R}^N)$ as

$$\left\langle (-\Delta)_p^s u, \varphi \right\rangle = \int_{\mathbb{R}^{2N}} \frac{|u(x)-u(y)|^{p-2}\left(u(x)-u(y)\right)\left(\varphi(x)-\varphi(y)\right)}{|x-y|^{N+ps}} dxdy$$

In other words, the fractional p-Laplacian for $1 \leq p < \infty$ and $s \in (0,1)$ is an extension of the nonlocal operator $-\mathcal{L}_\Phi$. It is logical to work with Sobolev space $W^{s,p}\left(\mathbb{R}^N\right)$ and express the Dirichlet condition in $\mathbb{R}^N / \Omega$ rather than $\partial\Omega$ due to the nonlocal nature of $-\mathcal{L}_\Phi$ specified in (2). Although fractional Sobolev space are well known since the turn of the century, particularly in the field of harmonic analysis, and have gained popularity recently due to the immense work of (Caffarelli & Silvestre, 2007) and the references therein. Now that we have the variational setting on a suitable function space for (1), we can move to our problem. We also present some fundamental results for this topic. For all measurable functions $u : \mathbb{R}^N \to R$, we set

$$\|u\|_{L^p(\mathbb{R}^N)} = \left(\int_{\mathbb{R}^N} |u(x)|^p \, dx \right)^{\frac{1}{p}},$$

$$[u]_{s,p} = \left(\int_{\mathbb{R}^{2N}} \frac{|u(x)-u(y)|^p}{|x-y|^{N+sp}} dxdy \right)^{\frac{1}{p}},$$

where $p \in (1,\infty)$ and $s \in (0,1)$. The space of all function $u \in L^p\left(\mathbb{R}^N\right)$ such that $[u]_{s,p}$ is finite and endowed with the norm

$$\|u\|_{W^{s,p}(\mathbb{R}^N)} = \left(\|u\|^p_{L^p(\mathbb{R}^N)} + [u]^p_{s,p} \right)^{\frac{1}{p}}.$$

is defined as the fractional Sobolev space $W^{s,p}\left(\mathbb{R}^N\right)$. Further information on fractional Sobolev space can be found in (Nezza et al., 2012) and the references therein. We now define $W_0^{s,p}(\Omega)$ which is a closed linear subspace of $W^{s,p}\left(\mathbb{R}^N\right)$ as follows.

$$W_0^{s,p}(\Omega) = \left\{ u \in W^{s,p}\left(\mathbb{R}^N\right) : u = 0 \, a.e. \, in \, \mathbb{R}^N / \Omega \right\}.$$

It is easy to see that the norms $\|\cdot\|_{L^p(\mathbb{R}^N)}$ and $\|\cdot\|_{L^p(\Omega)}$ agree on $W_0^{s,p}(\Omega)$. The Poincaré type inequality is given by:

$\|u\|_{L^p(\Omega)} \leq C[u]_{s,p}$, for all $u \in W_0^{s,p}(\Omega)$.

Thus, we can equivalently renorm $W_0^{s,p}(\Omega)$, by setting $\|u\|_{W_0^{s,p}(\Omega)} = [u]_{s,p}$, for every $u \in W_0^{s,p}(\Omega)$. Let $p_s^* = \dfrac{Np}{N-sp}$, with the agreement that $p_s^* = \infty$ if $N \geq sp$. It is well known that $\left(W_0^{s,p}(\Omega), \|\cdot\|_{W_0^{s,p}(\Omega)}\right)$ is a uniformly convex reflexive Branch space. It is continuously embedded into $L^r(\Omega)$, for all $r \in \left[1, p_s^*\right]$ if $sp<N$, for all $1 \leq r < \infty$ if $N=sp$ and into $L^\infty(\Omega)$ if $N<sp$. It is also compactly embedded in $L^r(\Omega)$. for any $r \in \left[1, p_s^*\right)$ if $N^3 sp$ and in $L^\infty(\Omega)$. for $N<sp$. Furthermore, $C_c^\infty(\Omega)$ is a dense subspace of $W_0^{s,p}(\Omega)$ with respect to $\|\cdot\|_{W^{s,p}(\Omega)}$.

An associated energy functional to the problem (1) is defined as

$$I(u) = \int_{\mathbb{R}^{2N}} P\Phi\left(u(x) - u(y)\right) K(x,y)\,dxdy - \frac{1}{p_s^*}\int_{\Xi} |u|^{p_s^*}\,dx - \lambda \int_{\Omega} F\left(x, u(x)\right) dx,$$

Where $P\Phi(t) := \int_0^{|t|} \Phi(\tau)d\tau$ being the primitive of Φ. Thus by (3) we have

$$\Lambda^{-1}\frac{|t|^p}{p} \leq P\Phi(t) \leq \Lambda \frac{|t|^p}{p}, \tag{6}$$

for $t \neq 0$ and $P\Phi(0)=0$. The Frèchet derivative of I, which is in $W_0^{-s,p'}(\Omega)$, the dual space of $W_0^{s,p}(\Omega)$ where $p' = \dfrac{p}{p-1}$ is defined as

$$\begin{aligned}\left\langle I'(u), v\right\rangle = &\int_{\mathbb{R}^{2N}} \Phi\left(u(x) - u(y)\right)\left(v(x) - v(y)\right) K(x,y)\,dxdy \\ &- \int_{\Omega} |u|^{p_s^*-2} uv\,dx - \lambda \int_{\Omega} f(x,u)\,v\,dx \quad \forall v \in W_0^{s,p}(\Omega).\end{aligned} \tag{7}$$

We will now give a few definitions which will be used in this article.

Definition 2.1. (Cantizano & Silva, 2019) For i=1,2,3, let us define the manifold $M_i \subset W_0^{s,p}(\Omega)$ as

$$M_1 = \Big\{ u \in W_0^{s,p}(\Omega) : \int_\Omega u_+ > 0 \; and$$
$$\int_{\mathbb{R}^{2N}} \Phi(u_+(x) - u_+(y))(u_+(x) - u_+(y)) K(x,y)\, dx\, dy - \int_\Omega |u_+|^{p_s^*}\, dx = \lambda \int_{\Omega} f(x,u) u_+ dx \Big\}$$

$$M_2 = \Big\{ u \in W_0^{s,p}(\Omega) : \int_\Omega u_- > 0 \; and$$
$$\int_{\mathbb{R}^{2N}} \Phi(u_-(x) - u_-(y))(u_-(x) - u_-(y)) K(x,y)\, dx\, dy - \int_\Omega |u_-|^{p_s^*}\, dx = \lambda \int_{\Omega} f(x,u) u_- dx \Big\}$$

$$M_3 = \; M_1 \cap M_2$$

Where $u_+=\max\{u,0\}$ and $u_-=\max\{-u,0\}$.

Definition 2.2. (Cantizano & Silva, 2019) For i=1,2,3 let us define $K_i \subset W_0^{s,p}(\Omega)$ as $K_1=\{u \in M_1 : u \geq 0\}$, $K_2=\{u \in M_2 : u \leq 0\}$, $K_3=M_3$.

Definition 2.3. (Ekeland, 1974) Let Φ be a lower semicontinuous bounded below function from a Banach space X into $\mathbb{R} \cup \{+\infty\}$. For every $\epsilon>0$, there is $x_0 \in X$ such that $\Phi(x) \geq \Phi(x_0) - \epsilon \|x - x_0\|$ for every $x \in X$.

EXISTENCE RESULTS

We shall first prove a few lemmas before attempting to show the article's main result, which is expressed as Theorem (1.1).

Lemma 3.1. For every $w_0 \in W_0^{s,p}(\Omega), w_0 > 0 (w_0 < 0)$, there exists $t_\lambda > 0$ such that $t_\lambda w_0 \in M_1 (\in M_2)$. Moreover $\lim_{\lambda \to \infty} t_\lambda = 0$.

Consequently, given $w_0, w_1 \in W_0^{s,p}(\Omega), w_0 > 0, w_1 < 0$ with disjoint supports, there exist $\overline{t_\lambda}, \underline{t_\lambda}$ such that $\overline{t_\lambda} w_0 + \underline{t_\lambda} w_1 \in M_3$ and $\overline{t_\lambda}, \underline{t_\lambda} \to 0 \; as \; \lambda \to \infty$.

Proof. It is enough to prove the lemma for manifold M_1 since the other cases will follow analogously.

For $w \in W_0^{s,p}(\Omega)$ with $w \geq 0$, consider the functional

$$\varphi_1(w) = \int_{\mathbb{R}^{2N}} \Phi(w(x)-w(y))(w(x)-w(y))K(x,y)\,dxdy - \int_\Omega |w|^{p_s^*}dx - \lambda\int_\Omega f(x,w)w\,dx.$$

In order to prove the lemma we will show that for each $w_0 \in W_0^{s,p}(\Omega)$ there exists a 't_λ' such that $\varphi_1(t_\lambda w_0)=0$.

$$\varphi_1(tw_0) = \int_{\mathbb{R}^{2N}} \Phi(tw_0(x)-tw_0(y))(tw_0(x)-tw_0(y))K(x,y)\,dxdy - \int_\Omega |tw_0|^{p_s^*}dx - \lambda\int_\Omega f(x,tw_0)tw_0\,dx \tag{8}$$

$$\geq \Lambda^{-2}\int_{\mathbb{R}^{2N}} \frac{|tw_0(x)-tw_0(y)|^p}{|x-y|^{N+ps}}dxdy - t^{p_s^*}\int_\Omega |w_0|^{p_s^*}dx - \lambda\int_\Omega f(x,tw_0)tw_0\,dx$$
$$= \Lambda^{-2}t^p[w_0]_{s,p}^p - t^{p_s^*}\int_\Omega |w_0|^{p_s^*}dx - \lambda\int_\Omega f(x,tw_0)tw_0\,dx$$

Further on using (f_2), we get

$$\varphi_1(tw_0) \geq \Lambda^{-2}t^p[w_0]_{s,p}^p - t^{p_s^*}\int_\Omega |w_0|^{p_s^*}dx - \lambda c_4\int_\Omega |tw_0|^q dx \geq A_1t^p - Bt^{p_s^*} - E_1t^q \tag{9}$$

Where,

$$A_1 = \Lambda^{-2}[w_0]_{s,p}^p,\ B = \int_\Omega |w_0|^{p_s^*}dx,\ E_1 = \lambda c_4\int_\Omega |w_0|^q dx$$

and,

$$\varphi_1(tw_0) \leq \Lambda^2t^p[w_0]_{s,p}^p - t^{p_s^*}\int_\Omega |w_0|^{p_s^*}dx - \lambda c_3\int_\Omega |tw_0|^q dx \leq A_2t^p - Bt^{p_s^*} - E_2t^q \tag{10}$$

where, $A_2 = \Lambda^2[w_0]_{s,p}^p,\ B = \int_\Omega |w_0|^{p_s^*}dx,\ E_2 = \lambda c_3\int_\Omega |w_0|^q dx$

Since, $p<q<p_s^*$, it follows from (8) and (9) that $\varphi_1(tw_0)>0$ for a 't' small enough and $\varphi_1(tw_0)<0$ for some 't' big enough. Hence, there exists some 't_λ' such that $\varphi_1(t_\lambda w_0)=0$. We will now establish the existence of an upper bound of t_λ. For this it is enough to show the existence of some t_1 such that $\varphi_1(t_1w_0)<0$. From (9), it is easy to see that $\varphi_1(t_1w_0)<A_2t^p - E_2t^q$. On choosing t_1 such that $A_2t_1^p - E_2t_1^q = 0$

implying $t_1 = \left(\frac{A_2}{E_2}\right)^{\frac{1}{q-p}}$. Therefore, there exists $t_\lambda \in [0,t_1]$. The definition of φ_1 in (8) implies that $t_\lambda \to 0$ as $\lambda \to \infty$.

Lemma 3.2. There exist constants $\alpha_j > 0$ such that for every $u \in K_i$, $i=1,2,3$,

$$\alpha_1 [u]_{s,p}^p \leq \alpha_2 \left(\int_\Omega |u|^{p_s^*} dx + \int_\Omega \lambda f(x,u) u dx \right) \leq \alpha_3 I(u) \leq \alpha_4 [u]_{s,p}^p.$$

Proof. As $u \in K_i$ so

$$\int_{\mathbb{R}^{2N}} \Phi(u(x)-u(y))(u(x)-u(y)) K(x,y) dxdy = \int_\Omega |u|^{p_s^*} dx + \lambda \int_{\Omega'} f(x,u) u dx \tag{11}$$

Using (3), we get $\Lambda^{-2} [u]_{s,p}^p \leq \int_\Omega |u|^{p_s^*} dx + \lambda \int_{\Omega'} f(x,u) u dx$. On choosing $\alpha_2 = \Lambda^2 \alpha_1$, first inequality holds.

We now prove the last inequality i.e., $\alpha_3 I(u) \leq \alpha_4 [u]_{s,p}^p$. Using (f_2), (3) and (11) we get,

$$\begin{aligned}
\left|\lambda \int_\Omega F(x,u) dx\right| &= \lambda \int_\Omega F(x,u) dx \leq \lambda \frac{1}{c_2} \int_{\Omega'} f(x,u) u dx \\
&= \frac{1}{c_2} \left(\Phi(u(x)-u(y))(u(x)-u(y)) K(x,y) dxdy - \int_\Omega |u|^{p_s^*} dx \right) \\
&\leq \frac{\Lambda^2}{c_2} [u]_{s,p}^p - \frac{1}{c_2} \int_\Omega |u|^{p_s^*} dx
\end{aligned}$$

From here we obtain that,

$$-\lambda \int_\Omega F(x,u) dx \leq \frac{1}{c_2} [\Lambda^2 [u]_{s,p}^p - \int_\Omega |u|^{p_s^*} dx] \tag{12}$$

So, using (6) and (12)

$$I(u)=\int_{\mathbb{R}^{2N}}\mathcal{P}\Phi(u(x)-u(y))K(x,y)\,dxdy-\frac{1}{p_s^*}\int_\Omega |u|^{p_s^*}\,dx-\lambda\int_\Omega F(x,u)\,dx$$
$$\leq \frac{\Lambda^2}{p}[u]_{s,p}^p-\frac{1}{p_s^*}\int_\Omega |u|^{p_s^*}\,dx+\frac{1}{c_2}\left[\Lambda^2[u]_{s,p}^p-\int_\Omega |u|^{p_s^*}\,dx\right]$$
$$\leq \left(\frac{\Lambda^2}{p}+\frac{\Lambda^2}{c_2}\right)[u]_{s,p}^p=\Lambda^2\left(\frac{1}{p}+\frac{1}{c_2}\right)[u]_{s,p}^p$$

On choosing $\alpha_4=\Lambda^2\left(\frac{1}{p}+\frac{1}{c_2}\right)\alpha_3$, we obtain the inequality $\alpha_3 I(u)\leq \alpha_4[u]_{s,p}^p$.

We are now left to prove is the middle inequality i.e.

$$\alpha_2\left(\int_\Omega |u|^{p_s^*}\,dx+\int_\Omega \lambda f(x,u)u\,dx\right)\leq \alpha_3 I(u).$$

For this, we again use (f_2) and (6) to get

$$I(u)=\int_{\mathbb{R}^{2N}}\mathcal{P}\Phi(u(x)-u(y))K(x,y)\,dxdy-\frac{1}{p_s^*}\int_\Omega |u|^{p_s^*}\,dx-\lambda\int_\Omega F(x,u)\,dx$$
$$\geq \frac{\Lambda^{-2}}{p}[u]_{s,p}^p-\frac{1}{p_s^*}\int_\Omega |u|^{p_s^*}\,dx-\frac{1}{c_2}\lambda\int_\Omega f(x,u)u\,dx$$

Multiplying both sides by c2 and using (3) & (5),

$$c_2 I(u)\geq \frac{c_2\Lambda^{-2}}{p}[u]_{s,p}^p-\frac{c_2}{p_s^*}\int_\Omega |u|^{p_s^*}\,dx-\lambda\int_\Omega f(x,u)u\,dx$$
$$\geq \frac{c_2\Lambda^{-4}}{p}\int_{\mathbb{R}^{2N}}\Phi(u(x)-u(y))(u(x)-u(y))K(x,y)\,dxdy-\frac{c_2}{p_s^*}\int_\Omega |u|^{p_s^*}\,dx-\lambda\int_\Omega f(x,u)u\,dx$$
$$=\frac{c_2\Lambda^{-4}}{p}\left(\int_\Omega |u|^{p_s^*}\,dx+\lambda\int_\Omega f(x,u)u\,dx\right)-\frac{c_2}{p_s^*}\int_\Omega |u|^{p_s^*}\,dx-\lambda\int_\Omega f(x,u)u\,dx$$
$$=c_2\left(\frac{\Lambda^{-4}}{p}-\frac{1}{p_s^*}\right)\int_\Omega |u|^{p_s^*}\,dx+\lambda\left(\frac{c_2\Lambda^{-4}}{p}-1\right)\int_\Omega f(x,u)u\,dx$$

Since, from assumption (f_2) we know $c_2<p_s^*$, hence

$$c_2 I(u) \geq \left(\frac{c_2 \Lambda^{-4}}{p} - 1\right)\left(\int_\Omega |u|^{p_s^*} dx + \lambda \int_\Omega f(x,u) u dx\right)$$

Now, on choosing α_3=c_2 and $\alpha_2 = \left(\frac{c_2 \Lambda^{-4}}{p} - 1\right)$; where $\Lambda \in \left[1, \left(\frac{c_2}{p}\right)^{\frac{1}{4}}\right]$, the inequality follows.

Lemma 3.3. There exists a constant D>0 such that $[u_+]_{s,p}^p \geq D$ in K_1, $[u_-]_{s,p}^p \geq D$ in K_2, and $\left[u_+\right]_{s,p}^p, \left[u_-\right]_{s,p}^p \geq D$ in K_3.

Proof. From the definition of K_i, we have

$$\int_{\mathbb{R}^{2N}} \Phi(u_\pm(x) - u_\pm(y))(u_\pm(x) - u_\pm(y)) K(x,y) dxdy = \int_\Omega |u_\pm|^{p_s^*} dx + \lambda \int_\Omega f(x,u) u_\pm dx \tag{13}$$

Using (f2), (3), (5) and Sobolev embedding in (13) we get,

$$\Lambda^{-2}[u_\pm]_{s,p}^p \leq \|u_\pm\|_{p_s^*}^{p_s^*} + c_4 \|u_\pm\|_q^q \leq \tilde{C}[u_\pm]_{s,p}^r$$

Where r=q if $[u_\pm]_{s,p} < 1$ or $r = p_s^*$ if $[u_\pm]_{s,p} \geq 1$. Hence, using the fact that p<r, we get the result.

Lemma 3.4. For manifolds M_i and K_i the following holds:

(i) M_i is a C^1 sub-manifold of $W_0^{s,p}(\Omega)$ codimension 1, if i=1,2 and 2 if i=3 respectively.

(ii) The sets K_i are complete, and for every $u \in M_i$ we have

$$T_u W_0^{s,p}(\Omega) = T_u M_i \oplus span\{u_+, u_-\} \tag{14}$$

where $T_u M$ is the tangent space at u of the Banach manifold M.

(iii) The projection onto the first coordinate in (14) is uniformly continuous on bounded sets of M_i.

Proof. (i) Define $\widetilde{M_1} = \left\{u \in W_0^{s,p}(\Omega) : \int_\Omega u_+ > 0\right\}$, $\widetilde{M_2} = \left\{u \in W_0^{s,p}(\Omega) : \int_\Omega u_- > 0\right\}$ and $\widetilde{M_3} = \widetilde{M_1} \cap \widetilde{M_2}$. It is easy to see that $\widetilde{M_i}$ are open and $M_i \subset \widetilde{M_i}$ and hence it is sufficient to show that M_i is a regular submanifold of $W_0^{s,p}(\Omega)$. For this we will construct C[1] functions $\psi_i : \widetilde{M_i} \to \mathbb{R}^d$, where d denotes the codimension of M_i, such that M_i is the inverse of regular value of ψ_i. Define

$$\psi_1(u) = \int_{\mathbb{R}^{2N}} \Phi(u_+(x) - u_+(y))(u_+(x) - u_+(y))K(x,y)\,dxdy - \int_\Omega |u_+|^{p_s^*}\,dx - \lambda\int_{\Omega} f(x,u)u_+dx.$$

$$\psi_2(u) = \int_{\mathbb{R}^{2N}} \Phi(u_-(x) - u_-(y))(u_-(x) - u_-(y))K(x,y)\,dxdy - \int_\Omega |u_-|^{p_s^*}\,dx - \lambda\int_{\Omega} f(x,u)u_-dx.$$

$$\psi_3(u) = (\psi_1(u), \psi_2(u))$$

It can be observed that $M_i = \psi_i^{-1}(\vec{0})$. Hence, we need to show that $\vec{0}$ is a regular value of ψ_i. We will use $(u+\epsilon u_+)= u_+ + \epsilon u_+$ for proving this. For $u \in M_1$,

$$\langle \nabla\psi_1(u), u_+\rangle = \frac{d}{d\epsilon}\psi_1(u + \epsilon u_+)$$

$$= \frac{d}{d\int}\Big[\int_{\mathbb{R}^{2N}} \Phi((u + \int u_+)_+(x) - ((u + \int u_+)_+(y))((u + \int u_+)_+(x) - (u + \int u_+)_+(y))K(x,y)\,dxdy$$

$$-\int_\Omega |(u + \int u_+)_+|^{p_s^*}\,dx - \lambda\int_{\Omega} f(x, u + \int u_+)(u + \int u_+)_+dx\Big]$$

$$= \frac{d}{d\int}\Big[\int_{\mathbb{R}^{2N}} \Phi((u_+ + \int u_+)_+(x) - ((u_+ + \int u_+)_+(y))((u_+ + \int u_+)_+(x) - (u_+ + \int u_+)_+(y))K(x,y)\,dxdy$$

$$-\int_\Omega |(u_+ + \int u_+)_+|^{p_s^*}\,dx - \lambda\int_{\Omega} f(x, u_+ + \int u_+)(u_+ + \int u_+)_+dx\Big]$$

$$\int_{\mathbb{R}^{2N}}\Big[\Phi'((u_+ + \int u_+)(x) - ((u_+ + \int u_+)(y))(u_+(x) - u_+(y))((u_+ + \int u_+)(x)\,dx\,dy$$

$$-(u_+ + \int u_+)(y))K(x,y)\,dxdy + \Phi((u_+ + \int u_+)(x) - (u_+ + \int u_+)(y))((u_+(x) - u_+(y))K(x,y)\Big]$$

$$-p_s^*(1+\int)^{p_s^*}|u_+|^{p_s^*}dx - \lambda\int_\Omega f\left(x,u+\int u_+\right)u_+dx - \lambda\int_\Omega f_u\left(x,u+\int u_+\right)u_+^2dx$$

Since, $u \in M_1$ so

$$\frac{d}{d\int}\psi_1\left(u+\int u_+\right)|_{\int=0} = \int_{\mathbb{R}^{2N}}[\Phi'\left(u_+(x)-u_+(y)\right)(u_+(x)-u_+(y))^2K(x,y)$$
$$+\Phi\left(u_+(x)-u_+(y)\right)\left(u_+(x)-u_+(y)\right)K(x,y)]dxdy - \int_\Omega p_s^*|u_+|^{p_s^*}dx$$
$$-\lambda\int_\Omega f(x,u)u_+dx - \lambda\int_\Omega f_u(x,u)u_+^2dx$$

Hence, (4) and (5)

$$\frac{d}{d\int}\psi_1\left(u+\int u_+\right)|_{\int=0} \le a\Lambda[u_+]_{s,p}^p + \Lambda^2[u_+]_{s,p}^p - \int_\Omega p_s^*|u_+|^{p_s^*}dx - \lambda\int_\Omega f(x,u)u_+dx - \lambda\int_\Omega f_u(x,u)u_+^2dx$$
$$= \left(a\Lambda+\Lambda^2\right)[u_+]_{s,p}^p - \int_\Omega p_s^*|u_+|^{p_s^*}dx - \lambda\int_\Omega f(x,u)u_+dx - \lambda\int_\Omega f_u(x,u)u_+^2dx$$

We already have $\Lambda \in \left[1,\left(\frac{c_2}{p}\right)^{\frac{1}{4}}\right)$ from lemma 3.2. Choose 'a' such that $a\Lambda+\Lambda^2 < p_s^*$. Hence,

$$\frac{d}{d\int}\psi_1\left(u+\int u_+\right)|_{\int=0} \le p_s^*\left([u_+]_{s,p}^p - \int_\Omega p_s^*|u_+|^{p_s^*}dx\right) - \lambda\int_\Omega f(x,u)u_+dx - \lambda\int_\Omega f_u(x,u)u_+^2dx$$

$$p_s^*\lambda\int_\Omega f(x,u)u_+dx - \lambda\int_\Omega f(x,u)u_+dx$$
$$-\lambda\int_\Omega f_u(x,u)u_+^2dx = \left(p_s^*-1\right)\lambda\int_\Omega f(x,u)u_+dx - \lambda\int_\Omega f_u(x,u)u_+^2dx$$

From the assumption (f_2), there exist $c_1 \in \left(0,\frac{1}{p_s^*-1}\right)$ such that

$$\lambda\int_\Omega f(x,u)u_+dx \le c_1\lambda\int_\Omega f_u(x,u)u_+^2dx < \frac{1}{p_s^*-1}\lambda\int_\Omega f_u(x,u)u_+^2dx \qquad (15)$$

Using (15) we conclude that $\frac{d}{d\epsilon}\psi_1(u+\epsilon u_+)|_{\epsilon=0}<0$ which further implies $\langle\nabla\psi_1(u),u_+\rangle<0$. Consequently, $\nabla\psi_1(u)\neq 0$. Hence, M_1 is a regular sub-manifold of $W_0^{s,p}(\Omega)$.

Similarly it can be proved that $\langle\nabla\psi_2(u),u_-\rangle<0$ for $u\in M_2$ implying that M_2 is also a regular sub-manifold of $W_0^{s,p}(\Omega)$.

Now if $\langle\nabla\psi_1(u),u_-\rangle=\langle\nabla\psi_2(u),u_+\rangle=0$ for $u\in M_3$ then $\langle\nabla\psi_1(u),u\rangle<0$ and $\langle\nabla\psi_2(u),u\rangle<0$ implying that $\nabla\psi_3(u)\neq 0$ for $u\in M_3$. Therefore,

$$\langle\nabla\psi_1(u),u_-\rangle=\frac{d}{d\epsilon}\psi_1(u+\epsilon u_+)$$

$$=\frac{d}{d\epsilon}\Big[\int_{\mathbb{R}^{2N}}\Phi((u+\epsilon u_-)_+(x)-(u+\epsilon u_-)_+(y))((u+\epsilon u_-)_+(x)$$
$$-(u+\epsilon u_-)_+(y))K(x,y)\,dxdy-\int_\Omega|(u+\epsilon u_-)_+|^{p_s^*}\,dx-\lambda\int_\Omega f(x,u+\epsilon u_-)(u+\epsilon u_-)_+\,dx\Big]$$

$$=\frac{d}{d\epsilon}\Big[\int_{\mathbb{R}^N}\Phi(u_+(x)-u_+(y))(u_+(x)-u_+(y))K(x,y)\,dxdy$$
$$-\int_\Omega|u_+|^{p_s^*}\,dx-\lambda\int_\Omega f(x,u+\epsilon u_-)u_+dx=-\lambda\int_\Omega f_u(x,u+\epsilon u_-)u_+u_-dx \quad =0.$$

Similarly, $\langle\nabla\psi_2(u),u_+\rangle=0$ and hence M_3 is also regular sub-manifold of $W_0^{s,p}(\Omega)$.

(ii) To show that K_i is complete. Let (u_n) be a Cauchy sequence in K_i then $u_n\to$ in $W_0^{s,p}(\Omega)$ which further implies that $(u_n)_\pm\to u_\pm$. By using continuity of the functions, (f_2) and Lemma 3.3, we conclude that $u\in K_i$. Hence, K_i is complete.

Now, we show that

$$T_uW_0^{s,p}(\Omega)=T_uM_1\oplus span\{u_+\},$$

where $M_1=\{u:\psi_1(u)=0\}$ and

$$T_u M_1 = \left\{ v : \left\langle \nabla \psi_1(u), v \right\rangle = 0 \right\}.$$

Let $v \in T_u W_0^{s,p}(\Omega)$ be a unit tangential vector, then $v=v_1+v_2$ where $v_2=\alpha u_+$ and $v_1=v-v_2$. Choose $\alpha = \dfrac{\left\langle \nabla \psi_1(u), u \right\rangle}{\left\langle \nabla \psi_1(u), u_+ \right\rangle}$. From here we get $v_1 \in T_u M_1$. Hence, $\left\langle \nabla \psi_1(u), v_1 \right\rangle = 0$. Similarly, we can show that $T_u W_0^{s,p}(\Omega) = T_u M_2 \oplus span\{u_-\}$ and $T_u W_0^{s,p}(\Omega) = T_u M_3 \oplus span\{u_+, u_-\}$.

(iii) It is easy to see that the projection onto the first coordinate in (14) is uniformly continuous on bounded sets of M_i.

We next prove the following lemma.

Lemma 3.5. The unrestricted functional I verifies the Palais-Smale condition for energy level $c < \left(\dfrac{1}{c_2} - \dfrac{1}{p_s^*} \right) \dfrac{S^{\frac{N}{ps}}}{\Lambda^2}$, where $S = \inf\limits_{\phi \in C_c^\infty(\Omega)} \dfrac{\|\phi\|_{s,p}^p}{\|\phi\|_{p_s^*}^p}$ is the best Sobolev constant for the fractional Laplacian.

Proof. Let (u_n) be a Palais-Smale sequence. Then $I(u_n) \to c$ and $I'(u_n) \to 0$. Let us assume that $\|u_n\| \to \infty$ as $n \to \infty$. We have

$$c \geq \lim_{n\to\infty} I(u_n) - \frac{1}{c_2} \left\langle I'(u_n), u_n \right\rangle$$

$$= \lim_{n\to\infty} \int_{\mathbb{R}^{2N}} P\Phi(u_n(x) - u_n(y)) K(x,y)\, dxdy - \frac{1}{p_s^*} \int_\Omega |u_n|^{p_s^*}\, dx - \lambda \int_\Omega F(x, u_n)\, dx$$

$$-\frac{1}{c_2} \left(\int_{\mathbb{R}^N} \Phi(u_n(x) - u_n(y))(u_n(x) - u_n(y)) K(x,y)\, dxdy - \int_\Omega |u_n|^{p_s^*}\, dx - \lambda \int_\Omega f(x, u_n) u_n dx \right)$$

$$\geq \lim_{n\to\infty} \frac{1}{p\Lambda^2} \int_{\mathbb{R}^{2N}} \frac{|u_n(x) - u_n(y)|^p}{|x-y|^{N+ps}}\, dxdy - \frac{1}{p_s^*} \int_\Omega |u_n|^{p_s^*} - \lambda \int_\Omega F(x, u_n)\, dx + \frac{1}{c_2} \int_\Omega |u_n|^{p_s^*}\, dx$$

$$-\frac{\Lambda^2}{c_2}\int_{\mathbb{R}^{2N}}\frac{|u_n(x)-u_n(y)|^p}{|x-y|^{N+ps}}dxdy+\frac{\lambda}{c_2}\int_\Omega f(x,u_n)u_ndx$$

$$=\lim_{n\to\infty}\left(\frac{1}{\Lambda^2 p}-\frac{\Lambda^2}{c_2}\right)\|u_n\|_{s,p}^p+\left(\frac{1}{c_2}-\frac{1}{p_s^*}\right)\int_\Omega|u_n|^{p_s^*}dx+\frac{\lambda}{c_2}\int_\Omega\left[f(x,u_n)u_n-c_2F(x,u_n)\right]dx$$

Using (f_2) and the fact that $c_2 < p_s^*$ we get $c > \left(\frac{1}{\Lambda^2 p}-\frac{\Lambda^2}{c_2}\right)\|u_n\|_{s,p}^p$. Thus, we get a contradiction for $\left(\frac{1}{\Lambda^2 p}-\frac{\Lambda^2}{c_2}\right)>0$. Hence, (u_n) is bounded in $W_0^{s,p}(\Omega)$ whenever $\Lambda\in\left[1,\left(\frac{c_2}{p}\right)^{\frac{1}{4}}\right)$.

By the reflexive property of $W_0^{s,p}(\Omega)$ there exists a weakly convergent subsequence say $(u_n)\rightharpoonup u$ in $W_0^{s,p}(\Omega)$. This implies $u_n\to u$ in $L^q(\Omega); q\in\left[1,p_s^*\right)$ as $n\to\infty$.

In order to prove the strong convergence in $L^{p_s^*}(\Omega)$, we will use concentration compactness principle for the fractional Sobolev space which is as follows [refer [13]].

Theorem 3.6. Let (u_n) be a bounded sequence in $W_0^{s,p}(\Omega)$. Then upto a subsequence there exists $u\in W_0^{s,p}(\Omega)$, Borel measures μ and ν, J denumerable, $x_j\in\bar{\Omega}, \nu_j>0, \mu_j\geq 0$ with $\mu_j+\nu_j>0$ for $j\in J$ and $u_n\rightharpoonup u\, in W_0^{s,p}(\Omega)$ and is strongly convergent in $L^p(\Omega)$.

$$|D^s u_n|^p \overset{*}{\rightharpoonup} u, |u_n|^{p_s^*} \overset{*}{\rightharpoonup} \nu, du\geq|D^s u|^p+\sum_{j\in J}\mu_j\delta_{x_j}, d\nu=|u_n|^{p_s^*}+\sum_{j\in J}\nu_j\delta_{x_j}$$

and $\mu_j\geq S\nu_j^{\frac{p}{p_s^*}}$. Here, $\mu_j=\mu(x_j)$ and $\nu_j=\mu(x_j)$.

Hence, if we show that the index set J is empty then $u_n\to u$ in $L^{p_s^*}(\Omega)$. Let ξ be a unit ball such that $\xi_{j,\delta}(x)=\xi\left(\frac{|x-x_j|}{\delta}\right)$ where ξ has a compact support in B_2, $\xi=1$

in B_1 and $0\leq\xi\leq1$. Since (u_n) is a PS sequence corresponding to the functional I, hence $I'(u_n)\to 0$. Therefore, on considering test function as $\xi_{j,\delta}u_n$ we have $\langle I'(u_n),\xi_{j,\delta}u_n\rangle\to 0$. Now,

$$\langle I'(u_n),\xi_{j,\delta}u_n\rangle=\int_{\mathbb{R}^{2N}}\Phi(u_n(x)-u_n(y))(\xi_{j,\delta}u_n(x)-\xi_{j,\delta}u_n(y))K(x,y)dxdy$$
$$-\int_{\Omega}|u_n|^{p_s^*-2}u_n\xi_{j,\delta}dx-\lambda\int_{\Omega}f(x,u_n)\xi_{j,\delta}u_ndx \tag{16}$$

The first term of the RHS of equation (16) can be written as:

$$\int_{\mathbb{R}^{2N}}\Phi(u_n(x)-u_n(y))(\xi_{j,\delta}u_n(x)-\xi_{j,\delta}u_n(y))K(x,y)dxdy$$
$$=\int_{\mathbb{R}^{2N}}\Phi(u_n(x)-u_n(y))(u_n(x)-u_n(y))\xi_{j,\delta}(x)K(x,y)dxdy$$
$$+\int_{\mathbb{R}^{2N}}\Phi(u_n(x)-u_n(y))(\xi_{j,\delta}(x)-\xi_{j,\delta}(y))u_n(y)K(x,y)dxdy=I+II$$

We further have,

$$I=\int_{\mathbb{R}^{2N}}\Phi(u_n(x)-u_n(y))(u_n(x)-u_n(y))\xi_{j,\delta}(x)K(x,y)dxdy$$
$$\geq\Lambda^{-2}\int_{\mathbb{R}^{2N}}\frac{|u_n(x)-u_n(y)|^p}{|x-y|^{N+ps}}\xi_{j,\delta}(x)dxdy=\Lambda^{-2}\int_{\mathbb{R}^N}|D^su_n|^p\xi_{j,\delta}(x)dx$$

So, $\lim_{\delta\to0}\lim_{n\to\infty}I\geq\Lambda^{-2}\mu_j$. From [16], it can be seen that $\lim_{\delta\to0}\lim_{n\to\infty}II=0$. Hence, for the first term of RHS in (16) we have

$$\lim_{\delta\to0}\lim_{n\to\infty}\int_{\mathbb{R}^{2N}}\Phi(u_n(x)-u_n(y))(\xi_{j,\delta}u_n(x)-\xi_{j,\delta}u_n(y))K(x,y)dxdy\geq\Lambda^{-2}\mu_j$$

Also, $\lim_{\delta\to0}\lim_{n\to\infty}\int_{\Omega}|u_n|^{p_s^*}\xi_{j,\delta}dx=\nu_j$ and $\lim_{\delta\to0}\lim_{n\to\infty}\int_{\Omega}f(x,u_n)u_n\xi_{j,\delta}dx=0$.

Thus, from (16), we have $\Lambda^{-2}\mu_j-\nu_j\leq0$ which further implies $\mu_j\leq\frac{\nu_j}{›^{2}}$. On using the relation $\mu_j\geq S\nu_j^{\frac{p}{p_s^*}}$ we get either $\nu_j=0$ or $\nu_j\geq\frac{S^{\frac{N}{ps}}}{›^{2}}$. Let us assume that $\nu_j\geq\frac{S^{\frac{N}{ps}}}{›^{2}}$, then using the fact that norm is weakly lower semi-continuous we have

$$c = \lim_{n\to\infty} I(u_n)$$

$$= \lim_{n\to\infty} I(u_n) - \frac{1}{c_2}\left\langle I'(u_n), u_n \right\rangle$$

$$\geq \lim_{n\to\infty}\left(\frac{1}{c_2} - \frac{1}{p_s^*}\right)\int_{\Omega} |u_n|^{p_s^*}\, dx$$

$$\geq \liminf_{n\to\infty}\left(\frac{1}{c_2} - \frac{1}{p_s^*}\right)\int_{©} |u_n|^{p_s^*}\, dx$$

$$\geq \left(\frac{1}{c_2} - \frac{1}{p_s^*}\right)\left(\int_{A_\epsilon} |u|^{p_s^*} + \sum_{j\in J} \nu_j\right)$$

$$\geq \left(\frac{1}{c_2} - \frac{1}{p_s^*}\right)\nu_j$$

$$\geq \left(\frac{1}{c_2} - \frac{1}{p_s^*}\right)\frac{S^{\frac{N}{ps}}}{›^{2}}$$

Hence, for $c < \left(\frac{1}{c_2} - \frac{1}{p_s^*}\right)\frac{S^{\frac{N}{ps}}}{\Lambda^2}$, the index set J is empty as ν_j=0 and since $\mu_j \leq \frac{\nu_j}{\Lambda^2}$ so μ_j=0. Therefore, $\int_{\Omega} |u_n|^{p_s^*} \to \int_{\Omega} |u|^{p_s^*}$. We also know that $W_0^{s,p}(\Omega)$ is compactly embedded in $L^r(\Omega), r \in \left[1, p_s^*\right)$ and hence $u_n \to u$ in $L^r(\Omega)$, for $1 \leq r < p_s^*$. Consider $\widetilde{u_n} = u_n - u$. Then $\widetilde{u_n} \rightharpoonup 0$ in $W_0^{s,p}(\Omega)$. Consider,

$$\left\langle I'(\widetilde{u_n}), \widetilde{u_n} \right\rangle =$$

$$\int_{\mathbb{R}^{2N}} \Phi\left(\widetilde{u_n}(x) - \widetilde{u_n}(y)\right)\left(\widetilde{u_n}(x) - \widetilde{u_n}(y)\right) K(x,y)\, dxdy - \lambda \int_{\Omega} |\widetilde{u_n}|^{p_s^*}\, dx - \int_{\Omega} f\left(x, \widetilde{u_n}\right)\widetilde{u_n}\, dx \tag{17}$$

The second and the third term of the functional approach to 0 as $n \to \infty$ since $u_n \to u$ in $L^{p_s^*}(\Omega)$. Also by (f_2), we see that the third term in (17) will also converge to 0. Hence from (17), we have first term in RHS of (17) will converge to zero. But

$$0 \leq \frac{1}{\Lambda^2} \|\widetilde{u_n}\|^p_{W_0^{s,p}(\Omega)} \leq \int_{\mathbb{R}^{2N}} \Phi\left(\widetilde{u_n}(x) - \widetilde{u_n}(y)\right)\left(\widetilde{u_n}(x) - \widetilde{u_n}(y)\right) K(x,y)\,dxdy \tag{18}$$

From (18) it is clear that $\lim_{n\to\infty} \|\widetilde{u_n}\|_{W_0^{s,p}(\Omega)} = 0$, that is $\widetilde{u_n} \to 0$ in $W_0^{s,p}(\Omega)$. Hence, $u_n \to u$ in $W_0^{s,p}(\Omega)$ and the unrestricted functional I satisfies Palais-Smale condition.

We now give a lemma which is a consequence of the Lemma 3.6.

Lemma 3.7. The functional $I|_{K_i}$ satisfies the Palais-Smale condition for every energy level $c < \left(\frac{1}{c_2} - \frac{1}{p_s^*}\right)\frac{S^{\frac{N}{ps}}}{\Lambda^2}$. Then u is also a critical point of the functional I and hence a solution of the problem (1).

Proof of theorem 1.1. Since I is bounded below on $K1$. Therefore, by Ekeland's variational principle, there exists $v_k \in K_1$ such that $I(v_k) \to c := \inf_{K_1}$ and $\left(I|_{K_1}\right)'(v_k) \to 0$.

We have to check if we choose λ large, we have $c < \left(\frac{1}{c_2} - \frac{1}{p_s^*}\right)\frac{S^{\frac{N}{ps}}}{\Lambda^2}$. It follows from Lemma 3.1. For instance, for c we have that for choosing $w_0 \geq 0, c \leq I(t_\lambda w_0) \leq \frac{t_\lambda^p \Lambda^2}{p}[w_0]^p_{s,p}$. Also, it follows from the estimate of t_λ in Lemma 3.1 that $c \to 0$ as $\lambda \to \infty$. Hence, there exists $\lambda'(p.p_s^*, n, c_3)$ such that for all $\lambda > \lambda', c < \left(\frac{1}{c_2} - \frac{1}{p_s^*}\right)\frac{S^{\frac{N}{ps}}}{\Lambda^2}$. The other cases are analogous. From the construction of K_i it can be seen that one solution is negative, one is positive and one is sign changing. Hence the theorem is proved.

CONCLUSION

The problem involving an integro-differential operator and a critical exponent has been studied. Variational method along with manifold and submanifold technique has been applied. Concentration-compactness principle has been used to deal with the critical exponent term as compact embedding fails for critical exponent case. As a result of this study we conclude the existence of three nontrivial solutions.

REFERENCES

Averna, D., & Bonanno, G. (2003). A three critical points theorem and its applications to the ordinary Dirichlet problem. *Topological Methods in Nonlinear Analysis, 22*(1), 93–103.

Bonder, J. F., Saintier, N., & Silva, A. (2018). The concentration-compactness principle for fractional order Sobolev spaces in unbounded domains and applications to the generalized fractional Brezis–Nirenberg problem. *Nonlinear Differential Equations and Applications NoDEA, 25*(6), 1–25.

Bonder, J. F., & Silva, A. (2009). *The concentration-compactness principle for variable exponent spaces and applications*. arXiv preprint arXiv:0906.1922.

Caffarelli, L., & Silvestre, L. (2007). An extension problem related to the fractional Laplacian. *Communications in Partial Differential Equations, 32*(8), 1245–1260.

Cantizano, N. A., & Silva, A. (2019). Three solutions for a nonlocal problem with critical growth. *Journal of Mathematical Analysis and Applications, 469*(2), 841–851.

De Nápoli, P. L., Bonder, J. F., & Silva, A. (2009). Multiple solutions for the p-Laplace operator with critical growth. Nonlinear Analysis: Theory. *Methods & Applications, 71*(12), 6283–6289.

Di Nezza, E., Palatucci, G., & Valdinoci, E. (2012). Hitchhiker's guide to the fractional Sobolev spaces. *Bulletin des Sciences Mathématiques, 136*(5), 521–573.

Ekeland, I. (1974). On the variational principle. *Journal of Mathematical Analysis and Applications, 47*(2), 324–353.

Evans, C. L. (1998). *Partial Differential Equations*. American Mathem. Soc.

Garcia Azorero, J., & Peral Alonso, I. (1991). Multiplicity of solutions for elliptic problems with critical exponent or with a nonsymmetric term. *Transactions of the American Mathematical Society, 323*(2), 877–895.

Giri, R. K., Choudhuri, D., & Soni, A. (2018). A problem involving a nonlocal operator. *Frac. Diffe. Calc.*, *8*(1), 177–190.

Holden, H., & Karlsen, K. H. (Eds.). (2012). *Nonlinear Partial Differential Equations*. Springer.

Kesavan, S. (2003). *Functional Analysis and Applications*. New Age Inter. Pvt. Ltd.

Kratou, M. (n.d.). *Three solutions for a semilinear elliptic boundary value problem*. https://www.ias.ac.in/public/Volumes/pmsc/forthcoming/PMSC-D-17-00359.pdf

Kuusi, T., Mingione, G., & Sire, Y. (2015). Nonlocal equations with measure data. *Communications in Mathematical Physics*, *337*(3), 1317–1368.

Lions, P. L. (1985). The concentration-compactness principle in the calculus of variations. The limit case, part 1. *Revista Matemática Iberoamericana*, *1*(1), 145–201.

Mosconi, S., & Squassina, M. (2016). Nonlocal problems at nearly critical growth. *Nonlinear Analysis: Theory. Methods & Applications*, *136*, 84–101.

Silva, A. (2011). Multiple Solutions for the p (x)– Laplace Operator with Critical Growth. *Advanced Nonlinear Studies*, *11*(1), 63–75.

Struwe, M. (1991). Three non-trivial Solutions of anticoercive boundary value problems for the Pseudo-Laplace-operator. *Journal fr die reine und angewandte Mathematik, 325*, 68-74.

Chapter 10
Women Entrepreneurship in Industrial Revolution 4.0:
Opportunities and Challenges

Soumi Majumder
Sister Nivedita University, Kolkata, India

ABSTRACT

Women entrepreneurship development is indeed an essential part of human resource development. In India the growth of women entrepreneurship is very low mostly in rural areas. Entrepreneurship is one of the most important issues under business paradigm. Women entrepreneurship, under this phenomenon is the recent concern. Women are very much aware about their skill, knowledge, capacity, rights, and work situations. The progress is more visible in upper class families than middle class society. This paper focuses on understanding of capacity of Indian women and their role effectiveness in the business development area under industrial revolution 4.0. It is the current trend of automation, it provides data exchange in the field of manufacturing technologies. With the application of Artificial Intelligence the women entrepreurship has taken a place of role playing to expand country's economy.

INTRODUCTION

Entrepreneurship development is the process of enhancing the skill and knowledge of entrepreneurs. The process must be done through several structured training and classroom coaching programs. The main aim of this program is to increase the numbers of entrepreneurs and generation of employment in the country. Entrepreneurship development process guides the new firms and ventures to accomplish their business

DOI: 10.4018/978-1-6684-7343-6.ch010

goals and improvement of nation's economy (Yadav and Unni, 2016). Another important phenomenon is, to keep in mind the risk related factors this process improves the managing capacity, developing capacity of a business and make it efficient. Entrepreneurship development program always encourages the participants to make a fruitful judgment and take a sensible decision for all kinds of business activities. Fig: 1 has explained the conceptual framework on entrepreneurship development. As per the report of Global Entrepreneurship Monitor (GEM) India in the year 2016-2017, it has been said that only 11 percent of adult population of India is engaged in "total early –stage entrepreneurial activity (TEA). Only 5 percent of the country's population takes the challenges to establish their own business. According to the report of Sixth Economic Census (2014), only 14 percent of Indian women run their own businesses. More than 90 percent of micro enterprises are run by women and among them 79 percent are self- financed. As per the data from World Bank, women contribute just 17 percent of India's GDP. It is less than half of the global average (Tiwari, 2007). The World Bank's vice president for South Asia, Annette Dixon has said if it is possible to join half of Indian women in the labor force, the annual pace of economic growth will be raised by 1.5 percent to 9 percent (Tambunan, 2007). World Bank Enterprise Survey has also said (in 2014), the percentage of women in leadership position is very low in India about to 9 percent. In abroad like USA, 25 percent of all businesses are own by women. In Canada women entrepreneurship has a great influence also where women own one-third of small business. In France the proportion of women entrepreneurship is one fifth.

In this dynamic world like men, women entrepreneurs are the part of sustained economic development as well as social progress (Kumar, 2008). In India women are playing vital roles in the society. Due to the lower status in the male dominated society their entrepreneurial ability has not been recognized properly. It is necessary to understand the priority of development and empowerment of women entrepreneurship. In India for developing women entrepreneurship different policies and programs are being implemented. Our society should change the mindset towards women. There is a need to give equal rights as per the constitution to the women. It is seen that the progress towards gender equality is slow and it happens due to non-availability of funds to policy commitments (Fortunato & Alter, 2015). The prerequisite for creating a good nation is women empowerment. When women are totally empowered, it is assured that society is stabled. The thought and value system of women empowerment leads to the development of a good family, good society as well as good nation. If a woman is fully empowered it does not mean that another individual has less power or powerless. The decision making power of women is influenced by family's behavior (Obaji & Olugu, 2014, Tambunan, 2009). Around half of the total world population is constituted by women and it is in India also. They are regarded as the better half of the society. In traditional concept of societies they were stayed into the four walls of houses and did performing household duties. In the modern era of society they have come out of the four walls of houses. They are taking part in all sorts of activities apart from household duties. As per the global evidences it has been stated that women can do excellent performance in different spheres of activities i.e. politics, administration, academics and social work. At present time they have started to jump into industry also. They are quite capable to run their enterprises (Mathew, 2010). The current trend of automation and data exchange in manufacturing technologies is known as industry 4.0. It consists of cyber-physical systems, cloud computing and many more. Industry 4.0 (Vimal et al., 2020, Gupta et al. 2021, Dey & Tamane, 2018, Dey et al., 2018, Bhatt et al., 2017) has created a smart factory under the area of industrialization. Under this concept women entrepreneurship has also become an important part of discussion. The objective of women entrepreneurship development in the dynamic world of industrial revolution 4.0 is to elevate the knowledge and enhance the skills of women entrepreneurs. This will increase country's GDP and national income (Akehurst, Simarro & MasTur, 2012, Tiwari & Tiwari, 2007). The economic development of the country will be taking place faster. Industry 4.0 focuses on achieving gender equality to demonstrate progress of women and girls. It is including productive and decent work for male and female both. Therefore there is a need to focus on women entrepreneurship development under the juncture of industry 4.0, its opportunities and challenges in the country.

In section 2, related studies have been discussed, section 3 is reported about importance and influencing factors of entrepreneurship, Entrepreneurial knowledge, skills, competencies and performance has been mentioned under section 4, section 5 is stated about role of women in Indian economy, women empowerment and entrepreneurship for industrial revolution 4.0 are reported under section 6, section 7 is reported about push and pull model of women entrepreneurship through AHP, section 8 is reported on AI and Women Entrepreneurship and section 9 has mentioned discussion and finally section 10 is conclusions.

LITERATURE REVIEW

The topic women entrepreneurship and its effectiveness in the country have become very much significant under industry revolution 4.0 eras. Females are also competitive as male in the journey of business establishment and development. Few studies have been made by many researchers. In developing countries the growth of women entrepreneurship has been largely ignored by the government and the society both. The study revealed that rural women entrepreneurs are very informal, unorganized, unregistered, less supported and less empowered. Therefore majority of women in

Figure 1. Conceptual Framework on Entrepreneurship Development

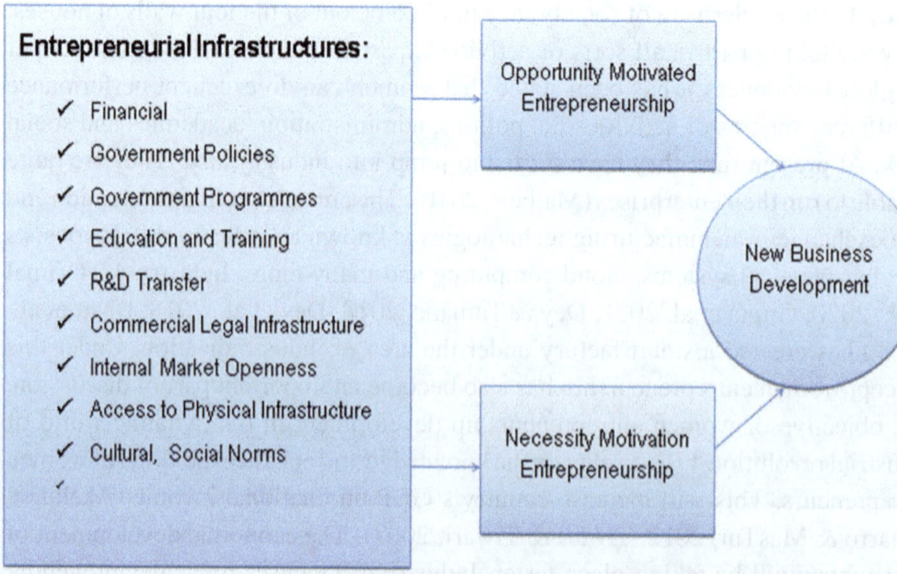

that area faced problems like lack of business skills, fund, and bargaining power within the market (Neithammer, 2013). In developing countries women entrepreneurs are playing a prime role in local economies and in micro- enterprises. This study said that economic status of women is not so much stable especially in rural areas (Fazalbhoy,2014).The earning opportunity is very less. In this scenario Self Help Groups (SHG's) has been generated for the purpose of economic independence (Eckhardt & Shane, 2003). In both the segments, rural and urban, there is a need to grow women entrepreneurship to make the country economically strong (Bruyat & Julien, 2001). It will create a high standard of living. Study has shown along with the household duties and farming the women is interested to take part in business. Another study discussed the issues regarding women entrepreneurship in relation with absence of direct ownership for property, poor degree of financial stability, absence of solid agenda in life, paradox of finance in economically poor as well as rich women, non-awareness of capabilities, low risk taking behaviour, lack of self-confidence and professional education, poor interaction with successful entrepreneurs, constraints in mobility and many more (Nawaz, 2009).

IMPORTANCE AND INFLUENCING FACTORS OF ENTREPRENEURSHIP

Importance of Entrepreneurship Development

Entrepreneurship is an important phenomenon in the country; so many reasons are involved under this concept to enhance its importance. It promotes social changes and drives innovation. Entrepreneurs are treated as national assets who are cultivated, motivated and remunerated to accomplish the goals. The most developed nations such as USA becomes one of the world leader for the reason of their forward thinking, research and development, innovation and invention and entrepreneurial abilities. Great entrepreneurs have the capacity to change the way of living and working (Wiklund et al., 2011). In addition their innovations establish entrepreneurial ventures with wealth creation. They are focusing on growing economy and sustainable development. Some necessities of entrepreneurship development have been stated below.

Entrepreneurs Add to National Income

The ventures of entrepreneurship generate new wealth. Existing business helps to maintain existing markets and tends to earn a high level of income. By generating entrepreneurship new products, services, technology can come and introduce in the new market. . Improvements of these components help to develop the existing market and wealth is generating in this way. Additionally there is a chance to increase employment and higher level of earnings. It leads to national income in terms of higher tax revenue and higher government spending (Baron & Shane, 2007). The government can use this revenue by investing in other struggling sectors and also on human capital development. Surplus wealth can be used by government to retrain the workers. If people are trained in entrepreneurial activities they can jump to establish their own business that very much concern with national income.

Entrepreneurs Spur Economic Growth

Entrepreneurship creates new products and services so that the process helps to bring economic growth. It stimulates other new ventures. An example can be given in this case, the IT industry in India has been formulated by many information technology companies during 1990s. Due to the expansion of IT industry many other sectors got benefitted from it, namely, call centres, network maintenance companies and hardware providers etc. It gives impacts on education and training institutes as well.

Jobs are created in different associated sector with IT industry. There is a spur of economic betterment in this way.

Entrepreneurs Create Social Change

Due to innovation of new products, services entrepreneurship breaks the tradition of business and offers unique kind of goods to the market. It reduces dependence on obsolete systems and technology. This improves quality of work life, morale and freedom to greater economy (Rumelt, 2005, Audretsch, 2012). The globalization of technology signifies that in developing countries the entrepreneurs have the accessibility of the same tools and techniques as it is existing in developed countries. Social changes can take place through this medium. Young entrepreneurs in developing countries can compete with a multimillion dollar existing product that has been already placed in a developed country.

Community Development

The entrepreneurs invest in community projects and support the charities financially. This provides development beyond their own ventures. They have the mind-set towards growing their communities. Few famous entrepreneurs invested money on education, public health and many more domains i.e., Bill Gates. The qualities of entrepreneurs always motivate others to enter into this philanthropy journey of entrepreneurship.

Factors Affecting Entrepreneurship Development

In the entrepreneurship development paradigm, there are so many factors exist that influence development of entrepreneurs very much. They are i)personal determinants, ii)aspiration and attitude, iii) family circumstances, iv) policies, incentives and facilities of government v) cultural and social determinants, vi) traditions and progeny and vii) business competition (Figure 2). The personal components always influence the spirit and motivate individuals to become entrepreneur. The individuals who have qualities like maturity, strong imaginations, optimism, ability to take risk, confidence on self, leadership abilities, dynamic personality, innovative thinking, hardworking can have the high chance to be succeed in the field of entrepreneurship (Sarfaraz et al., 2014). Aspiration and attitude says about the person's high ambition in career. Every person acts and behaves based on their aspiration and attitude. The mind of entrepreneurs comes from strong attitude and aspiration. Family circumstances are different for every individual; if the family members are not co-operative then it is a problem for the individual to set up career as entrepreneurs. There are several government policy exist like economic policy, import-export policy, taxation policy and licensing policy .These also have influence on entrepreneurship development programme. It becomes easy when government provides the support in terms of basic infrastructural facilities, i.e. electricity, raw-materials, transport, communication, banking, insurance, water and warehousing facilities. Cultural and social determinants include levels of education, personal behaviour, situation-based circumstances, traditions, social organizations, class and caste system, training programme for family etc. When there is a substantial impact of traditions of society and family on the working style of individuals then it comes under progeny and tradition factors. Due to change in time the concept of progeny and traditions are changing slowly under the area of entrepreneurial development. The development of entrepreneurships is obvious when a healthy competitive completion take place in the society. Unhealthy competition creates blocks in the path of entrepreneurial development programme.

ENTREPRENEURIAL SKILLS, COMPETENCIES AND PEFORMANCE

The Notion of Entrepreneurial Skills

Entrepreneurs are often taking up different roles to build and develop a business. A variety of skills set is required. Some important entrepreneurial skills are discussed below.

Skills on Business Management

To ensure the accomplishment of business goals business management skills and traits are required very much. These skills are needed for managing operations of different departments in the business organization as it contains a proper understanding of each function. Business management skills refer to the delegation of responsibility, multitasking, formulation of critical business decisions and many more (Yadav & Unni, 2016).

Skills on Communication and Active Listening

Effective communication is playing vital role to communicate with clients, team members and other stake holders. An entrepreneur requires superior written and verbal communication. On the other hand, along with communication they need to possess a good listening skills . They should be excellent listeners to make a clear

Figure 2. Factors affecting Entrepreneurship

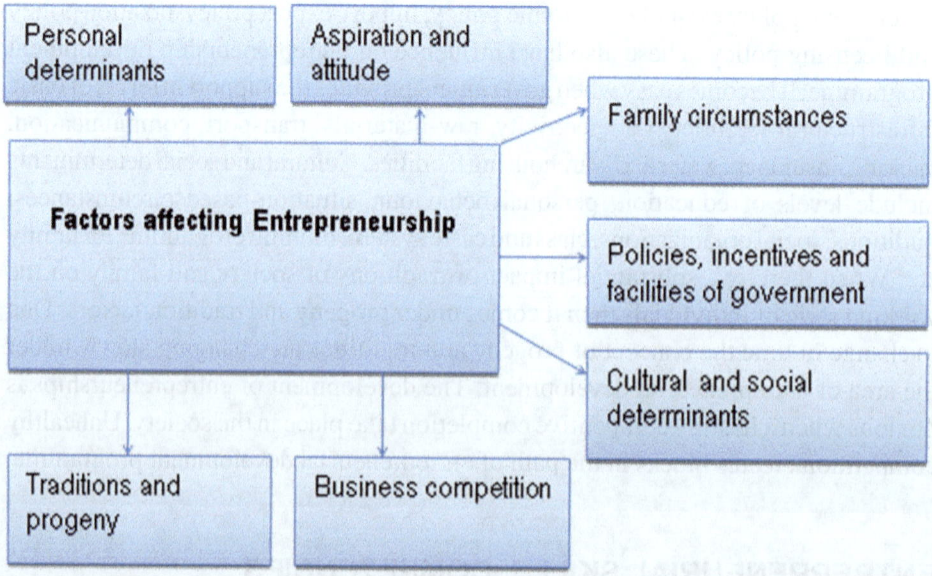

understanding about project requirements and necessary discussion during the time of project meetings.

Skills on Risk-taking

One of the essential skills of entrepreneur is risk taking behaviour and learns more on it. Tremendous success can come if they take calculated risk (Hani et al., 2012). A business can grow to the next level when they understand that risk is the opportunity to learn. Employers always prefer to take candidates who have risk taking potentiality.

Skills on Networking

To grow and promote a business networking skill is also needed for building and managing relationship with other professionals in the field. Future opportunities can be opened up through effective networking skills. It supports to build a solid brand. It helps to stay up to date with industry trends and make effective future teams. It is most desirable skills for entrepreneurs, by this they can fund their ideas from other professionals, can access with professional business expertise and earn feedback for their new ideas or ventures.

Technical Skills

By using digital tools and software technical skills can be gained. It is one kind of hard skills. Business owners should know about the usage of planning, marketing and budgeting software. If they have proper knowledge on software, they can easily manage the projects, track sales, and also can allocate a viable budget for the project work (Maden, 2015).

Apart from above skills some other skills are also essential for entrepreneurs, i.e. skills on problem solving, creative and critical thinking, leadership, time management and organizational skills as well. Often entrepreneurs face some unwanted challenging situations, to overcome that they must appeal their problem solving ability. Creative and critical thinking gives meaningful insights to the business and enhances level of profitability. Leadership is the process of influencing, motivating, inspiring, guiding the followers to achieve the business goals. Entrepreneurs with leadership quality can make success for the firm smoothly. Time management and organization skill tend to achieve productivity. It helps to understand different ways to prioritise tasks. It avoids procrastination. Managing time includes timely completion of entire project, analysing team's time, setting time limit for each task, delegating work for others, creating a to-do-list and using technology for organizing workspaces (Sharma, 2013., Agarwal & Lenka, 2018).

Entrepreneurial Competencies

Competency is a feature or characteristic of a person that gives result of effective performance. It is the summary of skills, knowledge, traits, self-image and motives that is very much needed for individual to perform a given task. In short, it can say that competencies is required for the entrepreneurs to start a new business venture and when it is carried out successfully, is known as entrepreneurial competencies. There are several kinds of competencies of entrepreneurs are listed below.

Personal Entrepreneurial Competencies

These competence abilities are required to perform the duties efficiently and effectively. This includes, i) initiative, ii)abilities to find and act on suitable opportunities, iii)persistence, iv)seeking for information, v)consciousness for high level qualitative work, vi) commitment towards work and efficiency, vii)assertiveness, viii)systematised ix)planning procedures) persuasion, xi)problem solving ability, xii)implementation of useful strategies. Figure 3 is the representation of personal entrepreneurial competencies. Initiative refers to the inner urge or motive of individual to do something new. Entrepreneurs are the people who always search for new opportunities and do act on such opportunities. Persistence means to overcome the obstacles requirement of capacity and skills. The successful entrepreneurs are very concern about to open their eyes and ears. New ideas and concepts can help them to realize goals. They are being ready all the time to consult experts for appropriate advices. High quality of work signifies the standard of excellence. In case of commitment to efficiency they focus on the ways to do things faster at a lower cost. They emphasize on new methods of work that make the work easy, simple, better and economical. Systematic planning refers to develop and use logical steps and sequences to accomplish goals. Problem solving ability is always required as businesses are facing uneven and uncertain situations. Assertiveness measures strong confidence, reliability and potentiality. Persuasion gives the thought to entrepreneurs to be competent in the domain of better performing and getting success in the business. Entrepreneurial competencies also states to use a variety of strategies to influence others or surroundings (Moses & Mordi, 2010). In this way they are able to develop business contracts and utilize influential persons to earn objectives.

Venture Initiation and Success

An entrepreneur should possess the competencies related to launching the enterprise and its growth and survival. This level of competency further be categorised into two ways. They are following.

I. Enterprise Launches Competencies

There are various enterprise launches competencies. these competencies are dealt with understanding the characteristics' of business, dealing with business, financing the business, complying with government rules and regulations, formulating marketing strategy, determining the location of business, selecting the type of ownership, obtaining technological assistance, developing and implementing of business planning and many more.

II. Enterprise Management Competencies

Enterprise management competencies are dealt with competency on protecting the business, managing customer credit and collections, managing and tracking the business records, coordinating sales efforts, promoting business products and services, managing human resources effectively and lastly overall management of businesses.

Figure 3. Personal Entrepreneurial Competencies

Entrepreneurial Performance

Entrepreneurial performance refers to the individual's ability that can be effective in many careers like medication, stores, marketing, suppliers etc. To obtain companies success entrepreneurship provides better outcomes in negotiations. Entrepreneurial performance establishes primary company concepts, develop new types of products, analyse market opportunities, formulation of modern environment, build up health and amicable relationship with investors (Ullah et al., 2013, Afrin et al. 2008). For the best possible performance, entrepreneurial commitment is the key indicator. The connection between performance and self- efficacy is the root for the entrepreneurial self-efficacy construct. The process of entrepreneurship becomes effective when business owners have the innovative techniques, abilities and actions. On the basis of entrepreneurial self- efficacy the owners are recognized in the market. This is performing a crucial part in generating their entrepreneurial objectives in relation with great performance of them. The theory of self -efficacy impacts on entrepreneurs' performance. It includes business owners' passions, level of determination and inspirations. It has been established that the entrepreneurial self -efficacy can predict greater performance of entrepreneurs even at the time of experiencing complicated and uncertainty.

ROLE OF WOMEN IN INDIA'S ECONOMY

In India it has been observed that 20.37% of women run medium scale enterprises on the basis of 23.3% labor force. At present they have been considered as backbone of the economy. As per the report of Mc. Kinsey global it is stated that by increasing women participation in the labor force India can add US$ 700 billion towards global GDP. The percentage of women employing under agriculture and manufacturing sector is greater than men, moreover it is noticed that literacy rate among women increased at 8.8% in FY21.this helps to make the bright prospects of the country. The research statistics states that India has 432 million working age women .the women entrepreneurs are providing direct employment to 22-27 million people through 13.5-15.7 women owned business. The Boston Consulting Group reported start-ups founded or co-founded by women entrepreneurs can generate 10% high cumulative revenue over a five year period time. These start ups consists of more inclusive work culture and employing thrice more women than men. As per Boston report It also has been said that in the next five years women –led businesses are estimated to grow 90%.The following report has been captured based on age on women entrepreneurship in India (Figure 4) (Source: https://www.womensweb.in/2019/07/). In X axis age is measured and in Y axis the percentage of generation of businesses.

From Figure 5 (Source: https://www.womensweb.in/2019/07/), it is observed that 72% women are under post graduate or diploma holder, 23% are graduate or diploma and 5% are under doctorate or post-doctorate.

In India, there are so many cities where women entrepreneurship has been developed, namely, Mumbai, Bangalore, NCR, Hyderabad, Goa, Pune, and others. From Figure 6 (Source: https://www.womensweb.in/2019/07/), it is noticed that Mumbai has highest development regarding women entrepreneurship(30%), Bangalore(15%), NCR (15%),Hyderabad (14%), Goa (11%),Pune (10%),and others (20%).

In India, women entrepreneurs are entering businesses in different fields. Figure 7 (Source: https://www.womensweb.in/2019/07/) has depicted the various sectors like

Figure 4. Age wise start up

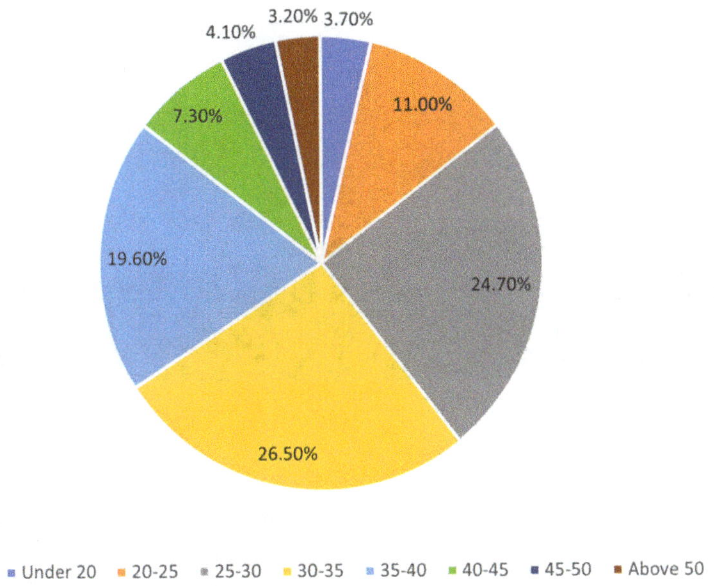

Figure 5. Education wise start up

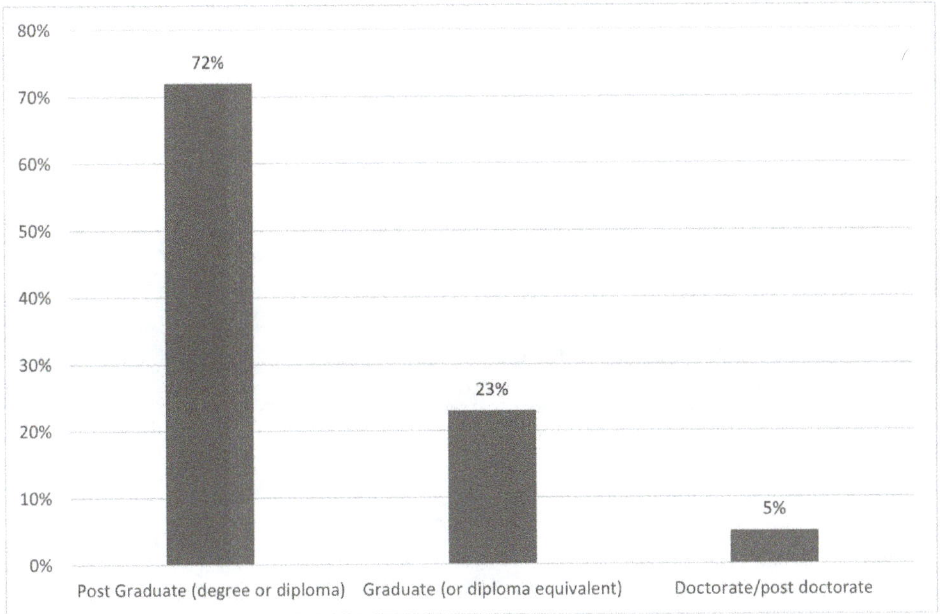

Figure 6. Location wise startup

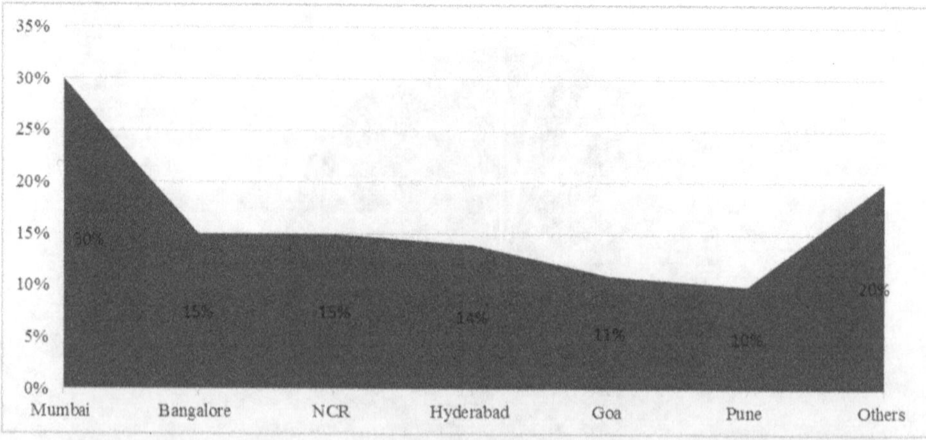

Apparel and accessories(10%), Food and Beverages (4%), Health and allied(12%), IT and Technology (5%), Professional services like law, architecture, CA etc(6%), Retail(5%), Publishing and Media(5%), Creative services like writing and design (16%), education(7%) and others (30%) as well.

The major businesses are run by women entrepreneurs in India, few examples have been given at this juncture. Around 45% of startups are owned by women entrepreneurs and among them, 50,000 are recognized by the Indian Government.

Figure 7. Industry wise Women Entrepreneurship

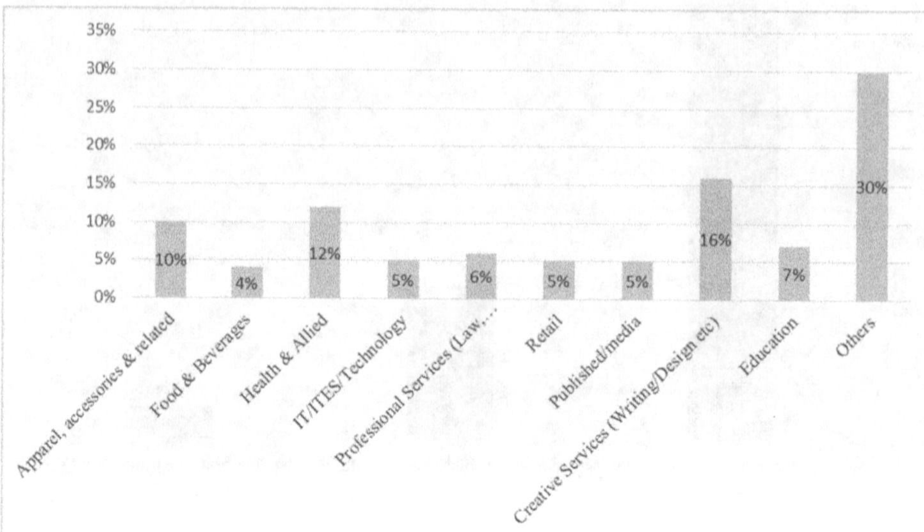

It is seen that the learning app BYJU'S founder is Divya Gokulnath (Mathew, R. V., & Panchanatham, N. 2011)., it was established in the year 2011, total fundraised US$ 8.5 billion and its market valuation is about US$ 18 billion. NYKAA, the fashion app is set up by Falguni Nayar in the year 2012. The total fundraising by the company is US$ 148.5 million and the market valuation is US$ 12.5 billion. Another company named ZOLO, the founder is Isha Choudhury (Osei, C. D., & Zhuang, J. 2020). It was established in the year 2015.ZOLO is a company that is offering to manage co-living spaces. It benefits basic and luxury living amenities including furnished rooms, food, and hygiene, housekeeping, internet, security facilities, etc. The fund raised by the company is US$ 90 million and the market valuation is US$ 100 million. MobiKwik, the company's payment gateway is offering a pre-paid type wallet that is fully digital. It is used for consumer payments, money transfer options; pay later options, loans, and various investments. Users enjoy wireless payments for recharging phones, online shopping and movie tickets without having a credit card or any bank account. Upasana Taku is the founder of MobiKwik, in the year 2009.The fund raised by the company is US4 380 million and the market valuation is US$ 750 million. On the other hand, another company, i.e. Thrillophila is a technology platform. It provides various activities and tours for travelers for different destinations. The founder is Chitra Daga, year of establishment is 2009. The Company's fundraising is US$1.24 million and its market valuation is US$ 4.48 million (Martinez Dy, A., & Jayawarna, D. 2020).

SUCCESSFUL WOMEN ENTREPRENEURSHIP IN INDUSTRIAL REVOLUTION 4.0

As per the report of World Economic Forum's, it is established that women entrepreneurs are facing trouble to set up the business due to lack of capital, credit, land and financial products. It can be said lack of care infrastructures are missing in case of developing women entrepreneurship. One most striking thing is, in the UK, about 350 billion of economic value is created for the agenda of establishment of female entrepreneurship as same as the rate on male entrepreneurship. In the global market there is a gap in case of social growth potential. Successful women entrepreneurs require crystal clear vision on their objectives. They should focus on future not only the present time and work accordingly. Therefore a clear vision must be made on what actually they want to accomplish despite of having roadblocks, distractions setbacks etc. Clear vision of entrepreneurs is helping to motivate their team members, potential investors and new clients and maintain the track to achieve short- term and long -term goals (Zhu et al., 2019). For women entrepreneurs there is need to identify big market opportunities and application of passion. This will add value and bring expertise. They should take training more on current technology. Like male entrepreneurs, the female also should take risk in case of generating profit or loss. All the time the upcoming situation is not predicted, they need to make decisions on the basis of incomplete information. In this juncture, keeping patience is very much required. It takes time to formulate a reputed company that's why patience is the key component in the journey of women entrepreneurship under 4.0 revolutions. It is essential to understand the financial position of the company from the side of women entrepreneurs. Cash on hand, way of profitability, growth projections, managing monthly burn rates, capital requirement for future, they should possess clear knowledge on these areas. In case of forecasting no one stay at level of 100 percent confidence, so that higher risk is involved in the process. Under the industrial revolution 4.0 eras, women entrepreneurship is creating good business with sound finance. Women entrepreneurs are always focusing on results. They are competing with males as well. They have made relationship with the people who provide motivation, challenge, inspire and support to them. The journey of female entrepreneurs is becoming more dynamic and full of high dedication than before. Learning, suitable opportunities and mentorship are the most discussed topic under the section of development of women entrepreneurship in India.

Women in Entrepreneurship —Challenges and Opportunities

In the recent years women entrepreneurship has become a matter of concern. For economic and social development, women entrepreneurship focuses on leadership skills, innovative ideas, decisiveness, problem solving ability etc. In the corporate sector over the past years, it has been notified that there is rise in leadership positions in terms of recognition and acceptance of women. As per the record it is said, in the USA more than 9 millions firms are owned by females, they have employed around 8 million of people and generated around $1.5 trillion in sales. At present across the globe women are accepting entrepreneurship. They are quite able to face and overcome various challenges. They are leading the business successfully. They utilize various opportunities and accomplish their objectives (Rajani & Sarada, 2008). Not only the opportunities they are acquiring also they face different challenges in the field of growth and development. The advantages and opportunities are discussed below.

I. A Diverse and Innovative Workforce

Innovation and creativity come from diversity in gender, age, culture and race. Across the world, top companies are aiming to prioritize and benefit from a diverse workforce. Men and women both are coming from different backgrounds with varied experiences that are helpful to shape their approach to business. Men and women both face challenge, by collaborating with each other they are able to perform creatively and take the company forward.

II. Strength in Soft Skills and Emotional Intelligence is an Advantage for Women

For getting success technical skills and knowledge are very much essential. Soft skill and emotional intelligence both are having impact on developing entrepreneurship. Emotional intelligence in leadership, it means ability to listen, empathy, self-awareness and accountability. To lead the companies, women can use their experiences and soft skill with emotional intelligence. It will help women to get forward.

III. Ability to Create a Woman-friendly Corporate Environment

The corporate culture of many companies is not suitable for women. On the other hand if a woman establishes her own company she can provide suitable and comfortable environment to other women who work in the company. The women have the capacity to create value towards their right and can enjoy a more authentic life.

In opposite case, there are different challenges that are faced by women entrepreneurs, are discussed below.

Gender Discrimination at Workplace

Industries are male dominated (Sharifabadi & Zadeh, 2020, Zhang, 2020). It becomes a challenge for the men when women come as leaders and direct men on their jobs. Though it is established, gender equality, in most of the corporate and they believe that the best person should deserve the job irrespective of gender. There are many instances where it is found unconscious gender biasness. Men are always encouraged to be bossy, in case of women there is a lack. They face problem in order to be assertive and aggressive.

Difficulty in Acquiring Funds

It is very difficult for woman to acquire fund from investors for start-ups. Women owned firms are facing more challenges regarding this. Many investors are in doubt that female cannot be a successful entrepreneur, that's why they avoid to invest in their businesses. For these thoughts of investors' women entrepreneurs are facing lots of challenges and incurring loss .Mostly they don't get a strong support from markets for establishing and smooth running of their businesses .

Building a Viable Support Network

Another major challenge and obstacle for women entrepreneurs is lack of support system in relation with efficient mentors and advisors. Around 48 percent of female founders are having the lack of able advisors and mentors, that stagnant their professional growth. The majority of the people under the world of business are highly mail dominated. It becomes so much difficult to generate own path in business establishment and development. Same problem is associated with introductions and connections into some of the top-notch business networking.

In this highly competitive business environment women have emerged as successful entrepreneurs. To accomplish the objectives of the organization they have proved that they are not less than their male counterparts (Zeidan & Bahrami, 2011). At presents every country brags of their successful women entrepreneurs who have achieved success in their life. They have faced many difficult situation and challenges in the journey of establishment of career and tactfully they have dealt with it. So many women are there, who hold key positions in the company. To accomplish their personal goals as well as organizational goals they are supported by their male counterparts. They are also supported by their family members, so that the journey of growth, success and prosperity become easy for them.

Women's Entrepreneurship Development (WED) Programme

International Labour Organization has set up the Women's Entrepreneurship Development Programme. It focuses to empower women who want to start, run and develop their business. ILO has taken the initiative for gender equality and sustainable development. The programme has been named as ILO-WED. It creates meaningful participation in entrepreneurship ecosystems. ILO-WED is using human-centred, systematic and integrated approach which is based on WED Strategy. They built a set of tested tools for implementing targeted actions and empowering women entrepreneurship. ILO-WED programme is helping women entrepreneurs to secure a decent job to themselves and others .In this way they are able to overcome barriers and develop productive and sustainable businesses (Manzanera-Román & Brändle, 2016). The ILO-WED and SME (small and medium enterprises) have established integrated approaches to support and build capacity for empowering women.

I. The Gender and Entrepreneurship Together (GET) Ahead

This is one of the most effective training programs under ILO-WED's core tools. GET Ahead has focused on key business management skill and soft skill for women. This program aims to address some of the barriers and challenges that women can face to start a business. It helps to bridge the gender gap (Idrus et al., 2014). It is totally different from conventional business training materials as it gives attention on entrepreneurial skills from a gender perspective. This training programme incorporates coaching and mentoring, also makes targets to the needs of women with lower literacy levels.

II. **The Improve Your Exhibiting Skills: A Training Guide (IYES)**

This kind of training program aims to increase women capacity to participate in trade fairs. The IYES program has adopted a three-pronged approach, namely trade fair participation, trade fair facilitation and trade show evaluation including follow-up.

III. The SME Unit's Start and Improve Your Own Business (SIYB)

One of the largest global man magnet training programmes is SIYB. It helps small -scale entrepreneurs for their start ups and growth. This program includes four training packages that work on different stages of business development.

EFFECTIVENESS OF PUSH AND PULL MODEL USING AHP

According to (Huda Al Matroushi et al., 2018) there are pull and push factors theory on women entrepreneurship, to identify and prioritize the factors that influence women entrepreneurs (Figure 8). This pull and push factors model is implemented in two cases, one is with business family and another one is with non- business family. This theory is based on analytic hierarchy process model. The data utilized by the researcher based on United Arab Emirates country. The analytical hierarchy process is a method that helps to organize and analyze complex decision through psychology and mathematics. AHP sets a rational framework for decision making to quantify its criteria and alternative options. It is relating those elements with the overall objective. The push and pull model of women entrepreneurship has six sub domains, i.e. education (Nakao & Nishide, 2020, Babo et al., 2021), training, skills, experience, motivation and knowledge. Education includes three components like primary, secondary and bachelors. Training includes personal development training and vocational training. Skills signify with technical skills, management skills (Iyer et al., 2015), entrepreneurial skills, decision making skills (Fong et al., 2020, Das et al., 2020, Ngan et al. 2016, Jayakumar et al., 2020), communication skills and leadership skills. Experience has been divided into two ways, namely entrepreneurial experience and work experience. Motivation and support come from

Figure 8. Push and Pull Model for Woman Entrepreneurship Development

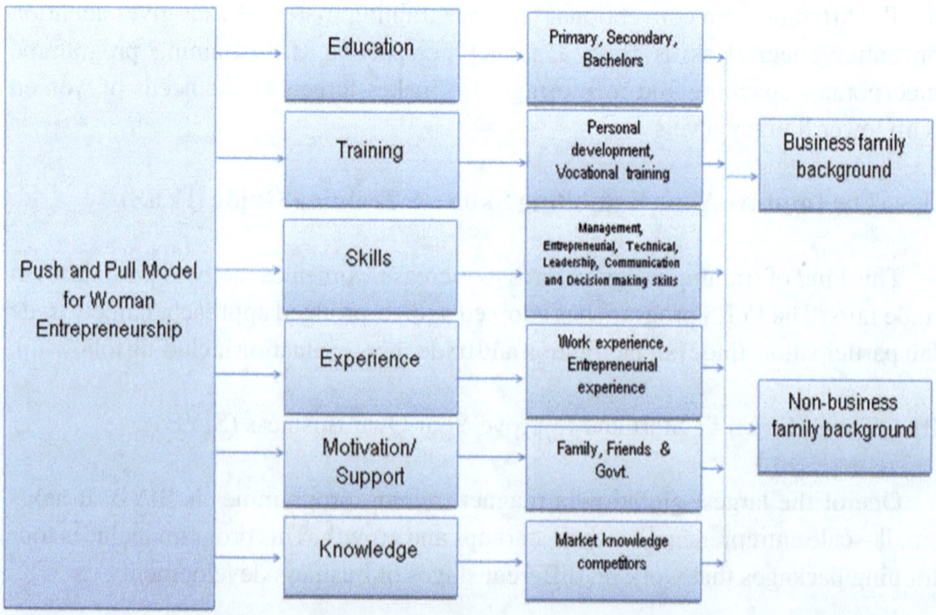

family members, friends and Government. Knowledge refers to market knowledge and knowledge about competitors.

AI and Women Entrepreneurship

Due to the change and advancements in technology, the meaning and concept of entrepreneurship have been modified for decades. It is a very clear picture in our country that businesses are always dominated by men. After that, when AI or artificial Intelligence boomed, women entrepreneurship and the rising numbers of women ventured into several businesses across the globe. AI is not only boosting women's entrepreneurship, rather, but it has also opened many doors with lots of opportunities for all (Sharma, A. (2019,). Data science and AI both have played a crucial role in the context of an entrepreneurial decision-making environment. Therefore, it is said that under industry 4.0 this field creates new paths for future entrepreneurial actions. One thing must say here, during the pandemic time when India witnessed massive job termination, jobs under AI were least affected. The corporate world is more focused on efficient virtual work, leading to high demand for AI (Majumder, S., & Mondal, A. (2021). AI helps to remove gender bias from the process of recruitment; this also addresses gender pay gaps. In the galaxy of women's entrepreneurship in business, one of the ways is to use the Natural Language

Processing Engine (NLP), where employers and candidates both can find matched skill sets, years of experience, and the number of finished projects. For further HR processes during the time of framing for best candidates in this area, gender and other factors will not be considered. There is a certain limitation in the field of AI that it can't change behavior, culture, or social norms. The work of human resources with the help of Artificial Intelligence in the workplace is getting faster but is not applicable to manpower's attitude, perception, and thinking habits. Women entrepreneurs are also taking challenges to sharpen up employees through the conduction of different AI-based training. This gives a potential and a more attractive workforce and a better fit for the organization. Women entrepreneurship is booming by the good quality of digital skills now a day. Women are much more responsible for the development of AI as tech is still a male-dominated industry. There exist brilliant female minds, and teams of researchers' in Google and Facebook too, they are running their own entrepreneurial ventures in the area of artificial intelligence. For example, Timnit Gebru is the efficient technical co-lead for ethical AI at Google. She is the leading advocate for accountability with advanced Artificial Intelligence technology (Majumder, S., & Dey, N. (2022). The former Uber director of engineering was Claire Delaumay, who works on the Isaac robots initiative.

Discussion

The increase in women entrepreneurship explores various field of education. It creates more aware society. People of the society become knowledgeable and it aids to develop the whole country. Presently women entrepreneurs are able to gather knowledge related to loans, taxes, short term and long term financial planning along with investments. They are creating a strong financial foundation for their family apart from contribution to the economy. To build different products and services they keep different mindset for development. They emphasize more on sustainability, empathy, inclusion, innovation and creativity tremendously. According to Mc Kinsey study it is established that, by 2025 a potentiality of adding 18% to India's GDP is exist if same work opportunities are also given to women. Against the 45% worldwide average, presently women contribute 22% of India's GDP. Though women entrepreneurship now combines with technologies but there is a thing that cannot be replaced by using information technology, it is empathy. Unfortunately women have this plentiful amount. This part of women makes them less likely to be benefitted from ASEAN Economic Community (AEC). Statistics has shown more than 70% of women stuck in business when it comes to the problems of complexity, i.e. connectivity and logistics related problems. Our society should encourage and support women towards flourish of women entrepreneurship at its apex level. Women entrepreneurship is most important issue to bring economic growth in the country.

If entrepreneurs can bring their own challenges, a favorable lucrative environment can be made and that will be significant to growth of the nation.

CONCLUSION

Women entrepreneurs are those entrepreneurs who contribute significantly towards economic growth with high level of dedication. From the study, it is said; women's entrepreneurship development program reduces poverty around the globe. There are so many challenges that a woman entrepreneur must face in her journey of establishment, though they are proving their best in form of every possible outcome. Encouragement is always required to boost women's motivation and make strong economic growth of the country. For the expansion of business, entrepreneurs need to introduce rewards; recognition for the sustainability of employees. The use of artificial intelligence has made the work of women entrepreneurs easier than the usage of modern technology. The opportunities have been increased and challenges have been decreased than previously by AI. Women entrepreneurs are not changing the society but also significantly contributing to the growth of any country.

REFERENCES

Afrin, S., Islam, N., & Ahmed, S. (2008). A multivariate model of micro credit and rural women entrepreneurship development in Bangladesh.

Afrin, S., Islam, N., & Ahmed, S. U. (2008). A Multivariate Model of Micro Credit and Rural Women Entrepreneurship Development in Bangladesh. *International Journal of Business and Management*, *3*(8), 169–185.

Agarwal, S., & Lenka, U. (2018). Why research is needed in women entrepreneurship in India: A viewpoint. *International Journal of Social Economics*.

Akehurst, G., & Simarro, E., & MasTur, A. (2012). Women entrepreneurship in small service firms: Motivations, barriers and performance. *Service Industries Journal*, *32*(15), 2489–2505.

Al Matroushi, H., Jabeen, F., Matloub, A., & Tehsin, M. (2020). Push and pull model of women entrepreneurship: empirical evidence from the UAE. *Journal of Islamic Accounting and Business Research*.

Audretsch, D. (2012). Entrepreneurship research. *Management Decision*.

Babo, R., Dey, N., & Ashour, A. S. (Eds.). (2021). *Workgroups E-Assessment: Planning, Implementing and Analysing Frameworks*. Springer Singapore.

Baron, R. A., & Shane, S. (2007). Entrepreneurship: A process perspective. *The psychology of entrepreneurship*, 19-39.

Bhatt, C., Dey, N., & Ashour, A. S. (Eds.). (2017). *Internet of things and big data technologies for next generation healthcare*.

Bruyat, C., & Julien, P. A. (2001). Defining the field of research in entrepreneurship. *Journal of Business Venturing*, *16*(2), 165–180.

Cunningham, J. B., & Lischeron, J. (1991). Defining entrepreneurship. *Journal of Small Business Management*, *29*(1), 45–61.

Das, H., Rout, J. K., Moharana, S. C., & Dey, N. (Eds.). (2020). *Applied Intelligent Decision Making in Machine Learning*. CRC Press.

Dey, N., Hassanien, A. E., Bhatt, C., Ashour, A., & Satapathy, S. C. (Eds.). (2018). *Internet of things and big data analytics toward next-generation intelligence,* (Vol. 35). Springer.

Dey, N., & Tamane, S. (Eds.). (2018). *Big data analytics for smart and connected cities*. IGI Global.

Eckhardt, J. T., & Shane, S. A. (2003). Opportunities and entrepreneurship. *Journal of Management*, *29*(3), 333–349.

Fazalbhoy, S. (2014). Women entrepreneurship as the way for economic development. *Annual Research Journal of Symbiosis Centre for Management Studies*, *2*(1), 117–127.

Fong, S. J., Li, G., Dey, N., Crespo, R. G., & Herrera-Viedma, E. (2020). Composite Monte Carlo decision making under high uncertainty of novel coronavirus epidemic using hybridized deep learning and fuzzy rule induction. *Applied Soft Computing*, *93*, 106282.

Fortunato, M. W., & Alter, T. (2015). Community entrepreneurship development: An introduction. *Community Development (Columbus, Ohio)*, *46*(5), 444–455.

Gupta, N., Gupta, S., Khosravy, M., Dey, N., Joshi, N., Crespo, R. G., & Patel, N. (2021). Economic IoT strategy: The future technology for health monitoring and diagnostic of agriculture vehicles. *Journal of Intelligent Manufacturing*, *32*(4), 1117–1128.

Hani, U., Rachmania, I. N., Setyaningsih, S., & Putri, R. C. (2012). Patterns of Indonesian women entrepreneurship. *Procedia Economics and Finance*, *4*, 274–285.

Idrus, S., Pauzi, N. M., & Munir, Z. A. (2014). The effectiveness of training model for women entrepreneurship program. *Procedia: Social and Behavioral Sciences, 129*, 82–89.

Iyer, V. R., Dey, N., & Chakraborty, S. (2015). Advent of Information Technology in the world of Tourism. In *Emerging innovative marketing strategies in the tourism industry* (pp. 44–53). IGI Global.

Jayakumar, L., Dumka, A., & Janakiraman, S. (2020). Energy aware dynamic mode decision for cellular D2D communications by using integrated multi-criteria decision making model. [IJACI]. *International Journal of Ambient Computing and Intelligence, 11*(3), 131–151.

Kumar, S. A. (2008). *Entrepreneurship development*. New Age International.

Maden, C. (2015). A gendered lens on entrepreneurship: Women entrepreneurship in Turkey. *Gender in Management*.

Majumder, S., & Dey, N. (2022). Artificial Intelligence and Knowledge Management. In *AI-empowered Knowledge Management,* (pp. 85–100). Springer.

Majumder, S., & Mondal, A. (2021). Are chatbots really useful for human resource management? *International Journal of Speech Technology, 24*(4), 969–977.

Manzanera-Román, S., & Brändle, G. (2016). Abilities and skills as factors explaining the differences in women entrepreneurship. *Suma de negocios, 7*(15), 38-46.

Martinez Dy, A., & Jayawarna, D. (2020). Bios, mythoi and women entrepreneurs: A Wynterian analysis of the intersectional impacts of the COVID-19 pandemic on self-employed women and women-owned businesses. *International Small Business Journal, 38*(5), 391–403.

Mathew, R. V., & Panchanatham, N. (2011). An exploratory study on the work-life balance of women entrepreneurs in South India. *Asian academy of management journal, 16*(2).

Mathew, V. (2010). Women entrepreneurship in Middle East: Understanding barriers and use of ICT for entrepreneurship development. *The International Entrepreneurship and Management Journal, 6*(2), 163–181.

Moses, C., & Mordi, C. (2010). Women entrepreneurship development in Nigeria: The effect of environmental factors. *BULETINUL Universitatii Petrol-Gaze din Ploieti, 62*(4), 43-52.

Mukit, M. M. H., Nabila, N. J., Islam, M. S., & Abdel-Razzaq, A. I. (2020). Valuing women entrepreneurship in the informal economy in Bangladesh. *ECONOMICS-Innovative and Economic Research*, *8*(2), 37–54.

Nakao, K., & Nishide, Y. (2020). The development of social entrepreneurship education in Japan. *Entrepreneurship Education*, *3*(1), 95–117.

Nawaz, F. (2009). Critical factors of women entrepreneurship development in rural Bangladesh.

Ngan, T. T., Tuan, T. M., Son, L. H., Minh, N. H., & Dey, N. (2016). Decision making based on fuzzy aggregation operators for medical diagnosis from dental X-ray images. *Journal of Medical Systems*, *40*(12), 1–7.

Niethammer, C. (2013). Women, entrepreneurship and the opportunity to promote development and business. *Brookings blum roundtable policy brief, 37*, 1-10.

Obaji, N. O., & Olugu, M. U. (2014). The role of government policy in entrepreneurship development. *Science Journal of Business and Management*, *2*(4), 109–115.

Osei, C. D., & Zhuang, J. (2020). Rural poverty alleviation strategies and social capital link: The mediation role of women entrepreneurship and social innovation. *SAGE Open*, *10*(2), 2158244020925504.

Rajani, N., & Sarada, D. (2008). Women entrepreneurship and support systems. *Studies on Home and Community Science*, *2*(2), 107–112.

Rumelt, R. P. (2005). Theory, strategy, and entrepreneurship. In *Handbook of entrepreneurship research* (pp. 11–32). Springer.

Sadi, M. A., & Al-Ghazali, B. M. (2010). Doing business with impudence: A focus on women entrepreneurship in Saudi Arabia. *African Journal of Business Management, 4*(1), 001-011.

Sarfaraz, L., Faghih, N., & Majd, A. A. (2014). The relationship between women entrepreneurship and gender equality. *Journal of Global Entrepreneurship Research*, *4*(1), 1–11.

Sharifabadi, A. M., & Zadeh, E. M. (2020). The evolution of university entrepreneurship over the past 20 years: A bibliometric analysis. *Entrepreneurship Education*, *3*(4), 457–488.

Sharma, A. (2019, November). Entrepreneurship and Role of AI. In *Proceedings of the 2019 2nd International Conference on Signal Processing and Machine Learning*, (pp. 122-126).

Sharma, P. (2013). Women entrepreneurship development in India. *Global Journal of Management and Business Studies*, *3*(4), 371–376.

Shiyal, S. M., Garg, A., & Rohini, R. (2019). Usage and Implementation of Artificial Intelligence in Entrepreneurship: An Empirical Study. *Seshadripuram Journal of Social Sciences (SJSS)*, *4*.

Tambunan, T. (2007). Entrepreneurship development: SMES in Indonesia. *Journal of Developmental Entrepreneurship*, *12*(01), 95–118.

Tambunan, T. (2009). Women entrepreneurship in Asian developing countries: Their development and main constraints. *Journal of Development and Agricultural Economics*, *1*(2), 027-040.

Tiwari, A. (2007). *Entrepreneurship development in India*. Sarup & Sons.

Tiwari, S., & Tiwari, A. (2007). *Women entrepreneurship and economic Development. Sarup*. Sons.

Ullah, M. M., Mahmud, T. B., & Yousuf, F. (2013). Women entrepreneurship: Islamic perspective. *European Journal of Business and Management*, *5*(11), 44–52.

Vimal, S., Khari, M., Dey, N., Crespo, R. G., & Robinson, Y. H. (2020). Enhanced resource allocation in mobile edge computing using reinforcement learning based MOACO algorithm for IIOT. *Computer Communications*, *151*, 355–364.

Vinay, D., & Singh, D. (2015). Status and Scope of Women Entrepreneurship.

Wiklund, J., Davidsson, P., Audretsch, D. B., & Karlsson, C. (2011). The future of entrepreneurship research. *Entrepreneurship Theory and Practice*, *35*(1), 1–9.

Yadav, V., & Unni, J. (2016). Women entrepreneurship: Research review and future directions. *Journal of Global Entrepreneurship Research*, *6*(1), 1–18.

Yadav, V., & Unni, J. (2016). Women entrepreneurship: Research review and future directions. *Journal of Global Entrepreneurship Research*, *6*(1), 1–18.

Zeidan, S., & Bahrami, S. (2011). Women entrepreneurship in GCC: A framework to address challenges and promote participation in a regional context. *International Journal of Business and Social Science*, *2*(14), 100–107.

Zhang, J. (2020). Pedagogical alignment for entrepreneurial development. *Entrepreneurship Education, 3*(3).

Zhu, L., Kara, O., & Zhu, X. (2019). A comparative study of women entrepreneurship in transitional economies: The case of China and Vietnam. *Journal of Entrepreneurship in Emerging Economies.*

Chapter 11
IoT Security Enhancement System:
A Review Based on Fusion of Edge Computing and Blockchain

Santosh Kumar Das
https://orcid.org/0000-0001-6852-1361
Sarala Birla University, Ranchi, India

Vikash Kumar
Sarala Birla University, Ranchi, India

ABSTRACT

The internet of things (IoT) is one of the new technologies based on new requirements and innovations. It is a set of equipment that consists of several sensor nodes and devices to make efficient communication. It is an embedded system that contains several sensor nodes, software, and exchanging equipment. There are several applications of IoT such as smart homes, health monitoring systems, smart factories, wearable health monitoring systems, logistic tracking, etc. Although, it has several applications and advantages, but it has also some limitations regarding data privacy, data security, data volume and data complexity, etc. Hence, it requires innovating and modeling of security of IoT based on different techniques. Therefore, in this chapter, a details review is discussed based on the fusion of edge computing and blockchain.

INTRODUCTION

Internet of Things (IoT) is a technology that is based on different sensor networks

DOI: 10.4018/978-1-6684-7343-6.ch011

and devices to make efficient communication for modeling applications (Abadía et al., 2022). It is used for several purposes such as data sensing, data access and its related transportation, data connectivity and its mechanism, app and application design and its human value and experience analysis, data value connectivity and action maintenance, etc. It requires designing of several requirements of the customer based on modeling applications such as smart applications and tangible benefits (Wojdała, 2021; Das, 2021).

Architecture of IoT

IoT is based on four layers as application layer, data processing layer, network layer and sensing layer which are shown in Figure 1 (Lima et al., 2021). The application layer is used to manage different applications based on the services of the customer and societies which depend on different needs. It helps to manages several information for the purpose of smart application. Data processing layer that is used to process data based on services and management. It helps to process data based on different areas and application. Network layer is used to maintain a connection based on wireless network. It may be a wireless sensor network, wireless mobile network, wireless ad-hoc network, etc. But each of the networks contains some sensor nodes having capabilities of sensing data and its processing information. The sensing layer is used to manage data sensing information. There are several components that are associated with this such as gateway, thing or device, cloud, and user interface based on different data and services. Short descriptions are as follows.

(i) **Application layer:** This layer is used the smart application layer which is used to manage and control several sensing smart applications. It also helps to manage the operation of IoT based on different devices.
(ii) **Data processing layer:** This layer is used for processing information system that helps to manage several data analytics information. It also helps to manage several data analytics based on some decision making system.
(iii) **Network layer:** It helps to manage several data transmission systems for managing some technologies. Its main purpose is a data acquisition system that helps to manage network technologies based on two devices such as the internet and network gateways.
(iv) **Sensing layer:** This layer is used for data sensing after that helps to gather information. It helps to sense data and gathers it by using a sensor and actuator. It helps to manage physical objects for handling several information.

Figure 1. Architecture of IoT

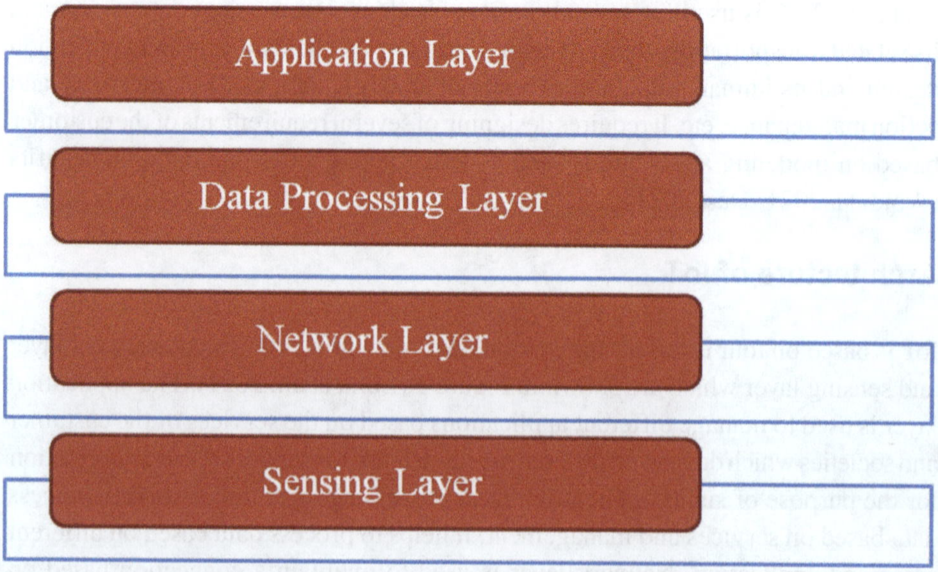

Application of IoT

IoT helps to make efficient communication based on services. It helps to establish a link between physical objects with software and some devices to make communication efficient. Some examples are given in this section as follows.

(a). Smart city maintenance
(b). Smart grid computing
(c). Smart home and building management
(d). Supply chain management
(e). Traffic management

The remaining part of this article is divided as: Section 2 describes details analysis of literature review for security related issues in the area of IoT. Section 3 discusses proposed methodology of secure IoT communication. Section 4 discusses conclusions and future works based on proposed work.

LITERATURE REVIEW

In the last few years, several works have been proposed for the purpose of IoT and its related application based on security purposes and its methodologies. Some of the works are discussed as Li et al. (2022) proposed multiple blockchain based authentication protocols for IIOT systems using PUF. In which first protocol is based on physical device compatible and low-cost SRAM PUF and second one is based on Arbiter PUF. Arbiter PUF is very difficult to clone physically but easy to clone mathematically which suit for authentication. Das et al. (2021) proposed a list of application in the area of wireless network and its related application. The main purpose of this book is to enhance the activity and application of wireless network. Each of the work in this book indicates architectural solution based on particular issues. It helps to design some related security issues based on different factors and its information system. It helps to manage efficiently several information. Wang et al. (2022) proposed lightweight and scalable protocol to control identity of IoT devices and access control of large scale of sensor data using decentralized ledger technology IOTA. In this protocol, to reduce energy demand IOTA technology is used and Inter Planetary file system (IPSF) is used for storage of large amount of data which generated by IOT devices. Yang et al. (2022) proposed lightweight authentication protocol mobile-edge computing. They designed two authentication phases based on modular square root technique. First authentication between the sensor node and edge node and second authentication is between edge node and application provider. Das et al. (2021) proposed a list of algorithms based on efficiency of machine learning methodologies. It helps to manage several machines learning information to deals with some problems. It helps to manage several operations and its application to solve real-life problem. Several machine learning algorithms are mentioned in this book that handle several areas and give new direction to the readers. Peng et al. (2022) proposed secure multidimensional data aggregation protocol for IOT devices using Chinese Remainder Theorem used to encrypt small integer vector of multiple dimensions. In order maintain nonrepudiation of data and efficient verification at edge devices, the signature process and batch verification process are combined. Yu et al. (2022) proposed a cryptographic protocol for data privacy for industrial internet of things using blockchain. For encryption symmetric key is used. Symmetric key is secured by private key. To prevent the loss of private key Samir secret sharing algorithm and blockchain mechanism is used. De et al. (2020) proposed a method that contains several nature inspired applications in the area of wireless sensor network. Several issues and problems are mentioned in this article to handle several operations of the network. The work is based on architecture system that handles several computation applications. Each of the application is inspired by nature with the context of algorithm. Vishwakarma and Das (2021) proposed an authentication scheme for IOT devices for secure communication using blockchain technology. In this scheme, Blockchain technology is combined with Advance Encryption Standard and Elliptic Curve digital signature algorithm. In this scheme, to reduce interruption in communication angular distance mechanism is used. Singh et al. (2021) proposed a centralized cross domain secure data sharing plate form for IIOT devises. Multiple security gateways using blockchain mechanism are used to store date in centralized cloud. To control malicious activity, centralized cloud verifies the activity using blockchain verification mechanism. Das et al. (2020) proposed a framework application for handling several operation of the network. It helps to manage several designed model of the network. Each of the models is based on some factors of application. It indicates solution prospective point of view operation in the area of wireless network. It helps to guide several operation and model of the network that helps to incorporate the system in smart application. Zhaofeng et al. (2021) proposed an authentication scheme in Edge and IoT system using blockchain mechanism. In this scheme edge devices considered as block to create blockchain network. In this model blockchain consensus, smart contract and complete blockchain authentication has been developed. Zhang et al. (2021) proposed privacy preserving authentication scheme for decentralized cloud edge IoT devices using combination of PUFs and blockchain technology. In this scheme some key has been addressed like 1) storage issue of each challenge response pairs (CRP) Generated by PUFs device. 2) Single point failure issue. 3) Synchronization of CRP. Vangala et al. (2021) proposed a smart contract-based authentication protocol for agriculture IoT devices using blockchain. In this scheme, mutual authentication and key agreement performed between two IoT devices and between IoT device and gateway node. In authentication process blockchain node is created by edge server. Block are verified and added to blockchain by smart contract-based consensus method. De et al. (2020) designed a survey model for wireless sensor network for application management. Each of the work is based on several algorithms and challenges system. But it helps to resolve several issues based of factors of information. Several solutions are mentioned in this proposed to resolve the challenges based on framework analysis of wireless sensor network. Ge et al. (2020) proposed a decentralized scheme to store data which generated by IoT devices using blockchain technique to ensure data privacy. In this scheme, Unspent Transaction Output (UTXO) verification method on RSA accumulator has been produced to reduce computational complexity of nodes. Guo et al. (2020) proposed a cross-domain authentication scheme for IoT devices using master- slave blockchain. In this scheme, trust issue of different IoT systems has been addressed. In this scheme, Byzantine fault tolerance using reputation value model (RIBFT) has been proposed for authentication and access control. Jangirala et al. (2020) proposed an RFID based light weight authentication scheme in edge computing system. In this scheme, bitwise exclusive or (XOR), cryptographic hash function and bit wise rotation operation are performed. Simulation based formal security verification is performed by AVISPA tool. Asheralieva et al. (2020) proposed decentralized, secure, scalable and self-organizing system architecture for IoT devices using blockchain in edge computing environment. Transaction detail is managed by peer of blockchain. New consensus mechanism is designed using vote of each peer. Reputation is considered to evaluate vote of each peer. Binh and Dey (2018) designed soft computing review system based on wireless sensor network and its application. The work is fully based on several techniques of soft computing technique that helps to model its related application. Each of the work helps to model and resolve the issues that are raised during the operation of wireless sensor network. Li et al. (2020) discussed the issue of knowledge sharing of distributed network edges, limited resource and mobility of edge devices. They also discussed different attack scenario like man in the middle attack and denial of service. They proposed client focused architecture for securing edge knowledge in IoT devices. Proof of popularity consensus method is used. Rathor et al. (2019) proposed a distributed security framework based on Software Defined Network (SDN) combined with blockchain technology for IoT devices. In this framework there is continuous monitoring and analysis of data flow of IoT system using SDN to secure the system. Blockchain provide decentralized attack detection to overcome single point of failure. Das et al. (2021) proposed a method for optimization in the area of conflicting environment. It is based on different types of optimization based on game optimization system. In this work, several types of rule matrices are considered as fuzzy nature and crisp nature. Both are optimize and model based on dataset and tune the parameters mathematically. Finally, it helps to model the application based on intelligent decision making system. Mohammad A and Issa T (2019) proposed a lightweight mutual authentication scheme and key agreement scheme to IoT devices based on short term identity and cumulative key hash chained. Dynamic identities and symmetric key are used to authenticate and maintain session between IoT nodes. Virtual domain segregation is used to ensure security between nodes. Pan et al. (2019) proposed an edge-IoT framework named as Edge-chain based on blockchain and smart contract. Blockchain is used to link edge severs with each IoT devices. Edge-chain used credit-based resource management system to controlled access of the resource based on priority, application category and past actions. Smart contracts are used to implement rules and policies in automated manner on IoT devices. Das et al. (2021) designed a fuzzy routing method for wireless ad-hoc network. The work is based on wireless ad-hoc network system. It is based on two parameters such as energy and distance. By combining both parameters helps to model the application. It also helps to optimize the network to make an efficient network lifetime based on fuzzification method. The work is evaluated in java programming for effectiveness system (Das & Giacalone, 2022).

Debe et al. (2019) proposed a distributed trust framework to maintain the reputation of public fog nodes using Ethereum Blockchain. To maintain of the reputation, user's opinion of their past interaction with the public fog nodes is considered. In this framework Ethereum blockchain and smart contract technologies are used to maintain trust between IoT devices and public fog nodes in the decentralized environment. Islam et al. (2019) proposed data acquisition scheme to gather information from IoT devices using unmanned arial vehicle (UAV) and store encrypted data in blockchain at mobile edge computing (MEC) server. Encryption of data is performed with the assistance of UAV. Validation of data and sender is performed at MEC server before storing the data in the blockchain. Thanh Binh et al. (2018) proposed a method for maximizing the area of coverage system in wireless sensor network. The work is based on optimization system that helps to model several algorithms and its analysis. The work is based on nature inspired optimization for handling several parameters. These parameters help to model and evaluate several computational times and help to manage adaptability of the system. Yang et al. (2018) proposed a data scheme based on Hyperledger Fabric v1.0 blockchain IoT edge computing environment. In this scheme, they focused to maintain temper registrant, reliable and distributed ledger by implementing smart contract. In this scheme, they also implemented point to point data share mechanism among remote IoT devices. Sharma et al. (2018) proposed decentralized cloud framework with a software defined network enabled fog nodes using blockchain technology. This framework provides low cost, secure, on demand access and high performance. All the computation is performed at edge of the IoT nodes that minimize the latency and maximize the security. Multiple computing recourses are created due to decentralized fog nodes. Brogan et al. (2018) discussed the use of distributed ledger technology to improve security in health care by ensuring authentication and integrity of data of health care IoT devices. In the paper they discussed use of authenticated messaging extension module (AMEM) of IOTA protocol to confidentiality, integrity and availability of data using temper proof distributed ledger. Thanh Binh and Hai Nam (2018) proposed an illustration for coverage purpose in wireless sensor network. The work is based on several coverage systems and its modeling based on optimization system. The work formulates several adapted and conventional problems by using soft computing methodologies. Finally, it helps to reduce the complexity of the system and helps to evaluate several performance analyses. Rahman et al. (2018) proposed a secure in- home therapy management architecture based on blockchain based distributed MEC model to provide minimum latency, security and availability on demand. Blockchain-Tor based distributed transaction is used to preserve privacy, storage, ownership and sharing of the data. Wang et al. (2018) discussed the privacy preserving technology in the field of intrusion detection and cryptanalysis of collaborative intrusion detection network (CIDN) scenario due to its distributed

Table 1. List of IoT security challenges

Sl. No.	Challenges	Sl. No.	Challenges
1	Access control	15	Lack of secure update system
2	Access control	16	Legacy liabilities
3	Authentication	17	Mobile security
4	Confidentiality	18	Partner and interface management
5	Confidentiality management	19	Policy enforcement
6	Data integrity	20	Prioritization of productivity
7	Identity management	21	Privacy maintenance
8	Insecure data storage and transfer	22	Secure middleware
9	Insecure network services	23	Supervision
10	Insecure or outdated components	24	System availability
11	Insufficient knowledge	25	Transport and communication management
12	Insufficient privacy protection	26	Trust management
13	Lack of physical barriers	27	Unclear job roles
14	Lack of resource management	28	Weak or hard coded password

nature. They proposed a privacy preserving architecture for signature-based intrusion detection in decentralized network based on fog devices.

METHODOLOGY BASED ON EDGE COMPUTING AND BLOCKCHAIN

In the last section, there are several works have been proposed for the purpose of IoT security enhancement. In this section, details illustrations are discussed for the purpose of security modeling of IoT based on the fusion of edge computing and blockchain. There are several security challenges of IoT which are illustrated in given Table 1.

There are several security challenges are discussed in this section for the purpose of security concerns of IoT. The basic component for maintenance of IoT security is based on the given three parameters shown in Figure 2.

(a) Privacy
(b) Authenticity
(c) Integrity

Figure 2. Basic components for IoT security maintenance

Privacy indicates maintaining privacy in the system and its model for the purpose of modeling in the system. It is used to secure the system based on services. Authenticity indicates managing the system authenticates based on different parameters. It helps to check its privacy based on authentication. Integrity is a mechanism that indicates to integrate of several systems based on different components. It helps to manage several components for the purpose of modeling.

To make efficient and secure communication in IoT requires two basic components such as edge computing and blockchain. Edge computing is an architecture design for distributed computing purposes. It helps to bring efficient data storage systems closer to the source of data. It is designed to enhance bandwidth and response time of the services. In other side, blockchain is a digital ledger based distributed system which is used for transaction purpose. The method of blockchain is decentralizing the design peer-to-peer methodology. It consists of different blocks where each block is linked one to another based on the method of cryptography. It contains hash functions that maintain the timestamp of data transactions. The combination of both helps to model the application and resolve several issues for the purpose of modeling and its management. In this section, some of the security attacks are mentioned which resolve or estimate the prone of error in IoT.

(a) Physical attack: It is based on directly device based attack. It influences directly IoT related devices.
(b) Communication attack: It is based on data transmission system that is based on different communication devices where each device is based on the communication system.
(c) Lifecycle attack: It is based on life cycle of IoT that indicates several device changes and its related maintenance system.
(d) Software attack: It is based on software related attacks that influence the different operations of IoT device systems.

Figure 3 shows the trust management of IoT that is based on the fusion of edge computing and blockchain. It helps to manage several information and application

Figure 3. Secure IoT System

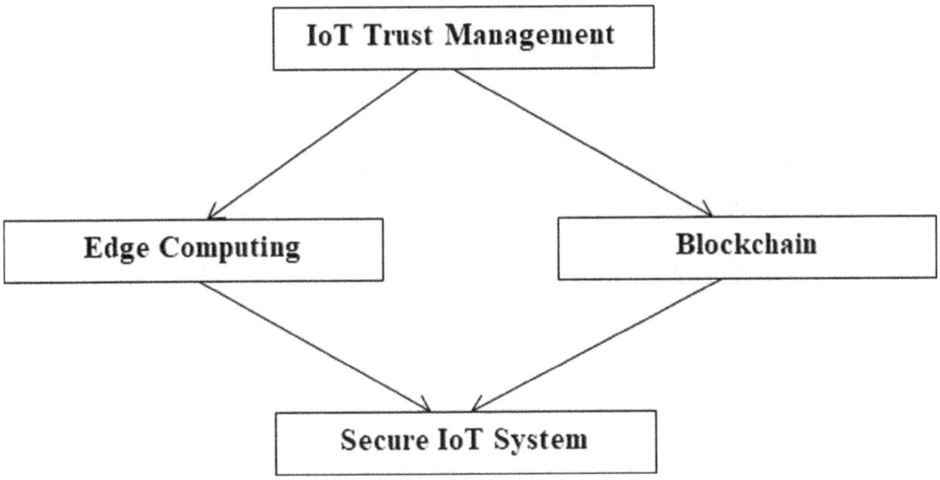

based on different services. Different features of edge computing and blockchain help to make the IoT system efficient.

Figure 4 shows the aiding benefits of IoT by fusion of edge computing. Architecture methodologies of edge computing help to model an efficient IoT for the purpose of model different service. There are several benefits such as computational efficiency, zero latency, cost and autonomous operation, lesser network load balancing system, privacy and security system and its modeling, reduced data exposure, etc. The combination of all benefits helps to make the application robust and secure.

CONCLUSIONS AND FUTURE WORK

IoT is one of the most popular innovations in the modern era which makes the application efficient and enable based on different services. There are several applications, issues and services are discussed in this chapter with the context of real-life applications. It has been observed that by fusion of edge computing in IoT helps to make an efficient and robust IoT application. It can also observe that by the fusion of blockchain, the services and transactions of IoT applications become more efficient and usable. Future work is to enhance this work with the help of simulation analysis based on different services.

Figure 4. Benefits of aiding Edge Computing in IoT

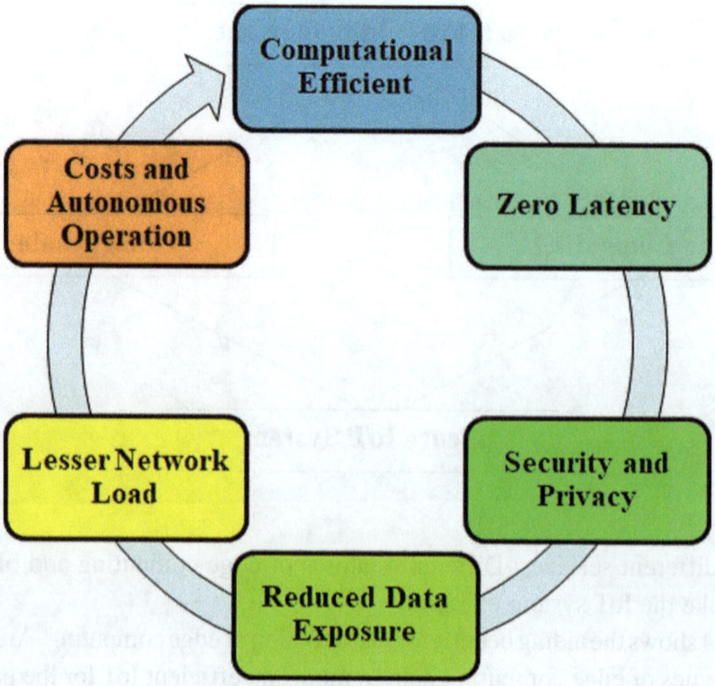

REFERENCES

Abadía, J. J. P., Walther, C., Osman, A., & Smarsly, K. (2022). A systematic survey of Internet of Things frameworks for smart city applications. *Sustainable Cities and Society*, 103949.

Alshahrani, M., & Traore, I. (2019). Secure mutual authentication and automated access control for IoT smart home using cumulative keyed-hash chain. *Journal of information security and applications, 45*, 156-175.

Asheralieva, A., & Niyato, D. (2020). Reputation-based coalition formation for secure self-organized and scalable sharding in iot blockchains with mobile-edge computing. *IEEE Internet of Things Journal*, *7*(12), 11830–11850.

Binh, H. T. T., & Dey, N. (Eds.). (2018). *Soft computing in wireless sensor networks*. CRC Press.

Binh, H. T. T., Hanh, N. T., Van Quan, L., & Dey, N. (2018). Improved cuckoo search and chaotic flower pollination optimization algorithm for maximizing area coverage in wireless sensor networks. *Neural Computing & Applications*, *30*(7), 2305–2317.

Binh, H. T. T., & Nam, N. H. (2018). Introduction to coverage optimization in wireless sensor networks. In *Soft computing in wireless sensor networks* (pp. 115–136). Chapman and Hall/CRC.

Brogan, J., Baskaran, I., & Ramachandran, N. (2018). Authenticating health activity data using distributed ledger technologies. *Computational and Structural Biotechnology Journal*, *16*, 257–266.

Das, S. K. (2021). Smart Design and Its Applications: Challenges and Techniques. In *Nature-Inspired Computing for Smart Application Design* (pp. 1–6). Springer.

Das, S. K., Das, A., Sinha, H. K., Das, S. P., & Nayak, S. (2021). Game theory based optimal decision-making system. In *Intelligent and Cloud Computing* (pp. 761–769). Springer.

Das, S. K., Das, S. P., Dey, N., & Hassanien, A. E. (Eds.). (2021). *Machine learning algorithms for industrial applications*. Springer.

Das, S. K., & Giacalone, M. (Eds.). (2022). *Fuzzy Optimization Techniques in the Areas of Science and Management*. CRC Press.

Das, S. K., Patra, N., Das, B. R., Das, S. P., & Nayak, S. (2021). Fuzzification-Based Routing Protocol for Wireless Ad Hoc Network. In *Intelligent and Cloud Computing* (pp. 237–245). Springer.

Das, S. K., Samanta, S., Dey, N., & Kumar, R. (Eds.). (2020). *Design frameworks for wireless networks*. Springer Singapore.

Das, S. K., Samanta, S., Dey, N., Patel, B. S., & Hassanien, A. E. (Eds.). (2021). Architectural Wireless Networks Solutions and Security Issues. Springer Singapore Pte. Limited. doi:10.1016/j.sysarc.2022.102638

De, D., Mukherjee, A., Das, S. K., & Dey, N. (2020). Wireless sensor network: applications, challenges, and algorithms. In *Nature inspired computing for wireless sensor networks* (pp. 1–18). Springer.

De, D., Mukherjee, A., Das, S. K., & Dey, N. (Eds.). (2020). Nature inspired computing for wireless sensor networks. Springer.

Debe, M., Salah, K., Rehman, M. H. U., & Svetinovic, D. (2019). IoT public fog nodes reputation system: A decentralized solution using Ethereum blockchain. *IEEE Access: Practical Innovations, Open Solutions*, *7*, 178082–178093.

Ge, C., Liu, Z., & Fang, L. (2020). A blockchain based decentralized data security mechanism for the Internet of Things. *Journal of Parallel and Distributed Computing*, *141*, 1–9.

Guo, S., Wang, F., Zhang, N., Qi, F., & Qiu, X. (2020). Master-slave chain based trusted cross-domain authentication mechanism in IoT. *Journal of Network and Computer Applications*, *172*, 102812.

Islam, A., & Shin, S. Y. (2019). BUAV: A blockchain based secure UAV-assisted data acquisition scheme in Internet of Things. *Journal of Communications and Networks (Seoul)*, *21*(5), 491–502.

Jangirala, S., Das, A. K., & Vasilakos, A. V. (2019). Designing secure lightweight blockchain-enabled RFID-based authentication protocol for supply chains in 5G mobile edge computing environment. *IEEE Transactions on Industrial Informatics*, *16*(11), 7081–7093.

Li, D., Chen, R., Liu, D., Song, Y., Ren, Y., Guan, Z., & Liu, J. (2022). Blockchain-based authentication for IIoT devices with PUF. *Journal of Systems Architecture*, *130*, 102638.

Li, G., Dong, M., Yang, L. T., Ota, K., Wu, J., & Li, J. (2020). Preserving edge knowledge sharing among IoT services: A blockchain-based approach. *IEEE Transactions on Emerging Topics in Computational Intelligence*, *4*(5), 653–665.

Lima, E., Moraes, J., Oliveira, H., Cerqueira, E., Zeadally, S., & Rosário, D. (2021). Adaptive priority-aware LoRaWAN resource allocation for Internet of Things applications. *Ad Hoc Networks*, *122*, 102598.

Pan, J., Wang, J., Hester, A., Alqerm, I., Liu, Y., & Zhao, Y. (2018). EdgeChain: An edge-IoT framework and prototype based on blockchain and smart contracts. *IEEE Internet of Things Journal*, *6*(3), 4719–4732.

Peng, C., Luo, M., Wang, H., Khan, M. K., & He, D. (2021). An Efficient Privacy-Preserving Aggregation Scheme for Multidimensional Data in IoT. *IEEE Internet of Things Journal*, *9*(1), 589–600.

Rahman, M. A., Hossain, M. S., Loukas, G., Hassanain, E., Rahman, S. S., Alhamid, M. F., & Guizani, M. (2018). Blockchain-based mobile edge computing framework for secure therapy applications. *IEEE Access: Practical Innovations, Open Solutions*, *6*, 72469–72478.

Rathore, S., Kwon, B. W., & Park, J. H. (2019). BlockSecIoTNet: Blockchain-based decentralized security architecture for IoT network. *Journal of Network and Computer Applications*, *143*, 167–177.

Sharma, P. K., Chen, M. Y., & Park, J. H. (2017). A software defined fog node based distributed blockchain cloud architecture for IoT. *IEEE Access: Practical Innovations, Open Solutions*, *6*, 115–124.

Singh, P., Masud, M., Hossain, M. S., & Kaur, A. (2021). Cross-domain secure data sharing using blockchain for industrial IoT. *Journal of Parallel and Distributed Computing*, *156*, 176–184.

Vangala, A., Sutrala, A. K., Das, A. K., & Jo, M. (2021). Smart contract-based blockchain-envisioned authentication scheme for smart farming. *IEEE Internet of Things Journal*, *8*(13), 10792–10806.

Vishwakarma, L., & Das, D. (2021). SCAB-IoTA: Secure communication and authentication for IoT applications using blockchain. *Journal of Parallel and Distributed Computing*, *154*, 94–105.

Wang, S., Li, H., Chen, J., Wang, J., & Deng, Y. (2022). DAG blockchain-based lightweight authentication and authorization scheme for IoT devices. *Journal of Information Security and Applications*, *66*, 103134.

Wang, Y., Meng, W., Li, W., Li, J., Liu, W. X., & Xiang, Y. (2018). A fog-based privacy-preserving approach for distributed signature-based intrusion detection. *Journal of Parallel and Distributed Computing*, *122*, 26–35.

Wojdała, M. (2021). The Internet of Things in carriage of goods by road-application, opportunities, problems. *Procedia Computer Science*, *192*, 3235–3242.

Yang, J., Lu, Z., & Wu, J. (2018). Smart-toy-edge-computing-oriented data exchange based on blockchain. *Journal of Systems Architecture*, *87*, 36–48.

Yang, X., Yi, X., Khalil, I., Luo, J., Bertino, E., Nepal, S., & Huang, X. (2021). Secure and Lightweight Authentication for Mobile Edge Computing Enabled WBANs. *IEEE Internet of Things Journal.*

Yu, K., Tan, L., Yang, C., Choo, K. K. R., Bashir, A. K., Rodrigues, J. J., & Sato, T. (2021). *A blockchain-based shamir's threshold cryptography scheme for data protection in industrial internet of things settings*. IEEE Internet of Things Journal.

Zhang, Y., Li, B., Liu, B., Hu, Y., & Zheng, H. (2021). A privacy-aware PUFs-based multiserver authentication protocol in cloud-edge IoT systems using blockchain. *IEEE Internet of Things Journal*, *8*(18), 13958–13974.

Zhaofeng, M., Jialin, M., Jihui, W., & Zhiguang, S. (2020). Blockchain-based decentralized authentication modeling scheme in edge and IoT environment. *IEEE Internet of Things Journal*, *8*(4), 2116–2123.

Chapter 12
Synthesis of Graphene:
Theory and Application

Arka Ghosh
University of Siegen, Germany

ABSTRACT

Single layer graphene is called monolayer graphene which exhibits some unique characteristics with a special arrangement of atoms in its crystal. In contrary, graphene can be sometimes modified for numerous application specific uses. This book chapter presents brief details of graphene structure, along with formation of sp2 hybridization, critical parameters of the unit cell, formation of σ and π bonds, different properties, synthesis processes and their advantages and disadvantages over one another, potential applications of graphene in different fields. Out of different synthesis processes, mechnical exfoliation, chemical exfoliation, reduced graphene oxide, chemical vapour deposition, and epitaxial growth are the most well-known processes, and are detailed and described here. Graphene application in integrated circuit, optical electronics generally in OLED, energy application generally for the storage purpose is elaborately discussed here in this chapter. Throughout the chapter the authors described everything with suitable diagrams as per the requirements.

INTRODUCTION

P. R. Wallace introduced single layer graphene theoretically in the year of 1947 (Wallace, 1947). In the year of 2004 i.e., after 57 years Geim and Novoselov discovered one atom thick crystallites from a piece of graphite and as a result of it in the year of 2010 they were awarded the Nobel Prize in Physics (Gerstner, 2010). As we require high speed and low power consumption devices in the field of electronics,

DOI: 10.4018/978-1-6684-7343-6.ch012

scaling of conventional silicon electronics in the order of nanometer is being done. Graphene based transistor known as Graphene Field Effect Transistor (GFET) is appeared to be potential replacement of conventional MOSFETs. International technology roadmap semi- conductors recognized graphene as a potential substitute of silicon based semi-conductor electronics. Graphene is a 2-D crystalline honeycomb lattice structure with sp2 hybridized carbon atoms having a molecular bond length of 0.142nm. Graphene- the thinnest material on earth has its exceptional electrical, mechanical, thermal and optical properties. The proposed work is based on constraint decision making system that contains several types of decision making. In the last decade, several types of decision-making works are done such as different architecture design and its methodologies of wireless network and its components (Das et al., 2021; Das et al., 2020; Das et al., 2021), metaheuristic and soft computing based work (Binh & Dey, 2018; Binh et al., 2020), machine learning based work (Das et al., 2021), smart application design based work (Das, 2021; Das & Giacalone, 2022), etc. A constraint of a system is responsible for slowing down of its performance or preventing it from the desired goal. In order to keep the flow of a project going identification, exploitation and elevation of constraints are very important.

In the subject like condensed matter physics as well as material science 2D graphene shows its unique electrical, mechanical, optical and chemical properties (Yang et al., 2018). In case of fundamental physics graphene has already shown its acceptance but in spite of having a lot of potential applications it needs further improvement to be commercialized and industrially acceptable. In this chapter different types of graphene synthesis processes have been de-scribed but commercial production of graphene is not so available till now. In the field of integrated circuit, optical electronics, energy application, graphene is already tested for some sort of application, but it need to be developed for future prospects. In the field of condensed matter physics graphene made itself an automatic choice due to its unusual as well as unique electronic spectrum. In a broad perspective, graphene has never been disappointed the researchers for its future potential applications and represents alone conceptually one atom thick material.

GRAPHENE STRUCTURE

Graphene is considered as thinnest as well as 2D allotrope of carbon which has an electronic configuration of $1s^2 2s^2 2p_x{}^1 2p_y{}^1 2p_z{}^0$ in its ground state. As far as electronic configuration of carbon is considered all the sub shells of the p orbital have same energy level i.e., p_x, p_y, p_z are having equivalent energy but there is no electron available in the p_z orbital in the ground state of the atom. The outer shell of an atom is called valence shell and electrons present in that shell is called valence electrons.

Figure 1. Electronic configuration and hybridization of carbon atom
(Spyrou & Rudolf, 2014)

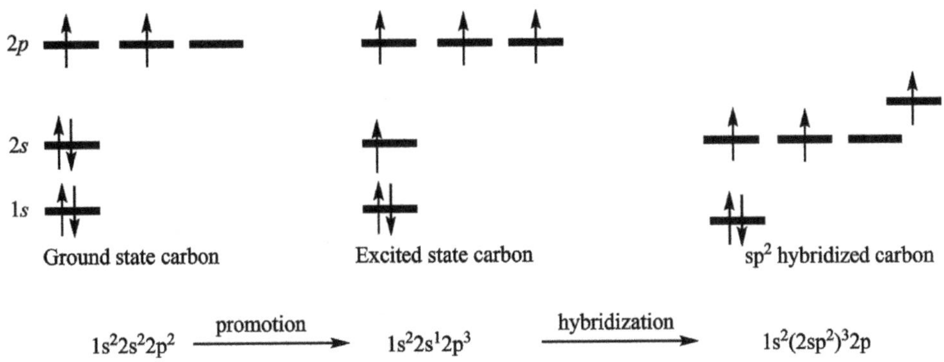

In case of carbon atom the number of valance electron is four. In the excited state these four valance electrons of carbon can form hybridized molecular orbital which are of three types namely sp, sp^2, and sp^3. In the figure 1 the electronic configuration of carbon atom has been shown. Where in the ground state carbon atom have 2 electrons in 1s orbital, 2 electrons in 2s and 2 electrons in 2p orbital. In its excited state one 1 electron from 2s orbital moves to higher energy state i.e. 2p orbital so, in this case 2s has 1 electron and 2p has 3 electrons. In the next step 3 electrons from 2s and 2p orbital make three sp^2 hybridized orbital (Singh et al., 2018). As we look at the figure 1 we can see that, in excited state one electron of 2s orbital jumps to $2p_z$ orbital which further transforms to three sp^2 hybridized orbital.

A sp^2 hybridized carbon atom shares electrons with three other carbon atoms adjacent to this which results in a planar structure of hexagonal honeycomb lattice which in turns is named as monolayer graphene. From the Fig. 2 (a) it can be seen that the unit-cell vectors $\vec{a_1}$ *and* $\vec{a_2}$ have the lattice constant of 2.46 Å where A and B are two carbon atoms. The planar ring which is formed here is very much stable due to resonance and delocalized electrons. In case of sp^2 hybridization any two carbon atoms of graphene participate to form in plane sigma (σ) bond which is formed by the hybridization of sp^2 (2s, $2p_x$, $2p_y$) molecular orbital apart from this an out of plane pi (π) bond is formed due to the hybridized $2p_z$ orbital which is perpendicular to the monolayer graphene's planar structure. The hybridized in plane σ bond has a very small interatomic bond length of ~ 1.42 Å, which makes the bond stronger than that of the sp^3 hybridized carbon–carbon (C-C) bonds in a diamond crystal; this gives monolayer graphene crystal unique mechanical properties (i.e. A tensile strength of 130.5 GPa and Young's modulus of 1 TPa). In monolayer graphene crystal, the half-filled π band is responsible for making the valance band and conduction band with zero band gap which permits free movement of electrons.

Figure 2. The crystal structure of monolayer graphene, where A and B are carbon atoms belonging to different sub-lattices, $\overrightarrow{a_1}$ and $\overrightarrow{a_2}$ are unit cell vectors (Pirzado et al., 2019)

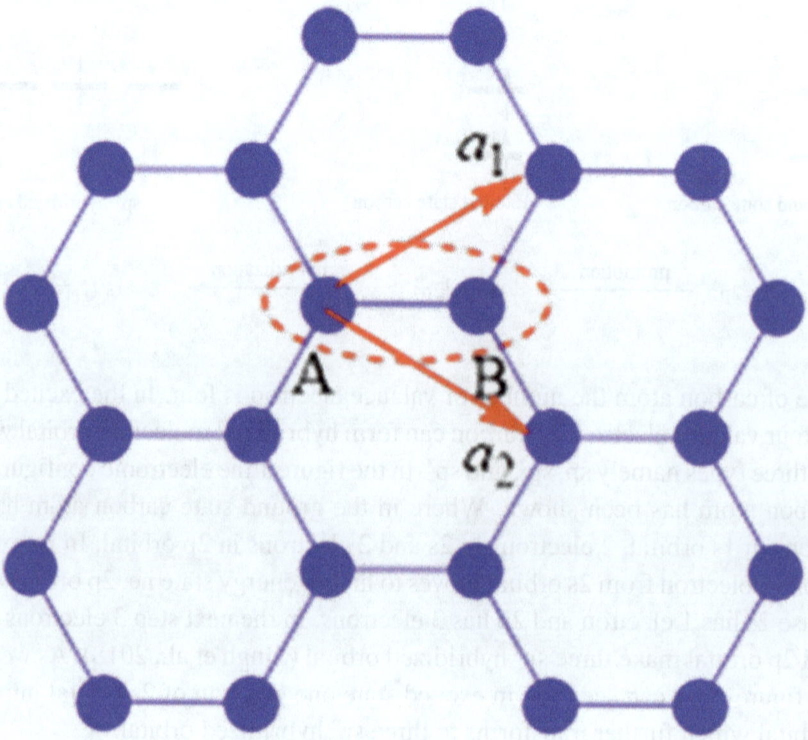

Furthermore, in case of bilayer and multilayer graphene the π-bonds which are out of plane provide a weak van der Waals interaction between adjacent graphene layers which makes multilayer graphene (graphite) brittle. From the figure 3 it can be seen that, head-to-head overlap of p orbital makes the in plane sigma bond and on the other hand pi bond is formed out of the plane which contributes significantly to the thermal and electrical conductivity properties of graphene.

In the hexagonal honeycomb lattice of monolayer graphene, two primitive unit cell vectors are written as (Bhuyan et al., 2016):

$$\overrightarrow{a_1} = \frac{a}{2}\left(1, \sqrt{3}\right) \tag{1.1}$$

and

Figure 3. Sigma bond and pi bond in the lattice formed by sp2 hybridization of carbon atom
(Irngartinger et al., 1992)

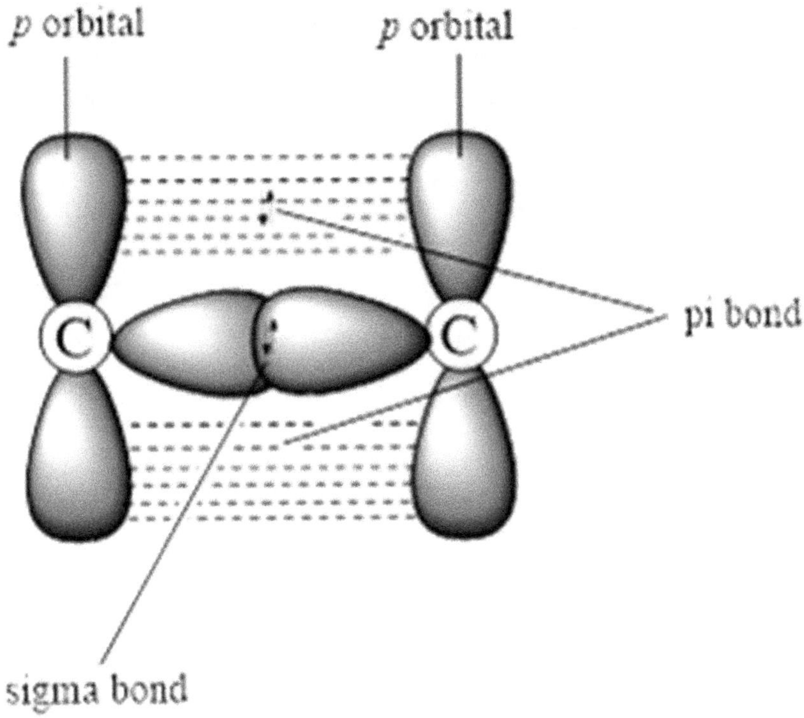

$$\overrightarrow{a_2} = \frac{a}{2}\left(1, -\sqrt{3}\right) \tag{1.2}$$

where

$$a = \sqrt{3}a_0 \approx \sqrt{3} \times 1.42\dot{A} = 2.46\dot{A} \tag{1.3}$$

is the lattice constant, which is the interatomic distance of unit cells.

The position vector of atom B_i, (l=1,2,3) relative to the atom A_i is termed as $\overrightarrow{\delta_i}$, and the three nearest neighbor vectors in real space are given by:

$$\overrightarrow{\delta_1} = \left(0, \frac{a}{\sqrt{3}}\right), \overrightarrow{\delta_2} = \left(\frac{a}{2}, -\frac{a}{2\sqrt{3}}\right) and\, \overrightarrow{\delta_3} = \left(-\frac{a}{2}, -\frac{a}{2\sqrt{3}}\right) \tag{1.4}$$

It is noted that, $\left|\overrightarrow{\delta_1}\right| = \left|\overrightarrow{\delta_2}\right| = \left|\overrightarrow{\delta_3}\right| = \frac{a}{\sqrt{3}}$, is the spacing between two nearest-neighbouring carbon atoms. From the figure 4 we can describe the reciprocal lattice of 2D honeycomb graphene lattice, where the crosses marked in the Figure 4 represent reciprocal lattice points, and the shaded hexagon represents the first Brillouin zone. The primitive reciprocal lattice vectors b_1 and b_2 satisfy the following conditions:

$$\overrightarrow{a_1}\overrightarrow{b_1} = \overrightarrow{a_2}\overrightarrow{b_2} = 2\pi \tag{1.5}$$

$$\overrightarrow{a_1}\overrightarrow{b_2} = \overrightarrow{a_2}\overrightarrow{b_1} = 0 \tag{1.6}$$

Therefore,

$$\overrightarrow{b_1} = \left(\frac{2\pi}{a}, \frac{2\pi}{\sqrt{3}a}\right) and\, \overrightarrow{b_2} = \left(\frac{2\pi}{a}, -\frac{2\pi}{\sqrt{3}a}\right) \tag{1.7}$$

PROPERTIES OF GRAPHENE

Electronic Properties

Some attractive and unique electronic properties of graphene have inspired scientists for the further research on it. A detailed approximation of the band structure of monolayer graphene can be derived from a single nearest neighbour tight binding calculation. Graphene is a semi metal because it has a zero (0) carrier density at the Dirac point that means its carrier density vanishes at Dirac point (Al-Shurman & Naseem, 2014), apart from this presence of pseudo-spin as well as the relativistic nature of the charge carriers (electrons and holes) are the three main electronic configuration of monolayer graphene which are the point of interest for further research on this unique material. Each and every carbon atom is connected to its three neighbouring atoms in the honeycomb graphene lattice by strong in-plane covalent bonds.

As graphene is a semi metal so its valence band is always occupying, and conduction band is empty in normal condition. This valence and conduction band meet at a point called as Dirac point where carrier densities are vanished, and density of states (DOS) becomes zero. Hence graphene is considered as zero-band gap

Figure 4. (a) Honeycomb lattice of 2D single atom thick graphene, where white and black circles indicate carbon atoms on A and B sites respectively, and (b) the reciprocal lattice of 2D graphene, where the shaded hexagon represents the corresponding Brillouin zone
(Ibrahim et al., 2014)

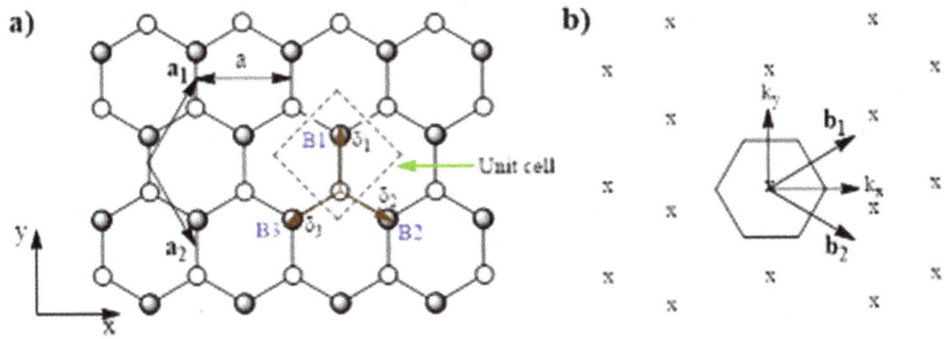

semiconductor with vanishing DOS at the Dirac Point but there is no energy gap between valence and conduction bands. Conduction band and valence band with majority charge carriers of holes and electrons respectively meet at a point. This point is named as Dirac point and as valence band and conduction band meet at a point that is why graphene is called as zero band gap semimetal (Gu et al., 2019). Near to the Fermi energy level band structure of graphene can be described in terms of two equivalent Dirac cones situated at K and K'. In order to scatter from K to K' an electron requires a large momentum change. Corresponding to two graphene sub lattices in parallel through the k and k' Dirac cones it is expected that electron transportation takes place in the honeycomb lattice structure of graphene. Now, pseudo spin and real spin is completely different and independent to each other, but it has an analogy with the real spin of the charge carriers. The basic Dirac equation i.e. relativistic Dirac equation described the linear dispersion relation of charge carrier i.e. electrons and holes where charge carriers are considered as Dirac Fermions (massless) which are capable of travelling with a group velocity of $v_F=1\times10^6 ms^{-1}$. As per the analysis of graphene's electronic properties based on relativistic Dirac equation it can be seen that dispersion relation of graphene is linear in nature close to the Dirac point. So, π orbitals together make π bands and π* bands which are responsible for exceptional electronic characteristics of graphene.

Mechanical Properties

The interatomic distance i.e. the length of the C-C sigma (σ) bond is 1.42Å which gives graphene an inherent strength and makes it the strongest material of the world with an tensile strength of 130 GPa (i.e., 13×10^{10} Pa), where A36 structural steel has a tensile strength of 400 MPa (i.e., 4×10^{8} Pa) and Kevlar (Yang et al., 2018), a synthetic fiber has a tensile strength of 375 MPa (i.e., 3.75×10^{8} Pa). Beside this graphene is lightest material ever discovered with a surface density of $0.77 mg/m^2$, that means paper of dimension $1m^2$ is 1000 times heavier than that of graphene. It is assumed that in order to cover a whole football ground with a graphene sheet it weigh less than one gram. This unique property of graphene is responsible for its elastic nature so that it can retain its original size. Atomic Force Microscope (AFM) is a scanning probe microscope (SPM) having a very high resolution and with the help of this microscope it is experimentally proved that a graphene sheet having a thickness between 2 and 8 nm has a Young's modulus of 0.5 TPa and spring constant in the range between 1-5 N/m. All these values are theoretically derived along with the superlative comparisons with very well-known conventional materials used in modern day life and considered the ideal case of graphene crystal where graphene is considered as an unflawed perfect crystal with no impurities containing with it. Along with this the major difficulty for making graphene commercial is its very expensive production process but it is steadily improving and showing its potential to be readily available with a reasonable cost and less complexity.

Electrical Properties

Graphene is a very good electrical conductor. The causes of such exceptional electrical properties are as follows- zero band gap semimetal, existence of π and π^* bands, two-dimensional honeycomb lattice structure of graphene. Graphene is blessed with a lot of properties resembles with that of the metals, but graphene is not actually a metal rather it can be termed as semi metal due to its unique arrangements of band structure. In hexagonal graphene lattice it is found that, electrons move along with a little scattering though in general electrons move very fast through conductor especially solid conductors. As the electrons inside the honeycomb lattice of graphene experience a little scattering so electrons have a chance to be accelerated to a higher speed. In order to get a comparative study of conductivity with the well-known conventional conductor graphene has 105 'imes higher conductivity than that of copper. Because of its super high charge carrier mobility graphene can lead over copper in terms of conductivity even though metals like copper have a high electron density. It is theoretically estimated that in an ideal case graphene has a conductivity of 15000cm2V-1s-1 'o 20000cm2V-1s-1 (Yang et al., 2018). Under some ideal experimental conditions this value can be shown.

Optical Properties

Monolayer graphene has an ability to absorb a maximum of 2.3% of white light which is an interesting property in the field of nano optics (Saito & Kataura, 2001). The electronic configuration and band structure which is mentioned before is responsible for this property. The charge carriers (electrons), having very high mobility due to the presence of little scattering, behaves like massless charge carriers. In the recent times, it has been proven that the absorption of white light depends on the fine n Structure Constant or Sommerfeld's constant (fundamental physical constant used to characterize the strength of electromagnetic interaction in between basic charged particles). Saturation fluency and saturable absorption are two important parameters for description of optical properties of graphene where saturation fluence is known as a property due to which with increase in light intensity absorption of light decreases. As per the research it can be found that saturable absorption takes place only when optical intensity reaches at the saturation fluence. In case of graphene at a higher energy level before a sharp rise to a constant universal value a region of reduced conductivity can be seen which resembles with Drude-like interband conductivity and afterwards density of state is being represented by the allowed states inside the bands.

Thermal Properties

Being a two-dimensional (2D) material, graphene is experiencing a bi-directional heat flow namely (in plane and out-of plane direction). In-plane thermal conductivity is high because of presence of phonons and free electrons; in contrary due to weak van der Walls coupling out of plane heat flow is limited to some extent. Including its specific heat, up to ballistic limits starting from diffusive limits, thermal conductivity of monolayer graphene extends up to different atomic modifications along with different defects and influence of substrates (Tang et al., 2020). For instance, up to a limited channel length graphene interconnects and transistors have the advantage from high in-plane thermal conductivity of 2D graphene.

Apart from this, weak thermal coupling with a specific substrate suggests that contacts and interfaces makes significant dissipation. Flow of heat in monolayer graphene or in the compounds of graphene has possibilities to be tunable through different ways, which includes scattering of phonons from substrates, interfaces, or edges. At the end, the unique thermal properties of honeycomb graphene due to its two-dimensional (2D) structure, makes a promising domain for the further researches of heat-flow physics and it potentially upgradation to the future thermal management applications. It is theoretically estimated that, 2D thermal conductivity

Figure 5. Flowchart of synthesis of graphene

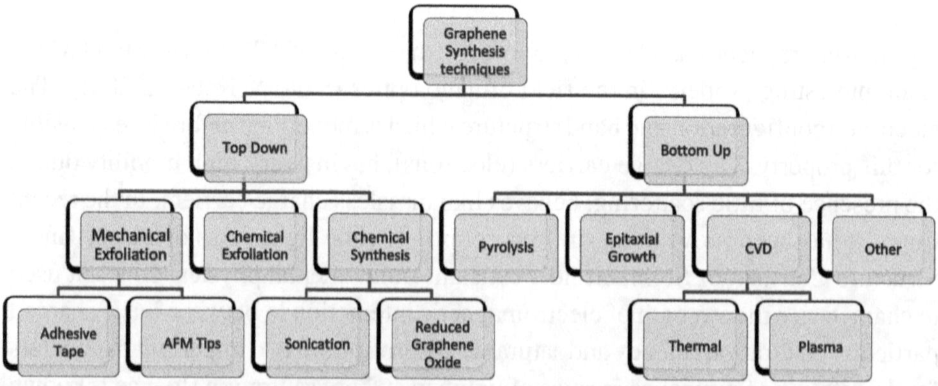

of graphene is 5000 $Wm^{-1}K^{-1}$ (Saito & Kataura, 2001) dominated by phonons, higher than for copper of 400 $Wm^{-1}K^{-1}$.

SYNTHESIS OF GRAPHENE

Synthesis of graphene means any process for extracting or fabricating graphene, as per the requirements of the desired size and purity of the specific product. The synthesis of graphene process can broadly be classified into two parts namely top-down process of graphene synthesis and bottom-up process of graphene synthesis (Syama & Mohanan, 2019). In top-down process, graphene sheets are produced by mechanical exfoliation, chemical exfoliation, and chemical synthesis process. In bottom up process, graphene sheets are produced by pyrolysis, epitaxial growth, chemical vapour deposition (CVD). A flowchart of synthesis of graphene is given below.

Mechanical Exfoliation

In order to have wide applications and viability of graphene commercial availability and mass production are very much required. The most promising way to produce large scale graphene is exfoliation of graphite and it is also very much cost effective. Preparation of graphene by exfoliating graphite is being described here. In this process from bulk graphite, graphene can be peeled off layer by layer in the ideal case. Meanwhile resistance between adjacent graphite layer flakes has to be overcome. Mechanical exfoliation means the process where graphene layers are separated from bulk graphite by using mechanical force. This basic requirement of this is using of an adhesive tape which can be attached to the graphite surface and by using mechanical force the tape can be peeled in addition with graphene layers attached with it. Another way is to rub one of the surfaces of bulk graphite with a different material to peel off graphene layers from the bulk graphite. This production process is so cost effective that mechanical exfoliation of graphene becomes the most effective top-down process to produce commercial graphene. Micromechanical exfoliation is associated in between two processes namely tip-based techniques of AFM or exfoliation by adhesive tape (Bhuyan et al., 2016). But AFM tip based approach is time consuming so it is very limited and apart from this, it provides a lower output, and moreover, there is much less control on layer number of the produced graphene. On the other hand, the micromechanical exfoliation by using adhesive tape is far easier as well as a quicker one and it provides large graphene sheet of high quality and these type of graphene sheets are desirable for preparation of device-grade graphene like highly oriented pyrolytic graphene (HOPG) and various experimental laboratory studies. When P.R. Wallace first proposed his theoretical view on free standing atomic plane i.e., graphene he assumed that due to thermo dynamical instability it is not possible to be physically existed and stable at nanometer scale. In order to overcome this difficulty Geim and Novoselov used repeatedly adhesive tape to prepare their first ever graphene and its dimension is up to 1 mm thick and in its nature, it was highly oriented pyrolytic graphite (HOPG). In their further progress of graphene synthesis, they made it into thinner pieces. The first step of mechanical exfoliation is to press the adhesive side of adhesive tape on the HOPG surface and then slowly peel away with a thick layer of graphite presents on the adhesive side of the tape. Then this tape with dark, shiny HOPG will undergoes repetitive folding and unfolding process on the clean section of same adhesive tape until the appearance of dark dull grey luster instead of shiny one. Now this adhesive tape with dark dull grey luster is optically transparent and is able to dissolve in acetone. Now this is graphene but in order to make it further use a Si wafer has to be dipped into the water followed by cleaning with water and propanol. Repetition of this process gives exfoliated flakes which stored on wafer surface. Initially it is having both thick films and thin films but ultrasound cleaning in propanol helps in removing thick flakes and left behind the thin ones (Nazarov). With the help of capillary forces and Van der Walls attraction force thin films remains attached with the Si wafer. So, in the first step of mechanical exfoliation the adhesive side of adhesive tape attached with the graphite and then it is gently peeled off then folding and unfolding of adhesive tape is being continued till the appearance of dark layer i.e., graphene.

Chemical Exfoliation

Graphite can be described as stacked layers of graphene and these layers can be separated by using chemical solution from the bulk. These thin layers are attached with each other due to weak van der Walls attraction force so in order to get graphene sheet, layer separation is needed by overcoming van der Walls attraction force (Nazarov). This usually needs some form of input energy (around 2 eV/nm^2) for the layer separation. The job is not finished just after the separation of layers from the bulk but restacking of these layers must be prevented. N-Methyl-2-pyrrolidone (NMP) (Saito & Kataura, 2001) is a polar solvent which is being used for the separation of graphene layer from a bulk graphite as well as it can isolate the layers by forming a colloidal solution. The classical process of chemical exfoliation suggests that it should be placed into a vial, after that a polar solvent like NMP should be added followed by a bath sonicator or a mild sonication for an hour. There may be large graphite chunks in the primary production but in order to get single and few layer graphene sheets of micron sized, centrifuging of the sample is required. This can finally give single or few layers graphene sheets. This finally produced monolayer or few layers graphene can be used for practical purposes like spray coating on specific substrate, printing (ink-jet), to make composites with polymers and electrophoresis deposition in some electronic devices. So, Chemical exfoliation is a process which is followed by two-steps. In order to increase the interlayer spacing, the van der Waals forces in between two layers has to be reduced. Thus it forms graphene-intercalated compounds (GICs). As a result of it, monolayer or few layers graphene will come out as a final product of this exfoliation process by sonication or rapid heating.

Reduced Graphene Oxide

When oxidization of graphene takes place then it is called oxidized graphene and it is a disrupted graphene lattice. For more than 150 years the nature, atomic structure, reaction with reducing agent, process to become chemically functionalized of graphene oxide has been studied but still there are a lot of uncertainties around this. But it is the potential process through which a huge amount of graphene sheets can be produced which can be of an amount of some grams to some kilograms (Nazarov). There are four well known chemical processes to chemically reduce the graphene oxide and get the pure graphene. These are as follows:

(a). Hummers method for graphite oxide (Hummers & Offeman, 1958)
(b). Chemical reduction of graphene oxide
(c). Heat treatment of graphene oxide
(d). Electrochemical reduction of graphene oxide

As far as application of reduced graphene oxide is concerned it can be used to produce ultra-strong but lightweight materials because it is ideal for application in those fields which is not strongly dependent on crystal quality (Saito & Kataura, 2001).

Chemical Vapour Deposition

Graphene can be fabricated over metals and the most popular way of doing this is chemical vapour deposition (CVD). The reasons for this are various and include the potential to scale up fabrication, the technique is already well established in industrial settings, and it is easy to set up in research laboratories among other attractive traits. As far as application of graphene is concerned it is required to have a graphene sheet of greater surface area and in CVD process graphene is being grown over a metal surface which results into a larger surface area. Transparent conducting electrodes for OLED as well as solar cell would be an ideal example for such application where an adjacent covering of graphene is essential. The fundamental difference of exfoliation and chemical vapour deposition is that in chemical vapour deposition there are no randomly scattered flakes over the substrate (Nazarov). At a higher temperature, the chemical vapour which deposits sp^2 carbon, ensures of passing a carbon feedstock over the surface of a catalytic substrate like transition metal. The catalyst used in this process then decomposes the feedstock by means of catalytic effect to provide a carbon supply. The catalyst used here has a potential which is called catalytic potential and it is well established in case of transition metals which is the outcome of partially filled d orbital of those transition metals. Another possibility is that there can be a formation of intermediate complex compounds those are responsible for absorbing and activating the reaction medium. In the end, by forming some intermediates or by providing a lower energy pathway for the reaction, the metals can change their oxidation states. In this step the main goal is to decompose the feedstock and provide a source of carbon as well (Gu et al., 2019). In case of Ni and Co the carbon is absorbed by metal followed by the formation of graphene. In case of Cu if carbon is less soluble then sp^2 hybridized carbon will form by the occurrence of surface process. This classical process of chemical vapour deposition (CVD) is very simple but the chemical dynamics of the reagents for carbon depositions as well as domain growth has its own limitation. The parameters of this chemical vapour deposition process which controls the production of graphene depends on some factors those are as follows:

(a). Cooling rate
(b). Carbon exposure time
(c). Concentration flow rate

(d). Carbon feedstock (source)

Apart from all these primary factors there are some secondary factors which can control the outcome these are as follows (Saito & Kataura, 2001):

(a). Oven reactor can affect the flow
(b). Deposition characteristics

The negative effects on the rate and chemical dynamics of this desired reaction is developed by the impurities. Depending on this factors and co-factors experimental balance must be vary from one reactor to another in the whole graphene manufacturing process through Chemical Vapour Deposition (CVD).

Epitaxial Growth

It is possible to grow graphene sheets epitaxially on silicon carbide (SiC) substrates. Transfer process is something complicated and without transfer process and only with the help of nano lithographic methods it is possible to patterned graphene. This is often the fundamental eye-catching advantage of this process over the standard technology available up to now. No metals furthermore as hydrocarbon involvement is there during this method which leads this process united of the cleanest graphene synthesis process till date because epitaxial matching is capable of providing its own carbon. The ultimate pure graphene which is that the outcome of this epitaxial growth is grown on carbide (SiC) process has an electronic mobility of 25000 cm^2/Vs together with the exhibition of long phase coherence lengths (Nazarov). That is why epitaxial growth process became so popular in the field of study of graphene synthesis. So, it is utilized an awfully wide range of graphene production in recent times. The control sublimation is the basic principle of extracting silicon from the carbide (SiC). The method through which a solid can directly converted to vapour phase is named sublimation process and, in this process, there is no intermediate liquid phase. This sublimation process does not preserve stoichiometry for binary compounds. Sublimation of a binary molecule depends on the energy (binary force) of the atoms which suggests that the atom which binds more tightly will sublimate at a later point of your time than the other atom. Just in case of SiC which is getting used here for the epitaxial process, Si sublimates first than that of the carbon and at the tip it leaves few layers of carbon behind. As a result of it, this residue of carbon layers rearranges on the respective surface, and they tried to minimize the energy and form graphene as the final product. In the figure 6 this epitaxial growth is shown where SiC substrate is going through controlled sublimation and as a result of its Si is sublimated from the substrate (Tang et al., 2020). As epitaxial growth fulfills

Figure 6. Epitaxial Growth of graphene
(Nguyen & Nguyen, 2016)

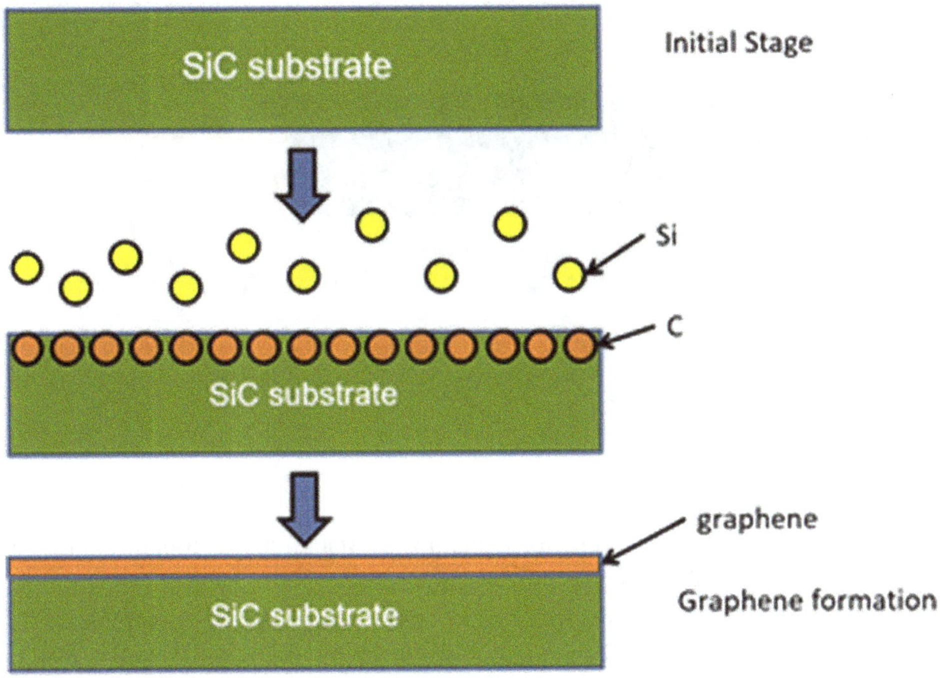

its demand of carbon by itself, so the carbon of the substrate SiC appears over the surface and from the required graphene layer. In order to create one layer graphene three bilayers of SiC is being required as per the calculation of molar densities.

APPLICATIONS OF GRAPHENE

Integrated Circuit

Complementary metal oxide semiconductor (CMOS) is the mother technology which makes itself irreplaceable in the field of silicon based integrated circuits and till date silicon is used as the channel in the transistor circuit (El-Showk, 2016). Where silicon channel is used for transportation of charge carriers through the integrated and logic gate circuits. As per the demand of current technology, it needs smallest as well as fastest device in the IC chip, but silicon-based industry is approaching towards a certain direction where it cannot be further shrunk as it came across a long way for over the decades. As per the research on the electronic properties of graphene it can be the potential replacement of silicon in the field of

Figure 7. Integrated Circuit (Top Gated)
(Coleman, 2012)

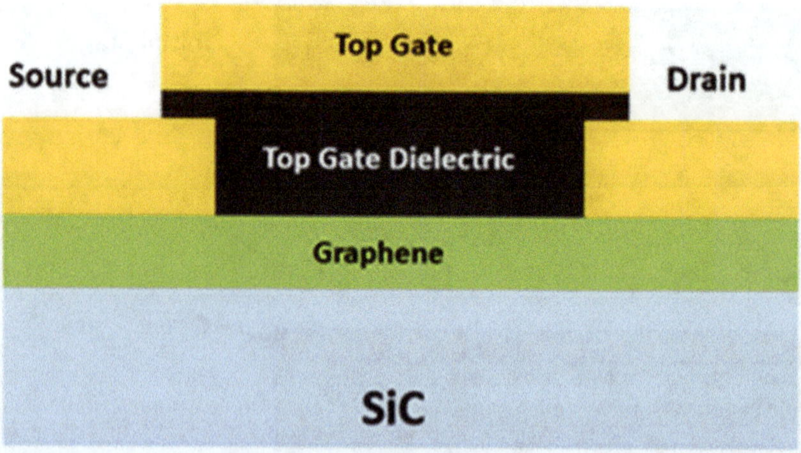

integrated circuit and semiconductor technology (Black, 2011). It can replace Si for the application in CMOS chip for carrying current as graphene has the highest electron mobility among the materials discovered so far. Fullerene film transistor is one of the newest discoveries of modern technology which is an expertise research based on the unique electronic characteristics of graphene. Silicon has limitations in terms of its carrier mobility as well as monolayer thickness and in regards with these properties of graphene, silicon will be potentially replaced in coming future days as far as semiconductor technology is concerned. Graphene is an allotrope of carbon so it is capable of making carbon nanotubes which is of zero (0) dimension as well as possess a planar structure so it can be used as the channel material in the semiconductor technology and in integrated circuit. In order to scale CMOS an important parameter is thickness of channel material which is ruled by silicon as of now but graphene is superior in this prospect as it is one atom thick monolayer crystal so graphene can be a substitution as technology goes to develop its further (Saito & Kataura, 2001).

Transparent Optical Electrodes (Optical Electronics)

Transparent conductors (TCs) are necessary in the field of electrochromic mirrors, transparent electromagnetic shields, touch screen technology, low emissivity window but the conventional conductors are not transparent in other words, transparent materials are not generally conducting as well. But as monolayer graphene is upto 97.7% transparent as well as a good conductor of electricity, so it is the unique one in these applications (El-Showk, 2016). In modern electronic products it becomes obvious to use conducting electrodes. The beginning and very well-known use of transparent conducting electrodes is liquid crystal display (LCD) and every household is very much familiar with this. As technology is heading towards the dependency on smart devices where use of transparent conductors for making touch screen, solid- state light devices, solar cells, Organic light emitting diode (OLEDs) is obvious then graphene presents itself as the automatic choice to be used in all those applications (Singh et al., 2018). Generally, in case of an OLED cathode and transparent anode is connected through a connecting wire and graphene presents at the anode side. Meanwhile in between anode and cathode there exists an organic light emitting layer (Saito & Kataura, 2001).

Graphene Based Energy Applications

The global energy demand is increasing sharply, and numerous efforts are being proposed in order to fulfil all these demands with as minimal damage as possible to the environment. The discovery of graphene and lots of exciting properties has global interest in the noble material, associating with electrochemical energy systems. Super capacitors store as well as deliver electrochemical energy along with high discharge rates. Their energy densities are exceptionally higher than that of the conventional dielectric capacitors (Chang & Hu, 2019). So as an obvious cause there is significant interest in super capacitors for the supplement of energy storage elements in coming future. As the conventional energy storage devices like battery and normal capacitor; battery can store less charge, but it takes less times to be fully charged and on the other hand normal capacitor stores more charge, but it takes too many times to be fully charged. Here comes super capacitor in rescue and merged both the advantages so that it can store more charges as well as takes less time to be charged additionally has a high longibility and high-power densities. Electrochemical double layer capacitors (EDLCs) performance parameters depend on conductivity of electrical charge carriers, surface area, surface area to mass ratio, pore size as well as distribution. Graphene has highest carrier mobility as a consequence having high electrical conductivity, attractive interlayer structure, high surface area to mass ratio obviously larger surface area make itself an automatic choice for EDLCs (Fu et al., 2020). So, in a graphene-based superconductor laser scribed graphene is used as electrodes namely cathode and anode. In between cathode and anode, a separator and electrolyte are used along with the plastic sheets to cover the graphene in both the sides (Saito & Kataura, 2001).

CONCLUSION

In this 21st century in the field of nanoelectronics, grapheme- the wonder materials have brought a revolutionary change as well as a promising future. Owing to its unique electrical, electronic, mechanical, thermal, chemical and optical characteristics than that of the conventional nanoparticles and nanostructured materials, two-dimensional mono layer graphene become a new branch for the further research on it. Graphene and its composites are capable of storing as well as transformation of energy. Apart from this, uses of graphene in integrated circuit design due to its high conductivity, Optical electronics is also a promising field for graphene to contribute. Its outstanding contribution towards human race is not only limited in the field of nanoelectronics but expanding in medical science, nanorobotics and many other fields also. In this chapter we covered a detailed of graphene starting from its discovery to properties, synthesis methods and different applications along with some potential advancement in different fields.

REFERENCE

Al-Shurman, K. M., & Naseem, H. (2014). CVD Graphene growth mechanism on nickel thin films. In *Proceedings of the 2014 COMSOL Conference in Boston*.

Bhuyan, M., Alam, S., Uddin, M., Islam, M., Bipasha, F. A., & Hossain, S. S. (2016). Synthesis of graphene. *International Nano Letters*, *6*(2), 65–83.

Bhuyan, M. S. A., Uddin, M. N., & Islam, M. M. (2016). Synthesis of graphene. *International Nano Letters*, *6*, 65–83.

Binh, H. T. T., & Dey, N. (Eds.). (2018). *Soft computing in wireless sensor networks*. CRC Press.

Binh, H. T. T., Hanh, N. T., Nghia, N. D., & Dey, N. (2020). Metaheuristics for maximization of obstacles constrained area coverage in heterogeneous wireless sensor networks. *Applied Soft Computing*, *86*, 105939.

Black, Charles. (2011). High-Performance Microelectronics. *Porous Polymers,* 359-385. Doi:. doi:10.1002/9780470929445.ch10

Chang, L., & Hu, Y. H. (2019). 3D Channel-structured graphene as efficient electrodes for capacitive deionization. *Journal of Colloid and Interface Science*, *538*, 420–425.

Coleman, J. (2012). Liquid Exfoliation of Defect-Free Graphene. *Accounts of Chemical Research*, *46*. Doi:10.1021/ar300009f

Das, S. K. (2021). Smart Design and Its Applications: Challenges and Techniques. In *Nature-Inspired Computing for Smart Application Design,* (pp. 1–6). Springer.

Das, S. K., Das, S. P., Dey, N., & Hassanien, A. E. (Eds.). (2021). *Machine learning algorithms for industrial applications*. Springer.

Das, S. K., & Giacalone, M. (Eds.). (2022). *Fuzzy Optimization Techniques in the Areas of Science and Management*. CRC Press.

Das, S. K., Maheswari, V., & Sharma, A. (2021). Wireless Networks: Applications, Challenges, and Security Issues. In *Architectural Wireless Networks Solutions and Security Issues* (pp. 1–10). Springer.

Das, S. K., Samanta, S., Dey, N., & Kumar, R. (Eds.). (2020). *Design frameworks for wireless networks*. Springer Singapore.

Das, S. K., Samanta, S., Dey, N., Patel, B. S., & Hassanien, A. E. (Eds.). (2021). *Architectural Wireless Networks Solutions and Security Issues*. Springer Singapore Pte. Limited.

El-Showk, S. (2016). *Making better and cheaper solar cells*. Nature Middle East. doi:10.1038/nmiddleeast.2016.4

El-Showk, S. (2016). *Making better and cheaper solar cells*. Nature Middle East. Doi:10.1038/nmiddleeast.2016.4

Fu, X. Y., Chen, Z. D., Han, D. D., Zhang, Y. L., Xia, H., & Sun, H. B. (2020). Laser fabrication of graphene-based supercapacitors. *Photonics Research*, *8*(4), 577–588.

Gerstner, E. (2010). Nobel prize 2010: Andre geim & konstantin novoselov. *Nature Physics*, *6*(11), 836–836. doi:10.1038/nphys1836

Gu, X., Zhao, Y., Sun, K., Vieira, C. L., Jia, Z., Cui, C., & Huang, S. (2019). Method of ultrasound-assisted liquid-phase exfoliation to prepare graphene. *Ultrasonics Sonochemistry*, *58*, 104630.

Hummers, W. S. Jr, & Offeman, R. E. (1958). Preparation of graphitic oxide. *Journal of the American Chemical Society*, *80*(6), 1339–1339.

Ibrahim, K. M., Irannejad, M., Ramadhan, A., Alayak, W., Sandeson, J., Cui, B., & Yavuz, M. (2014, August). Ultrafast light interaction with Graphene oxide aqueous solution. In *14th IEEE International Conference on Nanotechnology*, (pp. 830-831). IEEE.

Irngartinger, H., Oeser, T., Jahn, R., & Kallfaß, D. (1992). Pyramidalization of the sp2-Hybridized Carbon Atoms in 1, 4-Bridged 2, 5-Cyclohexadienes. *Chemische Berichte*, *125*(9), 2067–2073.

Kumar, K. (2018). An Overview on the Importance of Chemical Vapour Deposition Technique for Graphene Synthesis. Advanced Science. *Engineering in Medicine*, *10*. Doi:10.1166/asem.2018.2241

Nazarov, V. U. (2012). *Band-structure of graphene above the vacuum level.*

Nguyen, B. H., & Nguyen, V. H. (2016). Promising applications of graphene and graphene-based nanostructures. *Advances in Natural Sciences: Nanoscience and Nanotechnology*, *7*(2), 023002.

Pirzado, A. A., Le Normand, F., Romero, T., Paszkiewicz, S., Papaefthimiou, V., Ihiawakrim, D., & Janowska, I. (2019). Few-layer graphene from mechanical exfoliation of graphite-based materials: Structure-dependent characteristics. *ChemEngineering*, *3*(2), 37.

Saito, R., & Kataura, H. (2001). Optical properties and Raman spectroscopy of carbon nanotubes. In *Carbon nanotubes,* (pp. 213–247). Springer.

Singh, J., Rathi, A., Rawat, M., & Gupta, M. (2018). Graphene: From synthesis to engineering to biosensor applications. *Frontiers of Materials Science*, *12*(1), 1–20. Doi:10.100711706-018-0409-0

Singh, J., Rathi, A., Rawat, M., & Gupta, M. (2018). Graphene: From synthesis to engineering to biosensor applications. *Frontiers of Materials Science*, *12*(1), 1–20.

Spyrou, K., & Rudolf, P. (2014). An introduction to graphene. *Functionalization of graphene*, 1-20.

Syama, S., & Mohanan, P. V. (2019). Comprehensive application of graphene: Emphasis on biomedical concerns. *Nano-Micro Letters*, *11*(1), 6.

Tang, N., Tang, T., Pan, H., Sun, Y., Chen, J., & Du, Y. (2020). Magnetic properties of graphene. In Spintronic 2D Materials, (pp. 137-161). Elsevier.

Wallace, P. R. (1947). The band theory of graphite. *Physical Review*, *71*(9), 622–634. doi:10.1103/PhysRev.71.622

Yang, G., Li, L., Lee, W. B., & Ng, M. C. (2018). Structure of graphene and its disorders: A review. *Science and Technology of Advanced Materials*, *19*(1), 613–648.

Yang, G., Li, L., Lee, W. B., & Ng, M. C. (2018). Structure of graphene and its disorders: A review. *Science and Technology of Advanced Materials*, *19*(1), 613–648.

Chapter 13
Different Types of Compactness and Their Importance in Causality

Dwaipayan Mishra
Midnapore College (Autonomous), Vidyasagar University, India

Himadri Shekhar Mondal
Midnapore College (Autonomous), Vidyasagar University, India

ABSTRACT

The notions of different types of compactness in a spacetime manifold has reviewed in this article. Also, their relations with Cauchy hypersurfaces, which play very important role in globally hyperbolic spacetimes has been discussed. For example, A be a closed subset of a spacetime M having a compact intersection with all the Cauchy hypersurfaces of it, then $A \subset J(C)$ *for some compact set* $C \subset M$ *and conversely. Past and future compact sets in a spacetime and their interrelations with spacelike and timelike compactness are also discussed here by introducing necessary definitions, propositions, and diagrams wherever necessary, for the sake of understanding. Also, it is shown that, a closed advanced set is strictly future compact set. The relations among those compact sets themselves has been mentioned elaborately.*

INTRODUCTION

In mathematical general relativity, causality plays a very important role. Also, among the causal conditions, globally hyperbolicity is most important as it describes our reality more properly than others till now, see (Sanchez, 2006). Many real-life

DOI: 10.4018/978-1-6684-7343-6.ch013

problems can be modeled using decision constraints. It has immense significance in linear programming problems, optimization theory etc. This notion may also be useful in the study of causal theory of Spacetimes to explore some of its important aspects. (Das et al., 2015; Das et al., 2021a; Das et al., 2021b; Das et al., 2021c; De et al., 2020; Das et al., 2020). In 1953, Leray (Leray, 1953) introduced the concept of globally hyperbolicity. Since then, its application serves in various leading fields in mathematical physics as singularities of a gravitational collapse (Penrose, 1965; Witten, 2020), causal aspects of mathematical relativity (Wald, 1984; Penrose, 1969; Penrose, 1972), large scale cosmological structures (Hawking & Ellis, 1973; Beem et al., 1996) etc. Also, the existence of a Cauchy hypersurface in a spacetime is equivalent to globally hyperbolicity of the spacetime as one of the celebrated result due to Choquet-Bruhat (Choquet, 1971) and Geroch and Choquet-Bruhat (Choquet & Geroch, 1969) states that the metric tensor g of a globally hyperbolic spacetime M can be determined by Cauchy's data and a Cauchy hypersurface S. If in a globally hyperbolic spacetime M, a compact and acausal spacelike smooth submanifold C attains the boundary of it, then a smooth non-timelike Cauchy hypersurface S in M exists such that $C \subset S$. The strong cosmic censorship conjecture in general relativity is very much significant in regard to different aspects of causality theory, see (Geroch, 1970). There are many important interrelations among some type of compactness and Cauchy hypersurface. As for example, if $A \subset M$ is a closed set in M having a compact intersection with all the Cauchy hypersurfaces then $A \subset J(C)$ and converse for some compact set C. Hawking and Ellis (Hawking & Ellis, 1973) shows that if the Cauchy hypersurface Σ become compact, then the globally hyperbolicity condition can simply be replaced by assuming Σ as achronal only, where the timelike incompleteness occurs. Also, if a compact Cauchy surface in a spacetime M is given, we can find a constant mean curvature (CMC) Cauchy surface of same volume which has deep importance to initial value problem and also, these are very much stressed in general relativity, see (Muller & Sanchez, 2014). There is also compact chronology theorem which states that, a compact spacetime manifold can never be virtuous (Carter, 1971). The Hausdorff topology on compact set of a metric space agrees with the Vietoris topology which is defined on all the non-empty closed subsets of a certain topological space (Filippov, 1998). The act of using the Vietoris topology in the spaces of closed subsets in a compact set has been discussed in (Sorkin, 1990; Sorkin & Woolgar, 1999). In a globally hyperbolic spacetime, the space of causal curves connecting any two compact sets become compact, see (Sorkin & Woolgar, 1999). Also, in a compact space every Vietoris limit is a compact set, see (Sorkin & Woolgar, 1999). Muller and Sanchez (Muller & Sanchez, 2014) discussed mainly on globally hyperbolic spatially compact spacetimes. Hawking and Ellis (Hawking & Ellis, 1973) shown that a future inextendible causal curve Γ

either can enter into a compact set C and remain there forever or it may not remain within C but continually re-enter into C or it may not be remained in any such compact set not re-entering into such set for more than finite number of times. In first two cases Γ may be thought as if it has imprisoned in future directions and in last case Γ may be thought as if it is going to the edge (may either to a singularity or to infinity) of the spacetime. Actually, if every point of a compact set C is strongly causal, then no non-spacelike future-inextendible curve is imprisoned either totally or partially in C. Again, it is well-known that any compact spacetime contains closed timelike curve, as a result a compact spacetime cannot be chronological. Also, if a static spacetime is compact, then it is timelike geodesically connected and also admits a closed timelike geodesic (Samann, 2019; Seifert, 1967). In cosmology, the theory of gravitational lensing is indeed no other than a physical phenomenon which is closely related to the Lorentzian kind of Morse theory of causal curves (Perlick & Tausk, 2003). Navarro and Minguzzi (Navarro & Minguzzi, 2011) shown that any globally hyperbolic spacetime M always admit a sequence consisting of compact sets C_n satisfying $\bigcup_{n=0}^{\infty} C_n = M$; and for a given compact set C', there exists a $n>0$ such that $C' \subset C_n$. These types of sequences of compact sets satisfy more important conditions discussed in (Navarro & Minguzzi, 2011). Minguzzi and Sanchez (Minguzzi & Sanchez, 2008) shown that, every compact spacetime M can regularly be covered by another spacetime $\tilde{M}$ which endowed with a compact Cauchy hypersurface S containing a periodic timelike geodesic. It is very crucial for compactness of a static spacetime is that any two points *C^' ⊂ C_n* in a static spacetime M can be joined by a timelike geodesic. Also, any compact spacetime M endowed with a timelike conformal vector field is find to be totally vicious, that is in such case the chronological and hence causal future and past of any point $p \in M$ is the whole spacetime manifold M. The fact that any causal spacetime M satisfies the globally hyperbolicity conditions if f $J^-(C) \cap J^+(C)$ be compact for every compact set C, has been used several times in our study (Navarro & Minguzzi, 2011). Compactness of causal diamonds is a pre-requisite for global hyperbolicity condition. In (Hounnonkpe & Minguzzi, 2019), we see that compactness of causally convex hulls (Minguzzi, 2019; Hounnonkpe & Minguzzi, 2019) is also very important in this regard as the compactness of causally convex hulls of compact sets is equivalent to the compactness of causal diamonds. Henceforth there are some useful characterizations of different spacetimes on the basis of compactness of causally convex hulls. Also, the applications of compact sets lay very vital role in showing K-causality stable causality conditions are indeed equivalent. As in (Minguzzi, 2009) if M with respect to metric tensor g be K-causal spacetime, then for all compact set C, a metric g^C is found such that $g \leq g^C$ satisfying $g < g^C$ on C and M with respect to g^C will be K-causal. This statement actually leads to the result that, K- causal spacetimes are stably causal, see (Minguzzi, 2009). On the other hand, a physical field configuration ϕ defined over a globally hyperbolic spacetime M is simply a section of a vector bundle over it. If this ϕ holds a normal hyperbolic equation of motion endowed with the compactly supported initial conditions (Choquet & Geroch, 1969), then $supp\phi \subset J(C)$ for some compact set $C \subset M$ and thus it intersects every Cauchy hypersurface in a compact region (Bar et al., 2007; Friedlander, 1976). In this paper, we discussed the above-mentioned notions of compactness and their relevance with the causality theory, especially with the Cauchy surfaces. Also, the future scope of work has been mentioned. Also, we have proved some important results explicitly unmatched to any previous literatures. Along with this, we have established a relation between advanced sets with strictly future compact sets.

BASIC DEFINITIONS

Manifolds

Manifold is nothing other than a topological space having the geometry of neighborhood of each of its points similar to that of a Euclidean space. If manifold *M* is *n*-dimensional then the neighborhood of each of its points are homeomorphic to that of $\mathbb{R}^n$, see (Lee, 2011).

Spacetime

Taking account of all the criterion of modern physics a spacetime can be defined as a Housdorff, connected, paracompact (Hausdorff & second countable), four-dimensional $\mathbb{C}^n$-topological space ($n \geq 2$) endowed with a differentiable atlas having time direction, also carrying a torsion free connection consistent with a Lorentzian kind of metric. Also, spacetime is designed to satisfy the Einstein's equations. Although, the causality behavior of a spacetime is very important point to be explained later.

Tangent Space

The tangent space $T_p(M)$ at any point *p* of any differentiable manifold *M* is indeed a linear space of dimensions as manifold *M*. Informally, tangent space is the collection of all vectors of *M* at *p*. Any tangent vector at $p \in$ can be put as a linear combination of the co-ordinate derivatives $\left(\partial/\partial x^1\right)_p, \ldots\ldots, \left(\partial/\partial x^n\right)_p$, where M is an n-dimensional manifold, see (Spivak, 1979).

Causality

Thus, a non-zero tangent vector $v \in T_p(M)$ can be characterized as either timelike or lightlike or spacelike according as $g(v,v)$ is <0, $=0$ but $v \neq 0$, >0 respectively. The lightlike and zero vectors are usually called null vectors. causality is defined using the notion of tangent space through a Lorentzian metric. (Choudhury & Mondal, 2014).

The tangent spaces $T_p(M)$ at each $p \in M$ in any spacetime M has Minkowski structure with the light cones which is used to categorize all the non-zero vectors which is belonging to $T_p(M)$, into timelike, null (also known as lightlike) and spacelike vectors depending whether $g(u,u)$ is negative, zero and positive resp. Again, any timelike vector ζ is past or future directed depending upon whether $g(\zeta,u)$ is positive or negative resp. If these classifications are possible continuously, over the entire spacetime, we characterize the spacetime as a time orientable one (Hawking & Ellis, 1973; Beem et al., 1996; Mohajan, 2016).

Causal Relations

Any point $p \in M$ in a spacetime is usually known as an event. It is very important in mathematical relativity that how two events in a spacetime are related. In any spacetime an event p is chronologically related to another event q, if there exists a future directed timelike curve joining p and q. The point p is said to be causally related to q if there exists a future directed causal or constant curve connecting p with q. These relations are denoted by $p \leq q$, $p < q$ respectively. Obviously, from the elementary set theory the relations $\ll, <$ may be thought as a subset of $M \times M$ as $I^+ = \{(p,q): p \ll q\}$, $J^+ = \{(p,q): p \leq q\}$; I^-, J^- can also be defined analogously.

Now for an event $r \in M$, the chronological future $I^+(r)$ and the causal future $J^+(r)$ of r can be defined as $I^+(r) = \{s \in M: r \ll s\}$ and $J^+(r) = \{s \in M: r \leq s\}$ resp. Also, the chronological past $I^-(r)$ and causal past $J^-(r)$ of r can be defined time-dually. We also define $I^+(S) = \bigcup_{x \in S} I^+(x)$ and $J^+(S) = \bigcup_{x \in S} J^+(x)$, where S is subset of M. Also $J(S) = J^+(S) \bigcup J^-(S)$, see (Hawking & Ellis, 1973; Beem et al., 1996; Minguzzi, 2019).

Cauchy Hypersurface

In a spacetime *M*, a Cauchy hypersurface is a subset $S \subseteq M$, which is met only once by all inextendible timelike curves in *M*, see (Wald, 1984).

A subset *S* of *M* is called achronal (resp. acausal) if there does not exist two points *q,p*∈*S* such that *q*∈$I^+(p)$ (respectively *q*∈$J^+(p)$). That is an achronal (resp. acausal) set never be crossed by any timelike (resp. causal) curve more than once, see (Beem et al., 1996).

Cauchy hypersurfaces are thus achronal but not the converse. In a Minkowski spacetime $\mathbb{R}_1^n$, all the hyperplanes at a constant time t gives rise to Cauchy hypersurfaces. So, Cauchy hypersurface are closed achronal topological hypersurface which is met by each and every inextendible causal curves just once, see (Hawking & Ellis, 1973; Minguzzi & Sanchez, 2008).

DEFINITION OF DIFFERENT TYPE OF SPACETIMES

Chronological Spacetimes

A spacetime *M* is said to be chronological spacetime if one of the following equivalent conditions hold:

(i) Timelike loop do not exists.
(ii) d(p,p)<∞ (thus equas 0) for all p∈M, see (Hawking & Ellis, 1973; Beem et al., 1996).

Causal Spacetimes

A spacetime *M* is called causal if there exists no causal loop in*M* or equivalently if the strict causal relation is irreflexive which means, $p<q \Rightarrow p \neq q$, for all *p,q*∈*M*, see (Beem et al., 1996; Minguzzi & Sanchez, 2008).

Strongly Causal Spacetimes

A spacetime *M* is called strongly causal spacetime if for any neighbourhood V of *p*∈*M*, there is another neighborhood $U \subseteq V$, *p*∈*U* such that, any future directed causal curve $\Gamma:[0,1] \to M$ having end points in *U* is totally be contained in *V*, see (Hawking & Ellis, 1973; Minguzzi & Sanchez, 2008).

Globally Hyperbolic Spacetimes

A spacetime M is said to be globally hyperbolic spacetime if (i) M is strongly causal, and (ii) the sets $\{I^+(p)\cap I^-(q): p,q \in M\}$ are compact. The sets $I^+(p)\cap I^-(q)$ are said to be causal diamond. Although there are number of definitions available for global hyperbolicity, this definition perhaps is the most potent definition, see (Minguzzi & Sanchez, 2008; Bernal & Sanchez, 2007).

DEFINITION REGARDING DIFFERENT TYPE OF COMPACTNESS

Compact Space

If every open cover of a space M has finite subcover the M is said to be compact space, see (Hatcher, 2000; Munkres, 2000).

A spacetime is supposed to be paracompact but a compact spacetime is of least interest as causality conditions get violated in such spacetimes. Still compactness of spacetime characterizes no reasonable spacetimes. Thus, the study of compact spacetime do not lost its importance still now.

Paracompact Space

In a spacetime M, an open set $\mathcal{O}$ exists containing every point $p \in M$, such that $\mathcal{O}$ intersects with any finite number of members u_α of the atlas ($u\alpha$,$_\phi\alpha$), $_w$hich is said to be locally finite atlas. Now if a locally finite atlas $\{\mathcal{F}_\alpha, \psi_\beta\}$ exists for each and every atlas ($u\alpha$,$\phi_{\alpha)}$ of $_a$ given spacetime manifold such that $\mathcal{F}_\beta \subset u_\alpha$ for some α, then the space is called paracompact, see (Hawking & Ellis, 1973; Munkres, 2000). The notable fact is that a paracompact spacetime always attains a smooth positive definite Riemannian metric.

Now we define some other type of compact sets having importance in the study of globally hyperbolic spacetime, see (Bar, 2014; Bar et al., 2007):

Past Compact Sets

A closed set $A \subset M$ is said to be past compact if $J^-(p)\cap A$ is compact ∀p∈M. As an example, if Σ be a Cauchy surface, then J^+ (Σ) becomes past compact.

Strictly Past Compact Sets

A closed set $A \subset M$ is called Strictly past compact set if there is a compact subset $C \subset M$ such that $A \subset J^{+}(C)$.

Similarly, definitions and examples of future compact and strictly future compact sets can be made in a time dual manner.

Temporally Compact Sets

If a set $A \subset M$ be both future and past compact, then A is said to be temporally compact.

SOME RESULTS REGARDING THESE SETS

Proposition-1

In a globally hyperbolic spacetime, let a set A be strictly past compact, then $J^{+}(A)$ becomes so.

Proof: Clearly $J^{+}(A)$ is closed in a globally hyperbolic spacetime and also as A is strictly past compact so there is a compact set C such that $A \subset J^{+}(C)$. Thus $J^{+}(A) \subseteq J^{+}(J^{+}(C)) = J^{+}(C)$.

Proposition-2

In a globally hyperbolic spacetime, if $A \subset M$ be past compact then so is $J^{+}(A)$.

Proof: As A is past compact, so $A \cap J^{-}(p)$ is compact $\forall p \in M$. Now

$$J^{+}(A) \cap J^{-}(p) = J^{+}(A \cap J^{-}(p)) \cap J^{-}(p).$$

As $A \cap J^{-}(p)$ and {p} are compact so for a globally hyperbolic spacetime the set $J^{+}(A \cap J^{-}(p)) \cap J^{-}(p)$ is compact. Hence the result.

Proposition-3

A In a globally hyperbolic spacetime, a strictly past compact set is a past compact set.

Figure 1. This example shows that a future compact set may not be strictly future compact

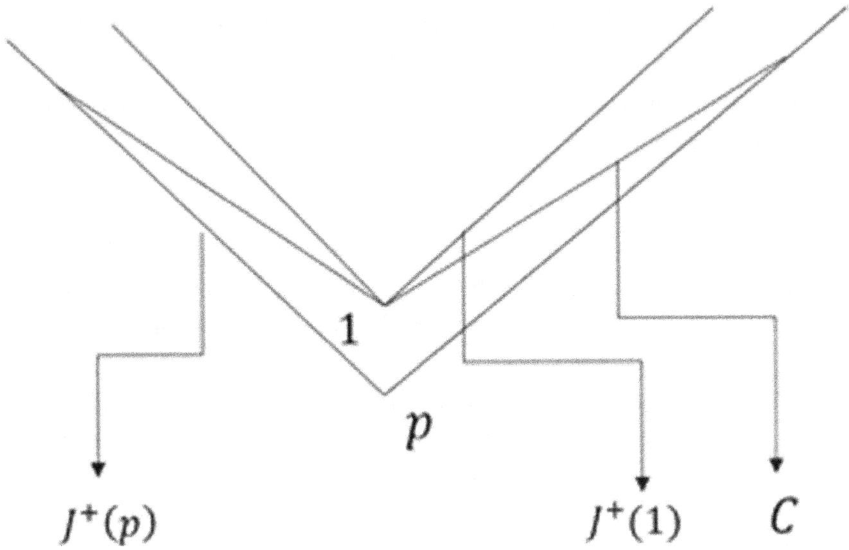

Proof: Let $A \subset M$ be a strictly past compact set. Thus, A is closed and there is a compact set C such that $\subset J^{+}(C)$. Then we are to show that, $A \cap J^{-}(p)$ is compact $\forall p \in M$. Now,

$$A \cap J^{-}(p) \subseteq J^{+}(C) \cap J^{-}(p).$$

Now $A \cap J^{-}(p)$ being a closed subset of the compact set $J^{+}(C) \cap J^{-}(p)$, is compact in M, as M is Hausdorff.

The converse is not true in general, will be clear from the following example,

Example Let C_1 be an open cone with vertex at 1 in a Minkowski space M containing $J^{+}(1)\setminus\{1\}$ which is closed in itself. Now define $A = M \setminus C_1$ which is clearly future compact set but this is not strict. As for each $p \in M$, the set $A \cap J^{+}(p) = J^{+}(p) \setminus C_1$ is compact but here the intersection of spacelike hyperplanes and A is not compact.

Here are some very useful connections between such types of compact sets and Cauchy hypersurfaces and hence with globally hyperbolic spacetimes, see (Choquet, 1971; Choquet & Geroch, 1969).

If *S* be a Cauchy hypersurface in a globally hyperbolic spacetime *M*, then the set of al causal path from r to *S* itself is compact in *M* and this is often denoted by C_r^S. An important fact is that the set D_r^S of points in *M* which can be visited by a causal curve from r to *S* also become compact, see (Wald, 1984; Beem et al., 1996).

Also, in a globally hyperbolic spacetime, there is a very important characterization of a compact set stated as follows:

Proposition-4

Let K ⊂M be compact in a given globally hyperbolic spacetime M. Also let Γ be a future inextendible causal curve which starts from p= Γ(0) ∈K, then there must be a t_(0)>0 satisfying Γ(t) ∉K ∀ t > t_(0),

Proof: Let Γ be defined on [a,b). On contrary let (t_n)_n be a sequence in [a,b) converging to b such that t_2n∈K and t_(2n+1) ∉K ∀ n ∈ N. Which means that Γ ⊂J(K), which is compact in a globally hyperbolic spacetime. Since there is a t_2n >t for every t ∈ [a,b) and so Γ(t) ∈ J^- (Γ(t_2n)) ⊂ J^- (K) which contradicts the fact that a globally hyperbolic spacetime is non- totally imprisoning. Hence the result.

Proposition-5

Let A ⊂M be closed. Then the following statments are equivalent (i) A is past (resp. future) compact. (ii) a smooth spacelike Cauchy surface Σ ⊂M exists satisfying A ⊂ J^+ (Σ) (resp. A ⊂ J^- (Σ)). (iii) a onto Cauchy temporal function t_c :M →R exists which will be bounded below (resp. bounded above) on the set A.

Now we state a necessary and sufficient condition for a set in a spacetime to become strictly past and future compact.

Proposition-6

Let $A \subset M$ be closed. Then, one of the necessary and sufficient condition for the set *A* to be strictly past compact is $A \subset J^+\left(K_S\right)$ for a compact set K_S of a smooth non-timelike Cauchy hypersurface $S \subset M$.

Proof: The given condition is sufficient follows from the definition trivially. Now conversely, since the set *A* is strictly past compact, then there is a compact subset $C \subset M$ satisfying $A \subset J^+\left(C\right)$. Now we take a smooth non-timelike Cauchy hypersurface $S \subset M$ satisfying $C \subset J^+\left(S\right)$ and we set $K_S = S \cap J^-\left(C\right)$. Then K_S is compact and also

$$A \subset J^{+}(C) \subset J^{+}\left(J^{+}(S) \cap J^{-}(C)\right) = J^{+}\left(S \cap J^{-}(C)\right) = J^{+}(K_S).$$

Hence the condition.

Similar condition for strictly future compact set can be formulated analogously by time-dual manner.

Proposition-7

Let Σ be a Cauchy surface in a globally hyperbolic spacetime and q∈D^+ (Σ), then J^+ (Σ)∩J^- (q) is compact, see (Wald, 1984; Beem et al.,1996; Geroch, 1970).

ADVANCED AND RETARDED SET

We name a set $A \subset M$ as advanced (respectively retarded) if $\overline{A} \subset J^{-}(K)$ (respectively $\overline{A} \subset J^{+}(K)$) for a compact $K \subset M$.

Note A closed set is both retarded and advanced if and only if it is compact.

Proposition-8

A closed advanced set is strictly future compact and hence future compact.

Proof: Let $A \subset M$ be closed and an advanced set $\overline{A} \subset J^{-}(K)$ for some compact set $K \subset M$. Now as *A* is closed, thus $A \subset J^{-}(K)$. As a result, *A* is strictly future compact.

Let *C(M)* and $C_S(M)$ stands for the set of all the Cauchy surfaces and set of all smooth spacelike Cauchy surface in a spacetime *M* respectively.

An important consequence of the above discussion can be stated as:

Proposition-9

If A⊂M be a closed set then the following statements are equivalent. (1) A is an advanced (resp. retarded) set. (2) J^+ (Σ)∩A (resp. J^- (Σ)∩A) is compact for all Σ∈C(M). (3) J^+ (Σ)∩A (resp. J^- (Σ)∩A) is compact for all Σ∈C_S (M).

Proof: As the causal future of every Cauchy hypersurface is past compact and a closed advanced set is future compact, thus the second condition is simply implied by the first condition. As C_S (M)⊂C(M) the third is also implied by the second condition trivially. From the third condition we have Σ∩A is compact for all Σ∈C_S

(M), then A⊂J(C) for a compact set C⊂M. Again if we choose a foliation of M by spacelike Cauchy surface Σ_t and also we have for any real t the set J^+ (Σ_t)∩A (resp. J^- (Σ_t)∩A) is compact (by the condition), then we can find a real t_1 such that A⊂J^- (Σ_t) (resp. A⊂J^+ (Σ_t)). Now setting K=Σ_(t_1)∩J(C), we have A⊂J^- (K) (resp. A⊂J^+ (K)). Thus A is advanced (resp. retarded) set. Hence the proposition.

Now we would like to define spacelike compact sets, for which let us state the following proposition-

Proposition-10

Let M be any globally hyperbolic spacetime, then the following conditions are equivalent for a closed set A⊂M. (i) There exists a compact subset A⊂M satisfying A⊂J(K). (ii) For every Cauchy surface Σ∈C(M), Σ∩A is compact. (iii) For every smooth spacelike Cauchy surface Σ∈C_S (M) , Σ∩A is compact

SPACELIKE COMPACT SETS

Any subset $A \subseteq M$ is called spacelike compact if $\overline{A}$ (which is essentially closed) satisfy any of the above three equivalent conditions. Advanced and retarded sets are simple and obvious example of spacelike compact sets.

Note A spacelike compact set in a spacetime *M* in general may not be compact. Whereas the intersection of it with any Cauchy hypersurface always become compact, see (Bar & Ginoux, 2011).

Here is a useful relation among future compact, spacelike compact and advanced sets.

Proposition-11

Any set *A* is advanced (respectively retarded) if and only if A is future (respectively past) compact as well as spacelike compact.

Proof: using above definitions we can simply prove it.

Proposition-12

In a globally hyperbolic spacetime *M*, let $\tilde{M}$ be a causally compatible open set in *M* containing a Cauchy surface of *M*. Also, if $A \subset M$ be a spacelike compact set with respect to *M*, then $\tilde{M} \cap A$ becomes spacelike compact set in $\widetilde{M}$, see (Bar & Ginoux, 2011).

TIMELIKE COMPACT SETS

Let, A ⊂M be a closed set and there exists a Cauchy hypersurface Σ ⊂M satisfying A ⊂ J^- (Σ) and A ⊂ J^+ (Σ). Then the set A is said to be timelike compact. It also is said to be temporally compact in some literature (Sanders, 2013; Bar et al., 2007).

Generally, if a set *A* becomes both future and past compact then equivalently we can say that *A* is timelike compact, see (Sanders, 2013).

SPACIALLY COMPACT SET

A closed set $A \subset M$ s called spacially compact if there is a compact subset $K \subset M$ such that $A \subset J(K)$.

Keeping in mind the statement that a spacetime is globally hyperbolic if and only if it contains a Cauchy surface, we state a very important theorem regarding spatially compactness-

Proposition-13

A globally hyperbolic spacetime *M* is spacially compact if and only if it contains compact Cauchy surfaces.

Proof Let *S* be the one of the compact Cauchy hypersurfaces. Then as *S* is Cauchy surface so, $M = J^{+}(S) \bigcup J^{-}(S)$. So, *M* is spacially compact. On the other hand, when *M* is spacially compact and $S \subset M$ be a Cauchy hypersurface, then $M \cap S = S$ is compact. Hence the result.

Proposition-14

One of the necessary and sufficient condition for $A \subset M$ to be spacially compact is that $A \subset J(K_S)$ for some compact set K_S of some smooth non-timelike Cauchy hypersurface $S \subset M$.

Figure 2. Possible relation among different types of compact sets discussed above

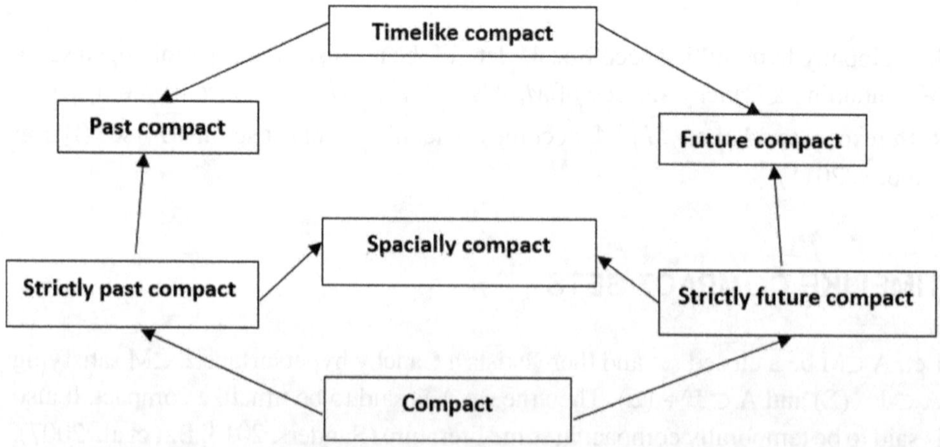

Proof Let A be spacially compact set and $S \subset M$ be a Cauchy hypersurface. Noe setting $K_S = S \cap J(C)$, we have the required compact subset. The converse part is very trivial.

Now we would like to state a condition under which the converse of the Proposition-**5** holds.

Proposition-15

In a spatially compact globally hyperbolic spacetime *M*, the notion of past compactness and that of strictly past compactness are equivalent.

Proof Let *A* be past compact in *M*. As *M* is spacially compact, so the Cauchy hypersurfaces are compact in *M*. As $A \subset M$ is past compact $\Longleftrightarrow$ a smooth non-timelike Cauchy surface $S \subset M$ exists satisfying $A \subset J^{+}(S) \Leftrightarrow A$ is strictly past compact (as *S* is compact here). Hence the proof.

Here is a possible relation among these sets which are followed from the above discussions quite obviously, the converse results of which are no longer valid.

CONCLUSION

We gave very brief overview of compactness (other types of compactness also) and their consequences and applications in globally hyperbolic spacetimes. Also, whether a compact Lorentzian manifold admits a closed geodesic or not can be verified in this regard. There are many spacetimes properties, which can also be verified for these types of compactness as well as their roles in characterizing the different type of spacetimes can be addressed. Specially, the relation between Cauchy hypersurfaces and these compact sets has been discussed here.

ACKNOWLEDGMENT

The author (D. Mishra) acknowledges the financial support of CSIR, Govt. of India. (File No. 08/702(0001)/2018-EMR-I).

REFERENCES

Bar, C. (2014). Green-hyperbolic operators on globally hyperbolic spacetimes.

Bar, C., & Ginoux, N. (2011). Classical and quantum fields on Lorentzian manifolds.

Bar, C., Ginoux, N., & Pfaffle, F. (2007). *Wave equations on Lorentzian manifolds and quantization. Mat. Soc*. EMS.

Beem, J. K., Ehrlich, P. E., & Easley, K. L. (1996). *Global Lorentzian Geometry*. Marcel Dekker Inc.

Bernal, A. N., & Sanchez, M. (2007). Globally hyperbolic spacetime can be defined as 'causal' instead of 'strongly causal'. *Classical and Quantum Gravity*, *24*(3), 745–749.

Carter, B. (1971). Causal structure in space-time. *General Relativity and Gravitation*, *1*(4), 349–391.

Choquet-Bruhat, Y. (1971). Problemede Cauchy pour le systeme integro differentiel d'Einstein-Liouville. *Annales de l'Institut Fourier*, *21*(3), 181–201. doi:10.5802/aif.385

Choquet-Bruhat, Y., & Geroch, R. (1969). Global aspects of the Cauchy problem in general relativity. *Communications in Mathematical Physics*, *14*(4), 329–335. doi:10.1007/BF01645389

Choudhury, B. S., & Mondal, H. S. (2014). *Continuous representation of a globally hyperbolic spacetime with non-compact Cauchy surfaces*. Anal. Math. Phys. Doi: doi:10.100713324-014-0093-x

Clarke, C. J. S. (1986). *Singularities: Global and local aspects in topological properties and global structure of spacetime*. Plenum Press.

Das, S. K., Dao, T. P., & Perumal, T. (Eds.). (2021). *Nature-Inspired Computing for Smart Application Design*. Springer Singapore.

Das, S. K., Das, S. P., Dey, N., & Hassanien, A. E. (Eds.). (2021). *Machine learning algorithms for industrial applications*. Springer.

Das, S. K., Samanta, S., Dey, N., & Kumar, R. (Eds.). (2020). *Design frameworks for wireless networks*. Springer Singapore.

Das, S. K., Samanta, S., Dey, N., Patel, B. S., & Hassanien, A. E. (Eds.). (2021). *Architectural Wireless Networks Solutions and Security Issues. Springer Singapore*. Pte. Limited.

Das, S. K., Tripathi, S., & Burnwal, A. P. (2015). *Intelligent energy competency multipath routing in wanet. Information systems design and intelligent applications*. Springer.

De, D., Mukherjee, A., Das, S. K., & Dey, N. (Eds.). (2020). *Nature inspired computing for wireless sensor networks*. Springer.

Filippov, V. (1998). Basic topological structures of ordinary differential equations, Mathematics and its Applications. (Volume 32). Kluwer Academic Publishers, Dordrecht.

Friedlander, F. G. (1976). *The wave equations on a curved space-time*. Cambridge University Press.

Geroch, R. (1970). Domain of dependence. *Journal of Mathematical Physics, 11*, 437–449.

Hatcher, A. (2000). *Algebraic topology*. Cambridge University Press.

Hawking, S. W., & Ellis, G. F. R. (1973). *The Large-Scale Structure of Space-Time*. Cambridge University Press. doi:10.1017/CBO9780511524646

Hawking, S. W., & Penrose, R. (1970). The singularities of gravitational collapse and cosmology. *Proceedings of the Royal Society of London. Series A, Mathematical and Physical Sciences*, *314*(1519), 529–548. doi:10.1098/rspa.1970.0021

Hounnonkpe, R. A., & Minguzzi, E. (2019). Globally hyperbolic spacetime can be defined without causal condition.

Kronheimer, E. H., & Penrose, R. (2008). *On the structure of causal spaces*. Cambridge University Press. doi:10.1017/S030500410004144X

Lee, J. (2011). *Introduction to topological manifolds. Graduate Texts in Mathematics*. Springer.

Leray, J. (1953). *Hyperbolic differential equations*. The Institute for Advanced Study.

Ling, E. (2020). Aspects of C^0 causal theory. *General Relativity and Gravitation, 52*(6). Doi: doi:10.100710714-020-02708-9

Minguzzi, E. (2009). K-Causality coincides with stable causality. *Communications in Mathematical Physics, 290*, 239–248.

Minguzzi, E. (2019). Lorentzian causality theory. *Living Rev. Relativ., 22*(3). DOI:10. 1007/s41114-019-0019-x.

Minguzzi, E. (2019). Lorentziaan causality theory. *Living Reviews in Relativity, 22*, 3. https://doi.org/10.1007/s41114-019-0019-x

Minguzzi, E., & Sanchez, M. (2008). In H. Baum & D. Alekseevsky (Eds.), The causal hierarchy of spacetimes. Recent developments in pseudo-Riemannian geometry, ESI Lect. EMS Pub. House, (pp. 299–358). doi:10.4171/051-1/9

Mohajan, H. K. (2016). Global hyperbolicity in spacetime manifold. *Int. J. Prof. Studies, 1*(1).

Muller, O., Sanchez, M. (2014). An invitation to Lorentzian geometry. *Jahresbericht der Deutschen Mathematiker-Vereinigung, 115*(3).

Munkres, J. R. (2000). Topology. Pearson Education.

Navarro, J. J. B., & Minguzzi, E. (2011). Global hyperbolicity is stable in the interval topology.

Navarro, J. J. B., & Minguzzi, E. (2011) Global hyperbolicity is stable in the interval topology.

O'Neill, B. (1983). *Semi-Riemannian Geometry*. Academic Press.

Penrose, R. (1965). Gravitational collapse and spacetime singularities. *Physical Review Letters, 14*(3), 57–59.

Penrose R (1969) Gravtational collapse: the role of general relativity. *Revista del Nuovo Cimento, 1*, 252-276.

Penrose, R. (1972). Techniques of Differential Topology in Relativity. CBMS-NSF Regional Conference Series in Applied Mathematics. SIAM.

Penrose, R. (1972). *Techniques of differential topology in relativity*. Society for Industrial and Applied Mathematics.

Penrose, R. (2002). Gravitational collapse – the role of general relativity. Gen. Rel. Grav, *34*, 1142–1164.

Perlick, V., & Tausk, D. V. (2003). Gravitational lensing from a spacetime perspective. *Living Reviews in Relativity*, *7*. www.livingreviews.org/Irr-2004-9

Ringstrom, H. (2009). The Cauchy problem in general relativity. ESI Lectures in Mathematics and Physics. EMS.

Samann, C. (2019). Global hyperbolicity for spacetimes with continuous metrics.

Sanchez, M. (2006). On causality and closed geodesics of compact Lorentzian manifolds and static spacetimes. *Differential Geometry and Its Applications*, *24*(1), 21–32. doi:10.1016/j.difgeo.2005.06.008

Sanchez, M. (2006) A note n causally simple and globally hyperbolic spacetimes.

Sanders, K. (2013). A note on spacelike and timelike compactness. *Clas. And Quan. Grav*, *30*(11), 115014. Doi: doi:10.1088/0264-9381/30/11/115014

Seifert, H. J. (1967). Global connectivity by timelike geodesics. *Z. Naturforsh*, *22a*, 1356–1360.

Sorkin, R. D. (1990) Spacetime and causal sets, Relativity and gravitation. Classical and quantum, 150-173.

Sorkin, R. D., & Woolgar, E. (1999) A causal order for spacetimes with C^0 Lorentzian metrics: Proof of compactness of the space of causal curves.

Spivak, M. (1979). *A comprehensive introduction to differential geometry*. Berkeley: Publish or perish, Inc.

Wald, R. M. (1984). *General Relativity*. The University of Chicago Press. doi:10.7208/chicago/9780226870373.001.0001

Witten, E. (2020) Light rays, singularities, and all that.

Chapter 14
On Characterization of b-Open Sets in Cofinite Topology on R

Dwaipayan Mishra
Midnapore College (Autonomous), Vidyasagar University, India

Himadri Shekhar Mondal
Midnapore College (Autonomous), Vidyasagar University, India

Motahar Reza
GITAM University (Deemed), India

Amalendu Rana
https://orcid.org/0000-0003-1457-7284
Jadavpur University, Kolkata, India

ABSTRACT

In this article, the characterization of different types of open sets in cofinite topology on the set of real numbers has been carved. Here specially b-open sets in the cofinite topology has been studied. The authors have characterized the b-open sets in the cofinite topology over R where one characterization facilitates to apprehend a group of sets to be b open and an other group of sets not to be b-open in the cofinite topology on the set of real numbers spontaneously. Furthermore, some important results on these sets are investigated and proved.

DOI: 10.4018/978-1-6684-7343-6.ch014

INTRODUCTION

The notion of open sets is one of the most important part in almost in every branch of mathematics. Actually, b-open sets are the generalization of the open sets. Thus, to find the b-open sets has some useful aspects in the study of topological spaces. These are also some other natural generalization of open sets as namely semi-open sets, β-open sets, pre-open sets and α-open sets. Andrijevic (Andrijevic, 1996) proposed the class of generalized open sets and named it as b-sets. He studied many interesting and useful fundamental properties of b-open sets. There are plenty of usefulness and applications of such sets (Ekici, 2005; Dvalishvili, 2004; Caldas et al., 2008; Ekici, 2008; Corson & Michael, 1964; Ekici & Noiri, 2009). The notion of decision constraints may be widely useful in determining various aspects of real analytic problems as well as topology optimization with decision constraints (Das et al., 2015; Das et al., 2021a; Das et al., 2021b; Das et al., 2021c; De et al., 2020; Das et al., 2020). Kumari and Raman (Kumari & Raman, 2013) discussed on characterization of b-open sets in a topological space.

The cofinite topology has very important role to find out several counter examples easily. Thus, the basic results regarding b-open sets can be founded strongly through the ideas and examples of cofinite topology on different sets (Steen et al., 1995). So, it is quite relevant and useful to study cofinite topology in the perspective of b-open sets.

The purpose of this investigation is to discuss some aspects of these sets in cofinite topology on $\mathbb{R}$ which may facilitate the introduction of such open (as a result closed in complemental manner) sets and allied concepts in this regards. We show that any infinite set in cofinite topology is essentially β-open as well as pre-open but not necessarily semi-open or α-open. Also, unlike many topologies on $\mathbb{R}$, cofinite topology restrict the closed sets to become b-open. These results can be explained in the view of topological properties of a finer and coarser topology on a particular space X, as $(\mathbb{R},\tau_u)$ is finer than $(\mathbb{R},\tau_f)$.

BASIC DEFINITIONS

Semi-Open Set

Let (U,τ) be a topological space. A subset X of the topological space U is said to be semi-open set (Andrijevic, 1986; Levine, 1963; Mashhour et al., 1982) if $X \subset cl(int(X))$.

Example-1: Let $(\mathbb{R}, \tau_u)$ is a topological space on the set of Real numbers $\mathbb{R}$ with the usual topology. Then,

(i) An open set in $(\mathbb{R}, \tau_u)$ is a semi open set.
(ii) A closed interval in $\mathbb{R}$ is semi open set in $(\mathbb{R}, \tau_u)$.

β-Open Set

Let (U, τ) be a topological space. A subset X of the topological space U is said to be β-open set (Andrijevic, 1996) if $X \subset cl(int(cl(X)))$.

Example-2: Let $(\mathbb{R}, \tau_u)$ is a topological space on the set of Real numbers $\mathbb{R}$ with the usual topology. Then, an open set in $(\mathbb{R}, \tau_u)$ is a β-open set.

Pre-Open Set

Let (U, τ) be a topological space. A subset X of the topological space U is said to be pre-open set (Andrijevic, 1986) if $X \subset int(cl(X))$.

Example-3: Let $(\mathbb{R}, \tau_u)$ is a topological space on the set of Real numbers $\mathbb{R}$ with the usual topology. Then, an open set in $(\mathbb{R}, \tau_u)$ is a pre-open set.

α-Open Set

Let (U, τ) be a topological space. A subset X of the topological space U is said to be semi-open set (Andrijevic, 1996) if $X \subset int(cl(int(X)))$.

Example-4: Let $(\mathbb{R}, \tau_u)$ is a topological space on the set of Real numbers $\mathbb{R}$ with the usual topology. Then, an open set in $(\mathbb{R}, \tau_u)$ is α-open set.

b-Open Set

Let (U, τ) be a topological space. A subset X of the topological space U is said to be b-open set (Abd El-Monsef et al., 1983) if $X \subset cl(int(X)) \cup int(cl(X))$. The complement of a b-open set of X is called b-close set.

Example-5: Let $(\mathbb{R},\tau_u)$ is a topological space on the set of Real numbers $\mathbb{R}$ with the usual topology. Then, a semi open set and pre-open set in $(\mathbb{R},\tau_u)$ are b-open sets, but this result does not hold good generally, as we can find an example easily.

Example-6: Let $X=\{a,b,c,d\}$ endowed with the topology $\tau=\{\phi, X, \{a\}, \{b\}, \{a,b\}\}$. Then, the collection of different types of open sets are as follows:

The collection of all semi-open sets on (X,τ) is $\{\phi$, X, {a}, {b}, {a,b}, {a,c}, {a,d}, {b,c}, {b,d}, {a,b,c}, {a,b,d}, {a,c,d}, {b,c,d}}, the collection of all β-open sets on (X,τ) is $\{\phi$, X, {a}, {b}, {a,b}, {a,c}, {a,d}, {b,c}, {b,d}, {a,b,c}, {a,b,d}, {a,c,d}, {b,c,d}},, the collection of all pre-open sets on (X,τ) is $\{\phi$, X, {a}, {b}, {a,b}, {a,b,c}, {a,b,d}}, the collection of all α-open sets on (X,τ) is $\{\phi$, X, {a}, {b}, {a,b}, {a,b,c}, {a,b,d}} and the collection of all b-open sets on (X,τ) is the union of the collection of all semi-open sets on (X,τ) and the collection of all pre-open sets on (X,τ). Also, here the collection of all b-open sets on (X,τ) is the collection of all semi-open sets on (X,τ).

BASIC DEFINITIONS ON COFINITE TOPOLOGY

Cofinite Topology

Let X be a non-empty set. The topology τf on X is said to be cofinite topology if τf contains all the subsets of X whose complement is finite or ϕ or X itself.

Therefore, the open set in cofinite topology are the sets whose complement is finite and closed sets are only the finite sets and ϕ and X. Thus, co-finite topology on a finite set is identical with discrete topology.

The usual topology τu on $\mathbb{R}$ is strictly finer than the cofinite topology on it i.e., all the open sets in cofinite topology on $\mathbb{R}$ remains still an open set in usual topology on $\mathbb{R}$.

Cofinite topology has no minimal basis if it is defined on an infinite set as in that case the topology is T1 and non-discrete.

In addition, unconditionally every set in cofinite topology is compact and it remains always connected if the underlying set is of infinite cardinality otherwise it is disconnected.

SOME RESULTS REGARDING THESE SETS

As, semi-open, β-open, pre-open and α-open and thus b-open sets are the generalization of open sets, (Hamzah & Alameer, 2020) then one can easily find the following lemma.

Lemma-1

Every open sets in $(\mathbb{R}, \tau_u)$ are semi-open, β-open, pre-open and α-open and thus b-open.

Lemma-2

Every closed interval (but not necessarily the closed sets) in $(\mathbb{R}, \tau_u)$ is semi-open, β-open (as a result b-open)

Proof: **Let I=**[a,b] be *a* closed interval in $(\mathbb{R}, \tau_u)$. Then as int[a,*b]=*(*a*,*b*), *cl[*a,b*]=*[*a*,*b*], *int*(a,*b)=*(*a*,*b*), *cl(*a,b*)=*[*a*,*b*]; *us*ing the definition we have the result.

But if we have a finite set A in $\mathbb{R}$, then A is closed but int(*A)=ϕ*. Thus, $A \subset cl(int(A))$ does not hold.

In fact a finite set can never be any of semi-open, β-open, pre-open and α-open or b-open sets, as the interior of every finite set in $(\mathbb{R}, \tau_u)$ turns to be void.

Note: **A** closed interval in $(\mathbb{R}, \tau_u)$ cannot be a pre-open or α-open set.

The following lemma justifies that b-open set is generalization of pre-open and semi-open sets.

Lemma-3

Every semi-open and pre-open sets in any topological space are b-open but not conversely.

Proof: This is follows directly from the definition of b-open set.

Also, in the example-6 the set $\{a,c\}$ is b-open but not pre-open.

In fact every open set is designed to be semi-open, α-open, pre-open, α-open, (and as a result) b-open in arbitrary topological space but the converse is not true. Each of these different open sets are the generalizations of open sets.

The following theorem gives a useful characteristics of open sets in co-finite topology.

Theorem-1

Let A be an infinite set in $(\mathbb{R},\tau_f)$. Then A is α-open and pre-open but not necessarily semi-open or α-open.

Proof: As $A \subseteq \mathbb{R}$ is infinite in $(\mathbb{R},\tau_f)$, so $cl(A)=\mathbb{R}$.

Thus $A \subset int(cl(A))$ and $A \subset cl(int(cl(A)))$. Hence the result.

Now, considering A=(0,¥) in $(\mathbb{R},\tau_f)$, one have $int(A)=\phi$, showing that, inspite of being infinite, the set A serves neither as semi-open nor as α-open there.

Corollary-1

Any infinite subset of $\mathbb{R}$ in cofinite topology is b-open.

Corollary-2

In $(\mathbb{R},\tau_u)$, there are infinite sets which are neither of the above types of open sets. The set of Natural number $(\mathbb{N})$ is an example of such infinite set in $(\mathbb{R},\tau_u)$. But $\mathbb{N}$ is β-open and pre-open in $(\mathbb{R},\tau_f)$.

Note: If we define the cofinite topology on a finite set then the cofinite topology reduce to the discrete topology and hence there is no distinction among open sets, semi-open sets, α-open sets, β-open sets and b-open sets. All of these sets have the same connotations.

The closed sets in $(\mathbb{R},\tau_f)$ are finite sets in $\mathbb{R}$ precisely. Now, the interior of any finite set in $(\mathbb{R},\tau_f)$ is ϕ. So, the interior of any closed set in $(\mathbb{R},\tau_f)$ is ϕ. Thus, $int(cl(A))$, $cl(int(A))$, $int(cl(int(A)))$, $cl(int(cl(A)))$ are turned to be empty in $(\mathbb{R},\tau_f)$.

Also, the closure of any finite set A is again the set A. So, $int(cl(A))=\phi$. Thus we have the following theorem:

Theorem-2

No closed sets in $(\mathbb{R},\tau_f)$ can be semi-open, β-open, pre-open or α-open and as a result no finite set in $(\mathbb{R},\tau_f)$ can be b-open.

Proof: The result is obvious from the above discussion.

Note-1: This result may violate for $(\mathbb{R},\tau_u)$. e.g. A=[2,3] is closed set in $(\mathbb{R},\tau_u)$. Now, *int*(A)=(2,3) and *cl*(*int*(A))=[2,3]. So, A being a closed set, it is also semi open.

Note-2: In $(\mathbb{R},\tau_f)$, any infinite subset of $\mathbb{R}$ is b-open and no closed set can be b-open.

CONCLUSION

In this article, the characterization of different types of open sets in cofinite topology has been delineated. One characterization helps to find out a group of sets to be b-open in cofinite topology in $\mathbb{R}$ easily. Any infinite set of $\mathbb{R}$ in cofinite topology is b-open whereas no closed set of $\mathbb{R}$ can be b-open there. In future the notion of b-connectedness can be incorporated.

ACKNOWLEDGMENT

The author (D. Mishra) acknowledges the financial support of CSIR, Govt. of India. (File No. 08/702(0001)/2018-EMR-I).

REFERENCES

Abd El-Monsef, M. E., El-Deeb, S. N., & Mahmoud, R. A. (1983). b-open sets and b-continuous mappings. *Bull. Fac. Assuit Univ.*, *12*, 77–90.

Andrijevic´, D. (1986). Semi preopen sets. *Mat. Vesnik*, *38*, 24–32.

Andrijevic´, D. (1996). On b-open sets. *Mat. Vesnik*, *48*, 59–64.

Caldas M, Jafari S, Latif RM, (2008). Applications of b-open sets and (b,s)-continuous functions. *King Fahd University of Petroleum & Minerals, Dept. of Math. Sci. Technical Report Series* TR 400.

Corson, H. H., & Michael, E. (1964). Metrizability of certain countable unions. *Illinois J. Math*, *8*, 351–360.

Das, S. K., Dao, T. P., & Perumal, T. (Eds.). (2021a). *Nature-Inspired Computing for Smart Application Design*. Springer Singapore.

Das, S. K., Das, S. P., Dey, N., & Hassanien, A. E. (Eds.). (2021c). *Machine learning algorithms for industrial applications*. Springer.

Das, S. K., Samanta, S., Dey, N., & Kumar, R. (Eds.). (2020). *Design frameworks for wireless networks*. Springer Singapore.

Das, S. K., Samanta, S., Dey, N., Patel, B. S., & Hassanien, A. E. (Eds.). (2021b). *Architectural Wireless Networks Solutions and Security Issues*. Springer Singapore Pte. Limited.

Das, S. K., Tripathi, S., & Burnwal, A. P. (2015). Intelligent energy competency multipath routing in wanet. In *Information systems design and intelligent applications* (pp. 535–543). Springer.

De, D., Mukherjee, A., Das, S. K., & Dey, N. (Eds.). (2020). *Nature inspired computing for wireless sensor networks*. Springer.

Dvalishvili, B. (2004). Connectedness of a fine topology and localization in bitopological spaces. *Georgian Mathematical Journal, 11*(4), 713–732. doi:10.1515/GMJ.2004.713

Ekici, E. (2005). On c-Urysohn spaces. *Adv. Stud. Contemp. Math, 11*(2), 219–226.

Ekici, E. (2008). Generalization of weakly clopen and strongly h-bcontinuous functions. *Chaos, Solitons, and Fractals*, *38*, 79–88.

Ekici, E., & Noiri, T. (2009). Decompositions of continuity, a-continuity and AB-continuity. *Chaos, Solitons, and Fractals*, *41*, 2055–2061.

Hamzah, H. A., & Alameer, N. T. A. (2020) β-sets in topological spaces and $\tau\beta$s topology, IOP *Conf. Series: Materials Science and Engineering,* 928, 042034, doi:10.1088/1757-899X/928/4/04203

Kumari V, Raman TCK, (2013) On characterization of b-open sets in a topological space. *International Journal of application or innovation in Engineering & Management*, 2(12), 229-235.

Levine, N. (1963). Semi-open sets and semi-continuity in topological space. *The American Mathematical Monthly*, *70*, 36–41.

Mashhour, A. S., Abd El-Monsef, M. E., & El-Deeb, S. N. (1982). On pre-continuous and weak pre-continuous mapping. *Proceedings of the Mathematical and Physical Society of Egypt*, *53*, 47–53.

Steen, L. A. Seebach, J. A. Jr. (1995) Counter examples in Topology. Berlin, New York: Springer-Verlag.

Chapter 15
Numerical Investigation of Turbulent Flow Transport Through a Corrugated Microtube

Amalendu Rana
https://orcid.org/0000-0003-1457-7284
Jadavpur University, Kolkata, India

Motahar Reza
GITAM University (Deemed), India

ABSTRACT

In this investigation, turbulent flow inside a corrugated micro-tube has been simulated numerically at low Reynolds number. The effect of irregular roughness of the wall of the microtube has been investigated and the turbulent flow model has been simulated to measure the state of turbulence. The results indicates that the irregular surface affects the flow pattern of the corrugated microtube. Turbulent kinetic energy and turbulent dissipation rate are augmented with respect to the position and fluctuated rapidly near to the outlet region of the corrugated microtube. Further, the Eigen values are calculated by taking the volume average of the Reynolds stress components and classified the state of the turbulence.

INTRODUCTION

Narrow confinement flow transport is generally challenged by laminar to turbulent

DOI: 10.4018/978-1-6684-7343-6.ch015

transition, flow instability and turbulent flow characteristics. In modern technology, the development of various small scale systems, micro electro mechanical systems (MEMS) have attracted researchers to augment their interest towards microfluidics which finds its applications in biomedical devices, electronics cooling, aerospace technology, micro liquid atomizers etc. (Weston et al., 2010; Stone et al., 2004; Laster & Santiago, 2004; West et al., 2002; Jang and Lee, 2000). Many real-life problems can be modeled using decision constraints. It has immense significance in linear programming problems, optimization theory etc. This notion may also be useful in the study of flow transport to explore some of its important aspects. The flow performance through fractal-like heat sinks, subject to both geometrical and flow constraints are assessed which are crucial in order to satisfy physical requirements of a design of microfluidic devices. Researchers are also very much inquisitive to know the different type of fluid transport patterns through microchannel. Therefore, they are interested to identify the flows through different roughened channels, structured surfaces, ribbed channels etc. They investigated the fluid flow transport phenomenon experimentally and numerically as well as analytically (Das et al., 2015; Das et al., 2021a; Das et al., 2021b; Reza & Rana, 2020; Wang et al., 2016; Reza et al., 2021; Rana et al., 2022; Das et al., 2021c; De et al., 2020; Das et al., 2020; Reza et al., 2020). Recently, there are so many theoretical and experimental observations are performed on the turbulent flow modeling (Sharp & Adrian, 2004; Kowalewski & Blonski, 2009; Kowalewski et al., 2006). By using the micro-PIV experiment, Li and Oslen (Li & Oslen, 2006) examined on the turbulent flow transport in large-scale structures through a microchannel. Natrajan and Christensen (Natrajan & Christensen, 2010) investigated the impact of the surface roughness on the flow through the rectangular microchannel. To investigate the heat transfer of turbulent flow through a triangular microchannel, a numerical experiment is performed by Rezaei et al. (Rezaei et al., 2017). Bazdar et al. (Bazdar et al., 2019) studied the turbulent flow and heat transfer through a wavy microchannel by considering different wavelengths.

The main objective of this study is to investigate the effect of anisotropy property of the turbulent flow of a Newtonian fluid and non-Newtonian fluid through a corrugated microtube. In this research work water is taken as a Newtonian fluid. A numerical simulation is done in ANSYS and the Reynolds stress components are calculated with proper boundary condition. Further, the turbulent kinetic energy and turbulent viscous dissipation rate are also examined in this observation.

Figure 1. Schematic sketch of the corrugated microtube

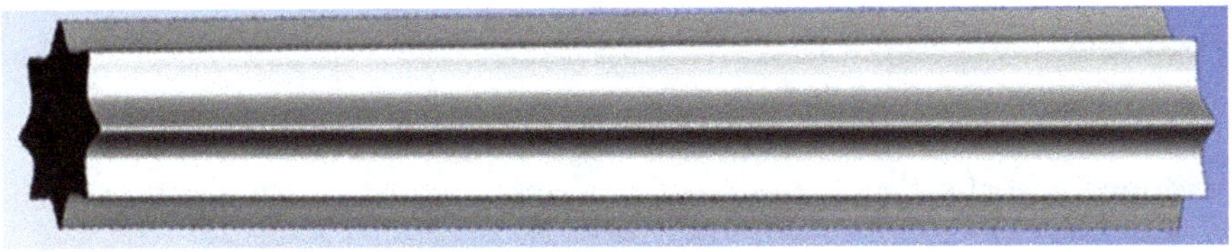

MATHEMATICAL MODEL AND PROBLEM FORMULATION

In this investigation, a three Dimensional, steady and incompressible flow is assumed through a corrugated walled microtube with hydraulic diameter 550μm and length 5mm. The microtube is considered with constant wall roughness (20μm) and constant wall heat flux 20000w/m^2 at the boundary. The flow is happened due to the standard pressure and the gravitational force. To make the simulation the velocity inlet and the pressure outlet boundary condition is used also the no-slip principle has been considered. The wall of the tube is under the constant heat flux boundary condition and the Newtonian fluid flow is simulated with the low Reynolds number.

Governing Equations of the flow

The governing equations of the flow are given by the continuity equation and the Reynolds Average Navior Stokes (RANS) equation, Reynolds Stress turbulent kinetic energy equations respectively which are as follows:

$$\frac{\partial \rho}{\partial t} + \nabla.\left(\rho U\right) = 0 \tag{1}$$

$$\frac{\partial u_i u_j}{\partial t} + U_k \frac{\partial u_i u_j}{\partial x_k} = P_{ij} - \varepsilon_{ij} + \Pi_{ij} - \frac{\partial u_i u_j u_k}{\partial x_k} + \upsilon \frac{\partial^2 u_i u_j}{\partial x_k^2} \tag{2}$$

$$\frac{\partial k}{\partial t} + U_k \frac{\partial k}{\partial x_k} = P - \varepsilon - \frac{\partial u_k p}{\partial x_k} - \frac{1}{2}\frac{\partial u_i u_j u_k}{\partial x_k} + v\frac{\partial^2 k}{\partial x_k^2} \tag{3}$$

Where,

$$\text{Production Tensor}\left(P_{ij}\right) = -\left(u_k.u_i \frac{\partial U_j}{\partial x_k} + u_k.u_j \frac{\partial U_i}{\partial x_k}\right),$$

$$\text{Dissipation Tensor}\left(\varepsilon_{ij}\right) = 2v \frac{\partial u_i u_j}{\partial x_k \partial x_k},$$

$$\text{Velocity-pressure gradient Tensor}\left(\Pi_{ij}\right) = R_{ij} - \frac{\partial}{\partial x_k} T_{kij},$$

$$\text{Pressure-rate-of-strain Tensor}\left(R_{ij}\right) = \frac{p}{\rho}\left(\frac{\partial u_j}{\partial x_i} + \frac{\partial u_i}{\partial x_j}\right)$$

$$\text{Pressure transport Tensor}\left(T_{kij}\right) = \left(\frac{pu_j}{\rho}\delta_{jk} + \frac{pu_i}{\rho}\delta_{ik}\right)$$

$$\varepsilon = \frac{1}{2}\varepsilon_{ii} = \frac{2}{2} v \frac{\partial u_i \partial u_i}{\partial x_k \partial x_k} = v \frac{\partial u_i \partial u_i}{\partial x_k \partial x_k},$$

$$P = \frac{1}{2} P_{ii} = -\frac{1}{2}\left(u_k u_i \frac{u_i}{\partial x_k} + u_k u_i \frac{\partial u_i}{\partial x_k}\right) = -u_k u_i \frac{\partial u_i}{\partial x_k},$$

$$\Pi_{ij} = R_{ij} - \frac{\partial}{\partial x_k} T_{kij} = 0 - 2\frac{\partial u_k p}{\partial x_k}.$$

Grid Study and Boundary conditions

Different sizes of the grid are tested to assure the independent result from the grid number. As it is seen that, when the grid number is more than 130000, then the archived results are independent from the grid number.

The flow has been investigated inside the corrugated microtube by using the numerical simulation at 37°C. To make the simulation the velocity inlet and the pressure outlet boundary condition is used also the no-slip principle has been

Figure 2. Meshing of the corrugated microtube

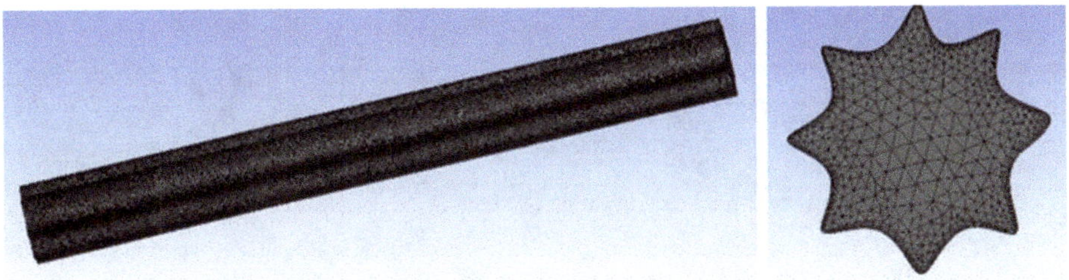

considered. The wall of the tube is under the constant heat flux boundary condition and the flow is simulated with Reynolds number less than 500.

NUMERICAL SIMULATION PROCEDURE

The governing equations are discretized by using the finite volume method and to simulate the flow directly the following setup conditions are used:

RESULTS AND DISCUSSIONS

In this article, the turbulent flow transport of water through the corrugated microtube has been examined. The turbulent flow model is simulated numerically in ANSYS. Using the simulation results for the turbulent flow performance through the microtube with corrugated wall, the results are presented graphically in figure

Table 1.

Used package and version	Fluent 19.2, double precision, segregated
Flow Type	Three Dimensional, steady and incompressible
Viscosity Model	Quadratic Pressure strain Reynolds stress model with scalable wall functions and Energy equation with viscous heating.
Medium	Newtonian fluid: water, constant density ρ= 998.2 kg/m^3 and viscosity μ=0.001003 kg/m.s (At 37°C)
Discretization Scheme	Pressure: standard Momentum: Second Order Upwind Turbulence Kinetic Energy: Second Order Upwind Turbulence Dissipation Rate: Second Order Upwind Reynolds Stress: Second Order Upwind Energy: Second Order Upwind

Figure 3. Velocity streamline for Re=301

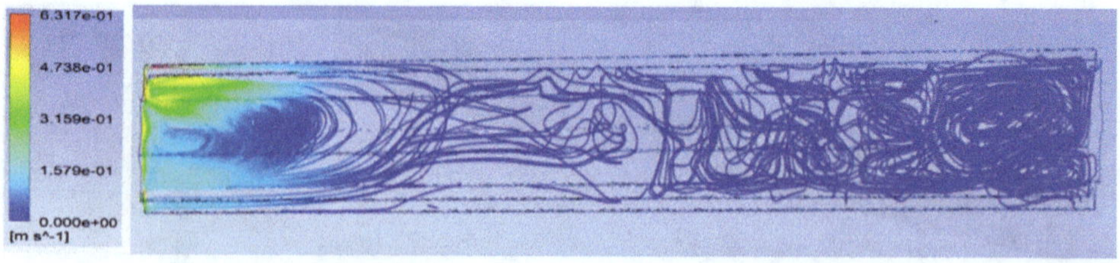

3-7. The fluctuations of the velocity are occurred due the irregular surface of the microtube. The velocity streamlines and the velocity contour are shown in figure 3 and figure 4. Due to the corrugated surface of the microtube, the velocity of the flow is very much fluctuated, which is displayed in figure 3. From figure 4 delineated the velocity contour in a cross-sectional plane of the corrugated microtube and the velocity is increased near to the boundary wall.

It is interesting to examine that the fluctuations of velocity magnitude concerning the position for *Re*=301 is displayed in figure 5. The velocity of the flow is enhanced at the inlet portion and after a certain position the velocity magnitude is fluctuated due to the corrugated surface of the microtube. Figure 6 and 7 display the nature of the turbulent kinetic energy and the turbulent dissipations rate concerning the position respectively. From figure 6, it is depicted that the turbulent kinetic energy fluctuation is increasing with the concerning position and it is more fluctuated in the outlet region. Almost similar performance is shown for the turbulent dissipation rate with respect to position in figure 7. But there fluctuations of turbulent dissipation rate is more rapid near to the outlet region of the corrugated microtube.

In this problem, the distribution of Reynolds stress ($u_i u_j$) is given by:

Figure 4. Velocity contour for Re=301

Figure 5. Velocity magnitude with respect to position for Re=301

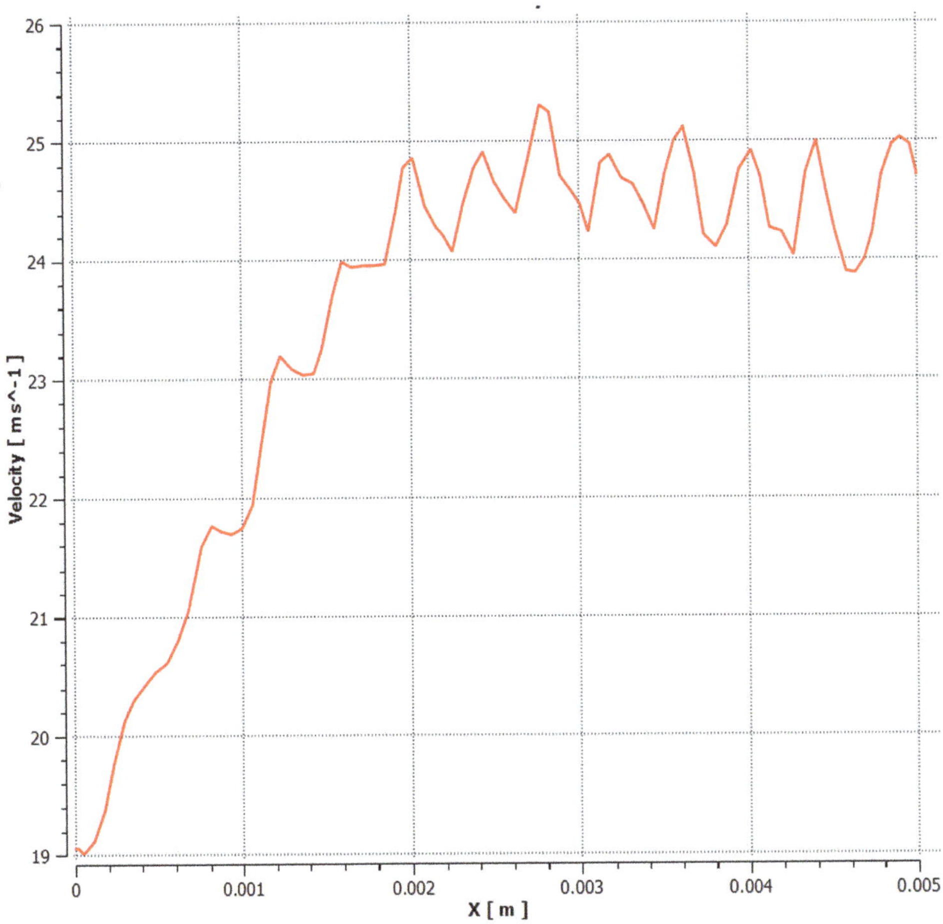

$$\begin{bmatrix} -0.002121 & 0.000023 & -0.000020 \\ 0.000023 & 0.002406 & -0.000107 \\ -0.000020 & -0.000107 & 0.003819 \end{bmatrix}.$$

The above all values are the volume average of the Reynolds stress and the volume average of the kinetic energy (k)=0.004173m^2/s^2.

Finally, one can easily calculate the anisotropic tensor and the Eigen values are given by0.1252, -0.0458, -0.0794. Then the canonical form of the anisotropic tensor is given by:

Figure 6. Turbulent kinetic energy with respect to position for Re=301

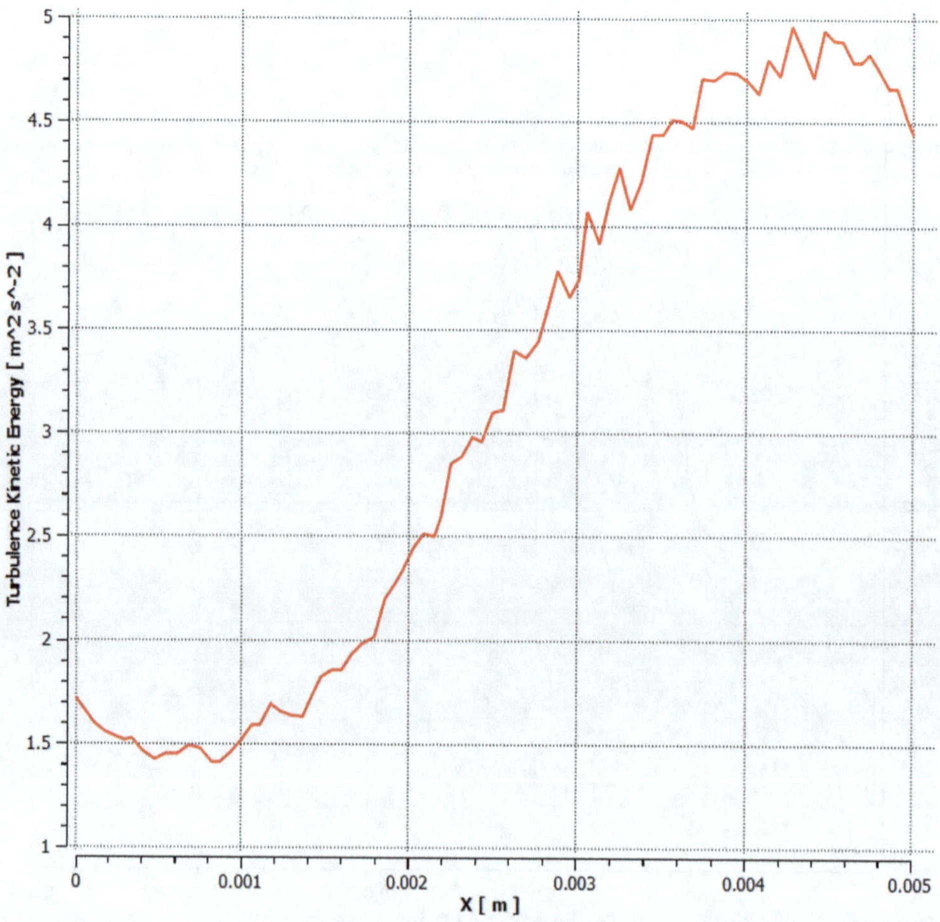

$$b_{ij}^{E} = \begin{bmatrix} 0.1252 & 0 & 0 \\ 0 & -0.0458 & 0 \\ 0 & 0 & -0.0794 \end{bmatrix}$$

Since all the Eigen values are non-zero and different. So, from the classification of turbulence property it is easy to conclude that the flow is anisotropic three component limiting state of turbulence.

Figure 7. Tubulent dissipation with respect to position for Re=301

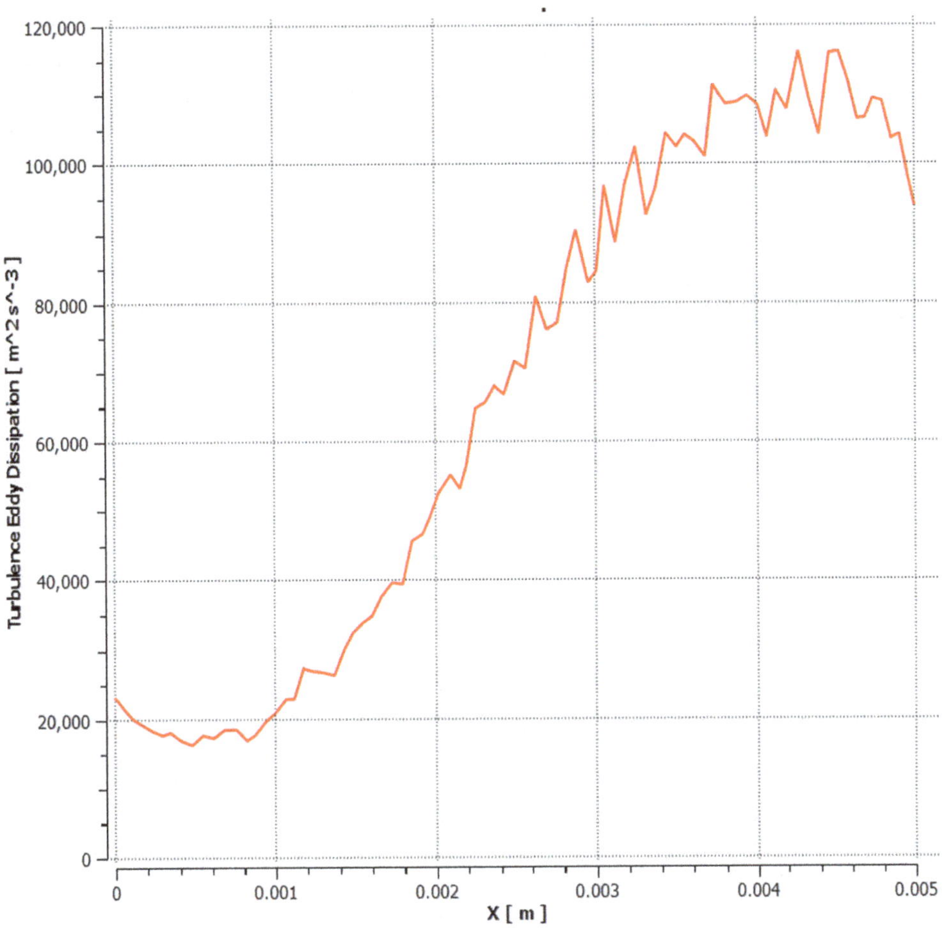

CONCLUSION

This numerical study deals with the turbulent flow transport phenomenon through a corrugated microtube. The flow velocity profiles and turbulent characteristics are simulated numerically. The computational work reveals with the following observations:

- Due to the corrugations of the microtbe the velocity profile is fluctuated and there is an enhancement in the velocity profile.
- The turbulent kinetic energy is more fluctuated in the outlet region of the corrugated microtube.

- Turbulent dissipation rate is improved with the concerning positions and it is rapidly fluctuated in the outlet region of the corrugated microtube.
- All the Eigen values of the Anisotropic tensor are real, non-zero and distinct. Therefore, from the classification property of the turbulence, it can be easily conclude that the state flow is anisotropic three component limiting state of turbulence.

ACKNOWLEDGMENT

This work was supported by SERB, Govt of India. (Grant File No. EMR/2016/006383). The authors would like to acknowledge this support.

REFERENCES

Bazdar, H., Toghraie, D., Pourfattah, F., Akbari, O. A., Nguyen, H. M., & Asadi, A. (2019). Numerical investigation of turbulent flow and heat transfer of nanofluid inside a wavy microchannel with different wave lengths. *Journal of Thermal Analysis and Calorimetry*. doi:10.1007/s10973-019-08637-3

Das, S. K., Dao, T. P., & Perumal, T. (Eds.). (2021a). *Nature-Inspired Computing for Smart Application Design*. Springer Singapore.

Das, S. K., Das, S. P., Dey, N., & Hassanien, A. E. (Eds.). (2021c). *Machine learning algorithms for industrial applications*. Springer.

Das, S. K., Samanta, S., Dey, N., & Kumar, R. (Eds.). (2020). *Design frameworks for wireless networks*. Springer Singapore.

Das, S. K., Samanta, S., Dey, N., Patel, B. S., & Hassanien, A. E. (Eds.). (2021b). *Architectural Wireless Networks Solutions and Security Issues. Springer Singapore*. Pte. Limited.

Das, S. K., Tripathi, S., & Burnwal, A. P. (2015). Intelligent energy competency multipath routing in wanet. In *Information systems design and intelligent applications* (pp. 535–543). Springer.

De, D., Mukherjee, A., Das, S. K., & Dey, N. (Eds.). (2020). *Nature inspired computing for wireless sensor networks*. Springer.

Jang, J., & Lee, S. S. (2000). Theoretical and experimental study of MHD (magnetohydrodynamic) micropump. *Sensors and Actuators. A, Physical, 80*, 84–89.

Kowalewski, T. A., & Blonski, S. (2009). Turbulence in micro-channels. *Fluid Mechanics and Thermodynamics, 7th World Conference on Experimental Heat Transfer*, Krakow, Poland.

Kowalewski, T. A., Blonski, S., & Korczyk, P. M. (2006). Turbulent flow in a micro-channel. *Microchannels and Minichannels, Fourth International Conference on Nanochannels*, Limerick, Ireland.

Laster, D. J., & Santiago, J. G. (2004). A review of micropumps. *Journal of Micromechanics and Microengineering, 14*, R35–R64.

Li, H., & Olsen, M. G. (2006). Examination of large-scale structures in turbulent microchannel flow. *Experiments in Fluids, 40*, 733–743.

Natrajan, V. K., & Christensen, K. T. (2010). The impact of surface roughness on flow through a rectangular microchannel from the laminar to turbulent regimes. *Microfluidics and Nanofluidics, 9*, 95–121.

Rana, A., Reza, M., & Shit, G. C. (2022). Electromagnetohydrodynamic thermo-fluidic transport in a porous microchannel with wall roughness. *Colloids and Surfaces. A, Physicochemical and Engineering Aspects, 650*, 129336.

Reza, M., & Rana, A. (2020). *Analysis of Exact Solutions of Electromagnetohydrodynamic Flow and Heat Transfer of Non-Newtonian Casson Fluid in Microchannel with Viscous Dissipation and Joule Heating*. Advances in Fluid Mechanics and Solid Mechanics. doi:10.1007/978-981-15-0772-4_11

Reza, M., Rana, A., & Patra, R. (2020). Thermal Characteristics of a casson fluid in microchannel through porous medium with an interfacial slip with electromagnetohydrodynamic effect. *ISHMT Digital Library, 2019*, 959–964. doi:10.1615/IHMTC-2019.1610

Reza, M., Rana, A., & Shit, G. C. (2021). Thermo-fluidic transport of electromagnetohydrodynamic flow in a corrugated porous medium microchannel. *The European Physical Journal Plus, 136*(5), 496.

Rezaei, O. Akbari, O.A. Marzban, A. Toghraie, D. Pourfattah, F. and Ramin Mashayekhi, R. (2017). The numerical investigation of heat transfer and pressure drop of turbulent flow in a triangular microchannel, *Physica E: Low-dimensional Systems and Nanostructures*. doi:10.1016/j.physe.2017.06.013

Sharp, K. V., & Adrian, R. J. (2004). Transition from laminar to turbulent flow in liquid filled microtubes. *Experiments in Fluids*, *36*, 741–747.

Stone, H. A., Stroock, A. D., & Ajdari, A. (2004). Engineering flows in small devices microfluidics toward a lab-on-a-chip. *Annual Review of Fluid Mechanics*, *36*(1), 381–411. doi:10.1146/annurev.fluid.36.050802.122124

Wang, L., Jian, Y., Liu, Q., Li, F., & Chang, L. (2016). Electromagnetohydrodynamic flow and heat transfer of third grade fluids between two micro-parallel plates. *Colloids and Surfaces. A, Physicochemical and Engineering Aspects*, *494*, 87–97.

West, J., Karamata, B., Lillis, B., Gleeson, J. P., & Alderman, J. K. (2002). Application of magnetohydrodynamic actuation to continuous flow chemistry. *Lab on a Chip*, 2, 224–230.

Weston, M. C., Gerner, M. D., & Fritsch, I. (2010). Magnetic Fields for Fluid Motion. *Analytical Chemistry*, *82*(9), 3411–3418. doi:10.1021/ac901783n PMID:20380431

Compilation of References

Abadía, J. J. P., Walther, C., Osman, A., & Smarsly, K. (2022). A systematic survey of Internet of Things frameworks for smart city applications. *Sustainable Cities and Society*, 103949.

Abbas, F., & Fan, P. (2018). Clustering-based reliable low-latency routing scheme using ACO method for vehicular networks. *Vehicular Communications*, *12*, 66–74.

Abd El-Monsef, M. E., El-Deeb, S. N., & Mahmoud, R. A. (1983). b-open sets and b-continuous mappings. *Bull. Fac. Assuit Univ.*, *12*, 77–90.

Abdalla, S., Al-Ameer, S. S., & Al-Magaishi, S. H. (2010). Electrical properties with relaxation through human blood. *Biomicrofluidics*, *4*, 034101.

Abdollahzadeh, S., & Navimipour, N. J. (2016). Deployment strategies in the wireless sensor network: A comprehensive review. *Computer Communications*, *91*, 1–16. doi:10.1016/j.comcom.2016.06.003

Abuassba, A. O., Zhang, D., Luo, X., Shaheryar, A., & Ali, H. (2017). Improving classification performance through an advanced ensemble based heterogeneous extreme learning machines. *Computational Intelligence and Neuroscience*.

Afrin, S., Islam, N., & Ahmed, S. (2008). A multivariate model of micro credit and rural women entrepreneurship development in Bangladesh.

Afrin, S., Islam, N., & Ahmed, S. U. (2008). A Multivariate Model of Micro Credit and Rural Women Entrepreneurship Development in Bangladesh. *International Journal of Business and Management*, *3*(8), 169–185.

Agarwal, S., & Lenka, U. (2018). Why research is needed in women entrepreneurship in India: A viewpoint. *International Journal of Social Economics*.

Ahmed, K., & Gregory, M. (2011, December). Integrating wireless sensor networks with cloud computing. In *2011 seventh international conference on mobile ad-hoc and sensor networks* (pp. 364-366). IEEE.

Akehurst, G., & Simarro, E., & MasTur, A. (2012). Women entrepreneurship in small service firms: Motivations, barriers and performance. *Service Industries Journal*, *32*(15), 2489–2505.

Akyildiz, I. F., Pompili, D., & Melodia, T. (2005). Underwater acoustic sensor networks: Research challenges. *Ad Hoc Networks*, *3*(3), 257–279. doi:10.1016/j.adhoc.2005.01.004

Akyildiz, I. F., Su, W., Sankarasubramaniam, Y., & Cayirci, E. (2002). A survey on sensor networks. *Communications Magazine, IEEE*, *40*(8), 102–114. doi:10.1109/MCOM.2002.1024422

Al Matroushi, H., Jabeen, F., Matloub, A., & Tehsin, M. (2020). Push and pull model of women entrepreneurship: empirical evidence from the UAE. *Journal of Islamic Accounting and Business Research.*

Al Qurashi, M., Angelopoulos, C. M., & Katos, V. (2020). An architecture for resilient intrusion detection in ad-hoc networks. *Journal of Information Security and Applications*, *53*, 102530.

Al-Jarrah, M. A., Al-Dweik, A., Kalil, M., & Ikki, S. S. (2018). Decision fusion in distributed cooperative wireless sensor networks. *IEEE Transactions on Vehicular Technology*, *68*(1), 797–811.

Alshahrani, M., & Traore, I. (2019). Secure mutual authentication and automated access control for IoT smart home using cumulative keyed-hash chain. *Journal of information security and applications, 45*, 156-175.

Alsheikh, M. A., Lin, S., Niyato, D., & Tan, H. P. (2014). Machine learning in wireless sensor networks: Algorithms, strategies, and applications. *IEEE Communications Surveys and Tutorials*, *16*(4), 1996–2018.

Al-Shurman, K. M., & Naseem, H. (2014). CVD Graphene growth mechanism on nickel thin films. In *Proceedings of the 2014 COMSOL Conference in Boston.*

Anand, M., Balaji, N., Bharathiraja, N., & Antonidoss, A. (2021). A controlled framework for reliable multicast routing protocol in mobile ad hoc network. *Materials Today: Proceedings.*

Andrijevic´, D. (1986). Semi preopen sets. *Mat. Vesnik*, *38*, 24–32.

Andrijevic´, D. (1996). On b-open sets. *Mat. Vesnik*, *48*, 59–64.

Asheralieva, A., & Niyato, D. (2020). Reputation-based coalition formation for secure self-organized and scalable sharding in iot blockchains with mobile-edge computing. *IEEE Internet of Things Journal*, *7*(12), 11830–11850.

Audretsch, D. (2012). Entrepreneurship research. *Management Decision.*

Averna, D., & Bonanno, G. (2003). A three critical points theorem and its applications to the ordinary Dirichlet problem. *Topological Methods in Nonlinear Analysis*, *22*(1), 93–103.

Azari, M., Sadeghi, A., & Chakraborty, S. (2020). Electroosmotic flow and heat transfer in a heterogeneous circular microchannel. *Applied Mathematical Modelling*. doi:10.1016/j.apm.2020.06.022

Babo, R., Dey, N., & Ashour, A. S. (Eds.). (2021). *Workgroups E-Assessment: Planning, Implementing and Analysing Frameworks*. Springer Singapore.

Babu, K. R., Suja, G. J., Samuel, P., & Jos, S. (2015, September). Performance Analysis of Big Data Gathering in Wireless Sensor Network Using an EM Based Clustering Scheme. In *2015 Fifth International Conference on Advances in Computing and Communications (ICACC)* (pp. 109-113). IEEE.

Bar, C. (2014). Green-hyperbolic operators on globally hyperbolic spacetimes.

Bar, C., & Ginoux, N. (2011). Classical and quantum fields on Lorentzian manifolds.

Bar, C., Ginoux, N., & Pfaffle, F. (2007). *Wave equations on Lorentzian manifolds and quantization. Mat. Soc.* EMS.

Baron, R. A., & Shane, S. (2007). Entrepreneurship: A process perspective. *The psychology of entrepreneurship*, 19-39.

Bazdar, H., Toghraie, D., Pourfattah, F., Akbari, O. A., Nguyen, H. M., & Asadi, A. (2019). Numerical investigation of turbulent flow and heat transfer of nanofluid inside a wavy microchannel with different wave lengths. *Journal of Thermal Analysis and Calorimetry*. doi:10.1007/s10973-019-08637-3

Beem, J. K., Ehrlich, P. E., & Easley, K. L. (1996). *Global Lorentzian Geometry*. Marcel Dekker Inc.

Bera, S., Chattopadhyay, M., & Dan, P. K. (2017). A two-stage novel approach using centre ordering of vectors on agglomerative hierarchical clustering for manufacturing cell formation. *Proceedings of the Institution of Mechanical Engineers. Part B, Journal of Engineering Manufacture*, *232*(14), 2651–2662.

Bernal, A. N., & Sanchez, M. (2007). Globally hyperbolic spacetime can be defined as 'causal' instead of 'strongly causal'. *Classical and Quantum Gravity*, *24*(3), 745–749.

Bhattacharya, A., & Sinha, K. (2017). An efficient protocol for load-balanced multipath routing in mobile ad hoc networks. *Ad Hoc Networks*, *63*, 104–114.

Bhatt, C., Dey, N., & Ashour, A. S. (Eds.). (2017). *Internet of things and big data technologies for next generation healthcare*.

Bhuyan, M., Alam, S., Uddin, M., Islam, M., Bipasha, F. A., & Hossain, S. S. (2016). Synthesis of graphene. *International Nano Letters*, *6*(2), 65–83.

Bimol, S., Saikia, M., & Devi, L. P. (2014, March). Achieving knowledge management through Cloud Computing: A case in higher education. In *2014 International Conference on Computing for Sustainable Global Development (INDIACom)* (pp. 222-227). IEEE.

Binh, H. T. T., & Dey, N. (Eds.). (2018). *Soft computing in wireless sensor networks*. CRC Press.

Binh, H. T. T., Hanh, N. T., Nghia, N. D., & Dey, N. (2020). Metaheuristics for maximization of obstacles constrained area coverage in heterogeneous wireless sensor networks. *Applied Soft Computing*, *86*, 105939.

Binh, H. T. T., Hanh, N. T., Van Quan, L., & Dey, N. (2018). Improved cuckoo search and chaotic flower pollination optimization algorithm for maximizing area coverage in wireless sensor networks. *Neural Computing & Applications*, *30*(7), 2305–2317.

Binh, H. T. T., & Nam, N. H. (2018). Introduction to coverage optimization in wireless sensor networks. In *Soft computing in wireless sensor networks* (pp. 115–136). Chapman and Hall/CRC.

Black, Charles. (2011). High-Performance Microelectronics. *Porous Polymers,* 359-385. Doi:. doi:10.1002/9780470929445.ch10

Bonder, J. F., & Silva, A. (2009). *The concentration-compactness principle for variable exponent spaces and applications*. arXiv preprint arXiv:0906.1922.

Bonder, J. F., Saintier, N., & Silva, A. (2018). The concentration-compactness principle for fractional order Sobolev spaces in unbounded domains and applications to the generalized fractional Brezis–Nirenberg problem. *Nonlinear Differential Equations and Applications NoDEA*, *25*(6), 1–25.

Borkar, G. M., Patil, L. H., Dalgade, D., & Hutke, A. (2019). A novel clustering approach and adaptive SVM classifier for intrusion detection in WSN: A data mining concept. *Sustainable Computing: Informatics and Systems*, *23*, 120–135. doi:10.1016/j.suscom.2019.06.002

Bosch, P., De Schepper, T., Zeljković, E., Famaey, J., & Latré, S. (2020). Orchestration of heterogeneous wireless networks: State of the art and remaining challenges. *Computer Communications*, *149*, 62–77. doi:10.1016/j.comcom.2019.10.008

Brogan, J., Baskaran, I., & Ramachandran, N. (2018). Authenticating health activity data using distributed ledger technologies. *Computational and Structural Biotechnology Journal*, *16*, 257–266.

Bruyat, C., & Julien, P. A. (2001). Defining the field of research in entrepreneurship. *Journal of Business Venturing*, *16*(2), 165–180.

Buren, M., Jian, Y., & Chang, L. (2014). Electromagnetohydrodynamic flow through a microparallel channel with corrugated walls. *Journal of Physics. D, Applied Physics*, *47*, 425–501.

Caffarelli, L., & Silvestre, L. (2007). An extension problem related to the fractional Laplacian. *Communications in Partial Differential Equations*, *32*(8), 1245–1260.

Caldas M, Jafari S, Latif RM, (2008). Applications of b-open sets and (b,s)-continuous functions. *King Fahd University of Petroleum & Minerals, Dept. of Math. Sci. Technical Report Series* TR 400.

Canada-Bago, J., Fernandez-Prieto, J. A., Gadeo-Martos, M. A., & Velasco, J. R. (2010). A new collaborative knowledge-based approach for wireless sensor networks. *Sensors (Basel)*, *10*(6), 6044–6062.

Cantizano, N. A., & Silva, A. (2019). Three solutions for a nonlocal problem with critical growth. *Journal of Mathematical Analysis and Applications*, *469*(2), 841–851.

Cao, B., Zhao, J., Lv, Z., Liu, X., Kang, X., & Yang, S. (2018). Deployment optimization for 3D industrial wireless sensor networks based on particle swarm optimizers with distributed parallelism. *Journal of Network and Computer Applications*, *103*, 225–238.

Carter, B. (1971). Causal structure in space-time. *General Relativity and Gravitation*, *1*(4), 349–391.

Chakraborty, R., Dey, R., & Chakraborty, S. (2013). Thermal characteristics of electromagnetohydrodynamic flows in narrow channels with viscous dissipation and Joule heating under constant wall heat flux. *International Journal of Heat and Mass Transfer*, *67*, 1151–1162.

Chakraborty, S., & Paul, D. (2006). Microchannel flow control through a combined electromagnetohydrodynamic transport. *Journal of Physics. D, Applied Physics*, *39*, 5364–5371.

Chander, B., Kumar, B., & Kumaravelan, P. (2018). Analysis of Machine Learning in Wireless Sensor Network. *IACSIT International Journal of Engineering and Technology*, *7*, 185–192.

Chandrakar, P. (2019). A secure remote user authentication protocol for healthcare monitoring using wireless medical sensor networks. *International Journal of Ambient Computing and Intelligence*, *10*(1), 96–116. doi:10.4018/IJACI.2019010106

Chang, L., & Hu, Y. H. (2019). 3D Channel-structured graphene as efficient electrodes for capacitive deionization. *Journal of Colloid and Interface Science*, *538*, 420–425.

Chatterjee, S., Hore, S., Dey, N., Chakraborty, S., & Ashour, A. S. (2017). Dengue fever classification using gene expression data: a PSO based artificial neural network approach. In *Proceedings of the 5th international conference on frontiers in intelligent computing: theory and applications* (pp. 331-341). Springer.

Chen, X., Makki, K., Yen, K., & Pissinou, N. (2009). Yen, Pissinou N. Sensor Network Security: A Survey. *IEEE Communications Surveys and Tutorials*, *11*(2), 52–73. doi:10.1109/SURV.2009.090205

Choquet-Bruhat, Y. (1971). Problemede Cauchy pour le systeme integro differentiel d'Einstein-Liouville. *Annales de l'Institut Fourier*, *21*(3), 181–201. doi:10.5802/aif.385

Choquet-Bruhat, Y., & Geroch, R. (1969). Global aspects of the Cauchy problem in general relativity. *Communications in Mathematical Physics*, *14*(4), 329–335. doi:10.1007/BF01645389

Choudhury, S., Majumder, A., & Sarkar, D. (2018). Enhanced-Ant-AODV for optimal route selection in mobile ad-hoc network. *Journal of King Saud University-Computer and Information Sciences*.

Choudhury, B. S., & Mondal, H. S. (2014). *Continuous representation of a globally hyperbolic spacetime with non-compact Cauchy surfaces*. Anal. Math. Phys. Doi:10.100713324-014-0093-x

Clarke, C. J. S. (1986). *Singularities: Global and local aspects in topological properties and global structure of spacetime*. Plenum Press.

Coleman, J. (2012). Liquid Exfoliation of Defect-Free Graphene. *Accounts of Chemical Research*, *46*. Doi:10.1021/ar300009f

Corson, H. H., & Michael, E. (1964). Metrizability of certain countable unions. *Illinois J. Math*, *8*, 351–360.

Cunningham, J. B., & Lischeron, J. (1991). Defining entrepreneurship. *Journal of Small Business Management*, *29*(1), 45–61.

Das, S. K., Samanta, S., Dey, N., & Kumar, R. (Eds.). (2020). Design frameworks for wireless networks. Springer Singapore.

Das, S. K., Samanta, S., Dey, N., Patel, B. S., & Hassanien, A. E. (Eds.). (2021). Architectural Wireless Networks Solutions and Security Issues. Springer Singapore Pte. Limited. doi:10.1016/j.sysarc.2022.102638

Das, H., Rout, J. K., Moharana, S. C., & Dey, N. (Eds.). (2020). *Applied Intelligent Decision Making in Machine Learning*. CRC Press.

Das, S. K. (2021). Smart Design and Its Applications: Challenges and Techniques. In *Nature-Inspired Computing for Smart Application Design* (pp. 1–6). Springer.

Das, S. K. (2021). Smart Design and Its Applications: Challenges and Techniques. In *Nature-Inspired Computing for Smart Application Design,* (pp. 1–6). Springer.

Das, S. K., Dao, T. P., & Perumal, T. (Eds.). (2021a). *Nature-Inspired Computing for Smart Application Design*. Springer Singapore.

Das, S. K., Das, A., Sinha, H. K., Das, S. P., & Nayak, S. (2021). Game theory based optimal decision-making system. In *Intelligent and Cloud Computing* (pp. 761–769). Springer.

Das, S. K., Das, B., & Burnawal, A. P. (2014). Intelligent energy competency routing scheme for wireless sensor network. *International Journal of Research in Computer Applications and Robotics*, *2*(3), 79–84.

Das, S. K., Das, S. P., Dey, N., & Hassanien, A. E. (Eds.). (2021). *Machine learning algorithms for industrial applications*. Springer.

Das, S. K., & Giacalone, M. (Eds.). (2022). *Fuzzy Optimization Techniques in the Areas of Science and Management*. CRC Press. doi:10.1201/b23138

Das, S. K., Kumar, A., Das, B., & Burnwal, A. P. (2013). Ethics of reducing power consumption in wireless sensor networks using soft computing techniques. *International Journal of Advanced Computer Research*, *3*(1), 301.

Das, S. K., Maheswari, V., & Sharma, A. (2021). Wireless Networks: Applications, Challenges, and Security Issues. In *Architectural Wireless Networks Solutions and Security Issues* (pp. 1–10). Springer.

Das, S. K., Patra, N., Das, B. R., Das, S. P., & Nayak, S. (2021). Fuzzification-Based Routing Protocol for Wireless Ad Hoc Network. In *Intelligent and Cloud Computing* (pp. 237–245). Springer.

Das, S. K., Samanta, S., Dey, N., & Kumar, R. (Eds.). (2020). *Design frameworks for wireless networks*. Springer Singapore. doi:10.1007/978-981-13-9574-1

Das, S. K., Samanta, S., Dey, N., Patel, B. S., & Hassanien, A. E. (Eds.). (2021). *Architectural Wireless Networks Solutions and Security Issues*. Springer Singapore Pte. Limited.

Das, S. K., Samanta, S., Dey, N., Patel, B. S., & Hassanien, A. E. (Eds.). (2021). *Architectural Wireless Networks Solutions and Security Issues. Springer Singapore*. Pte. Limited.

Das, S. K., & Tripathi, S. (2017). Energy efficient routing formation technique for hybrid ad hoc network using fusion of artificial intelligence techniques. *International Journal of Communication Systems*, *30*(16), e3340. doi:10.1002/dac.3340

Das, S. K., & Tripathi, S. (2018). Adaptive and intelligent energy efficient routing for transparent heterogeneous ad-hoc network by fusion of game theory and linear programming. *Applied Intelligence*, *48*(7), 1825–1845. doi:10.100710489-017-1061-6

Das, S. K., & Tripathi, S. (2018). Intelligent energy-aware efficient routing for MANET. *Wireless Networks*, *24*(4), 1139–1159. doi:10.100711276-016-1388-7

Das, S. K., & Tripathi, S. (2019). Energy efficient routing formation algorithm for hybrid ad-hoc network: A geometric programming approach. *Peer-to-Peer Networking and Applications*, *12*(1), 102–128. doi:10.100712083-018-0643-3

Das, S. K., Tripathi, S., & Burnwal, A. P. (2015). Intelligent energy competency multipath routing in wanet. In *Information systems design and intelligent applications* (pp. 535–543). Springer.

Das, S. K., Tripathi, S., & Burnwal, A. P. (2015). *Intelligent energy competency multipath routing in wanet. Information systems design and intelligent applications*. Springer.

De Cock, M., Dowsley, R., Horst, C., Katti, R., Nascimento, A. C., Poon, W. S., & Truex, S. (2017). Efficient and private scoring of decision trees, support vector machines and logistic regression models based on pre-computation. *IEEE Transactions on Dependable and Secure Computing*, *16*(2), 217–230.

De Nápoli, P. L., Bonder, J. F., & Silva, A. (2009). Multiple solutions for the p-Laplace operator with critical growth. Nonlinear Analysis: Theory. *Methods & Applications*, *71*(12), 6283–6289.

De, D., Mukherjee, A., Das, S. K., & Dey, N. (Eds.). (2020). Nature inspired computing for wireless sensor networks. Springer.

Debe, M., Salah, K., Rehman, M. H. U., & Svetinovic, D. (2019). IoT public fog nodes reputation system: A decentralized solution using Ethereum blockchain. *IEEE Access: Practical Innovations, Open Solutions*, *7*, 178082–178093.

De, D., Mukherjee, A., Das, S. K., & Dey, N. (2020). Wireless sensor network: applications, challenges, and algorithms. In *Nature inspired computing for wireless sensor networks* (pp. 1–18). Springer.

De, D., Mukherjee, A., Das, S. K., & Dey, N. (Eds.). (2020). *Nature inspired computing for wireless sensor networks*. Springer. doi:10.1007/978-981-15-2125-6

Delen, D., Zaim, H., Kuzey, C., & Zaim, S. (2013). A comparative analysis of machine learning systems for measuring the impact of knowledge management practices. *Decision Support Systems, 54*(2), 1150–1160.

Devi, R., Jha, R. K., Gupta, A., Jain, S., & Kumar, P. (2017). Implementation of intrusion detection system using adaptive neuro-fuzzy inference system for 5G wireless communication network. *AEÜ. International Journal of Electronics and Communications, 74*, 94–106. doi:10.1016/j.aeue.2017.01.025

Dey, N., Ashour, A. S., Beagum, S., Sifaki Pistola, D., Gospodinov, M., Peneva Gospodinova, E., Manuel, R. S., & Tavares, J. (2015). Parameter optimization for local polynomial approximation based intersection confidence interval filter using genetic algorithm: An application for brain MRI image de-noising. *Journal of Imaging, 1*(1), 60–84. doi:10.3390/jimaging1010060

Dey, N., Ashour, A. S., Shi, F., Fong, S. J., & Sherratt, R. S. (2017). Developing residential wireless sensor networks for ECG healthcare monitoring. *IEEE Transactions on Consumer Electronics, 63*(4), 442–449. doi:10.1109/TCE.2017.015063

Dey, N., Hassanien, A. E., Bhatt, C., Ashour, A., & Satapathy, S. C. (Eds.). (2018). *Internet of things and big data analytics toward next-generation intelligence,* (Vol. 35). Springer.

Dey, N., & Tamane, S. (Eds.). (2018). *Big data analytics for smart and connected cities*. IGI Global.

Di Nezza, E., Palatucci, G., & Valdinoci, E. (2012). Hitchhiker's guide to the fractional Sobolev spaces. *Bulletin des Sciences Mathématiques, 136*(5), 521–573.

Doherty, M. W., Michl, J., Dolde, F., Jakobi, I., Neumann, P., Manson, N. B., & Wrachtrup, J. (2014). Measuring the defect structure orientation of a single NV- centre in diamond. *New Journal of Physics, 16*. doi:10.1088/1367-2630/16/6/063067

Domínguez Gonzalez, R. V., & Martins, M. F. (2014). Knowledge management: An analysis from the organizational development. *Journal of Technology Management & Innovation, 9*(1), 131–147. doi:10.4067/S0718-27242014000100011

Dragonas, V., Tsoumanis, G., Koufoudakis, G., Papamichail, A., Oikonomou, K., & Stavrakakis, I. (2020). A Fairness-Aware topology independent TDMA MAC policy in time constrained wireless ad hoc networks. *Computer Networks, 171*, 107157.

Duque, S., & Omar, M. N. (2015). Using data mining algorithms for developing a model for intrusion detection system (IDS). *Procedia Computer Science, 61*, 46–51. doi:10.1016/j.procs.2015.09.145

Dvalishvili, B. (2004). Connectedness of a fine topology and localization in bitopological spaces. *Georgian Mathematical Journal, 11*(4), 713–732. doi:10.1515/GMJ.2004.713

Eckhardt, J. T., & Shane, S. A. (2003). Opportunities and entrepreneurship. *Journal of Management, 29*(3), 333–349.

Ekeland, I. (1974). On the variational principle. *Journal of Mathematical Analysis and Applications*, *47*(2), 324–353.

Ekici, E. (2005). On c-Urysohn spaces. *Adv. Stud. Contemp. Math*, *11*(2), 219–226.

Ekici, E. (2008). Generalization of weakly clopen and strongly h-bcontinuous functions. *Chaos, Solitons, and Fractals*, *38*, 79–88.

Ekici, E., & Noiri, T. (2009). Decompositions of continuity, a-continuity and AB-continuity. *Chaos, Solitons, and Fractals*, *41*, 2055–2061.

Eksert, M. L., Yücel, H., & Onur, E. (2021). Intra-and inter-cluster link scheduling in CUPS-based ad hoc networks. *Computer Networks*, *185*, 107659.

El Bekri, M., & Diouri, O. (2019). Pso based intrusion detection: A pre-implementation discussion. *Procedia Computer Science*, *160*, 837–842. doi:10.1016/j.procs.2019.11.002

El Hindi, K., AlSalman, H., Qasem, S., & Al Ahmadi, S. (2018). Building an ensemble of fine-tuned naive Bayesian classifiers for text classification. *Entropy (Basel, Switzerland)*, *20*(11), 857. doi:10.3390/e20110857 PMID:33266581

Elbasiony, R. M., Sallam, E. A., Eltobely, T. E., & Fahmy, M. M. (2013). A hybrid network intrusion detection framework based on random forests and weighted k-means. *Ain Shams Engineering Journal*, *4*(4), 753–762. doi:10.1016/j.asej.2013.01.003

El-Bendary, M. A. (2015). *Developing security tools of WSN and WBAN networks applications*. Springer Japan.

El-Showk, S. (2016). *Making better and cheaper solar cells*. Nature Middle East. doi:10.1038/nmiddleeast.2016.4

Evans, C. L. (1998). *Partial Differential Equations*. American Mathem. Soc.

Fang, W., Tan, X., & Wilbur, D. (2020). Application of intrusion detection technology in network safety based on machine learning. *Safety Science*, *124*, 104604. doi:10.1016/j.ssci.2020.104604

Fazalbhoy, S. (2014). Women entrepreneurship as the way for economic development. *Annual Research Journal of Symbiosis Centre for Management Studies*, *2*(1), 117–127.

Fescenko, I., Jarmola, A., Savukov, I., Kehayias, P., Smits, J., Damron, J., & Acosta, V. M. (2020). Diamond magnetometer enhanced by ferrite flux concentrators. *Physical Review Research*, *2*(2), 023394.

Filippov, V. (1998). Basic topological structures of ordinary differential equations, Mathematics and its Applications. (Volume 32). Kluwer Academic Publishers, Dordrecht.

Finke, I., & Will, M. (2003). Motivation for knowledge management. In *Knowledge Management* (pp. 66–91). Springer.

Fong, S. J., Li, G., Dey, N., Crespo, R. G., & Herrera-Viedma, E. (2020). Composite Monte Carlo decision making under high uncertainty of novel coronavirus epidemic using hybridized deep learning and fuzzy rule induction. *Applied Soft Computing*, *93*, 106282.

Fortunato, M. W., & Alter, T. (2015). Community entrepreneurship development: An introduction. *Community Development (Columbus, Ohio)*, *46*(5), 444–455.

Friedlander, F. G. (1976). *The wave equations on a curved space-time*. Cambridge University Press.

Fu, X. Y., Chen, Z. D., Han, D. D., Zhang, Y. L., Xia, H., & Sun, H. B. (2020). Laser fabrication of graphene-based supercapacitors. *Photonics Research*, *8*(4), 577–588.

Garcia Azorero, J., & Peral Alonso, I. (1991). Multiplicity of solutions for elliptic problems with critical exponent or with a nonsymmetric term. *Transactions of the American Mathematical Society*, *323*(2), 877–895.

Gawas, M. A., & Govekar, S. S. (2019). A novel selective cross layer based routing scheme using ACO method for vehicular networks. *Journal of Network and Computer Applications*, *143*, 34–46. doi:10.1016/j.jnca.2019.05.010

Ge, C., Liu, Z., & Fang, L. (2020). A blockchain based decentralized data security mechanism for the Internet of Things. *Journal of Parallel and Distributed Computing*, *141*, 1–9.

Geroch, R. (1970). Domain of dependence. *Journal of Mathematical Physics*, *11*, 437–449.

Gerstner, E. (2010). Nobel prize 2010: Andre geim & konstantin novoselov. *Nature Physics*, *6*(11), 836–836. doi:10.1038/nphys1836

Giri, R. K., Choudhuri, D., & Soni, A. (2018). A problem involving a nonlocal operator. *Frac. Diffe. Calc.*, *8*(1), 177–190.

González-Briones, A., Castellanos-Garzón, J. A., Mezquita Martín, Y., Prieto, J., & Corchado, J. M. (2018). A framework for knowledge discovery from wireless sensor networks in rural environments: A crop irrigation systems case study. *Wireless Communications and Mobile Computing*, 2018.

Gruber, A., Drabenstedt, A., Tietz, C., Fleury, L., Wrachtrup, J., & Borczyskowski, C. V. (1997). Scanning confocal optical microscopy and magnetic resonance on single defect centers. *Science*, *276*(5321), 2012–2014. doi:10.1126cience.276.5321.2012

Guo, S., Wang, F., Zhang, N., Qi, F., & Qiu, X. (2020). Master-slave chain based trusted cross-domain authentication mechanism in IoT. *Journal of Network and Computer Applications*, *172*, 102812.

Gupta, N., Gupta, S., Khosravy, M., Dey, N., Joshi, N., Crespo, R. G., & Patel, N. (2021). Economic IoT strategy: The future technology for health monitoring and diagnostic of agriculture vehicles. *Journal of Intelligent Manufacturing*, *32*(4), 1117–1128.

Gupta, S., Mittal, M., & Padha, A. (2017, December). Predictive analytics of sensor data based on supervised machine learning algorithms. In *2017 International Conference on Next Generation Computing and Information Systems (ICNGCIS)* (pp. 171-176). IEEE.

Gu, X., Zhao, Y., Sun, K., Vieira, C. L., Jia, Z., Cui, C., & Huang, S. (2019). Method of ultrasound-assisted liquid-phase exfoliation to prepare graphene. *Ultrasonics Sonochemistry*, *58*, 104630.

Hamzah, H. A., & Alameer, N. T. A. (2020) β-sets in topological spaces and $\tau\beta$s topology, IOP *Conf. Series: Materials Science and Engineering*, *928*, 042034, doi:10.1088/1757-899X/928/4/04203

Hani, U., Rachmania, I. N., Setyaningsih, S., & Putri, R. C. (2012). Patterns of Indonesian women entrepreneurship. P*rocedia Economics and Finance,* 4, 274–285.

Hao, R., Yang, H., & Zhou, Z. (2019). Driving Behavior Evaluation Model Base on Big Data from Internet of Vehicles. I*nternational Journal of Ambient Computing and Intelligence,* 1*0(*4), 78–95.

Hao, R., Yang, H., & Zhou, Z. (2019). Driving Behavior Evaluation Model Base on Big Data From Internet of Vehicles. *International Journal of Ambient Computing and Intelligence*, *10*(4), 78–95. doi:10.4018/IJACI.2019100105

Hassanien, A. E. (Ed.). (2021). *Architectural Wireless Networks Solutions and Security Issues*. Springer Singapore Pte. Limited.

Hatcher, A. (2000). *Algebraic topology*. Cambridge University Press.

Hawking, S. W., & Ellis, G. F. R. (1973). *The Large-Scale Structure of Space-Time*. Cambridge University Press. doi:10.1017/CBO9780511524646

Hawking, S. W., & Penrose, R. (1970). The singularities of gravitational collapse and cosmology. *Proceedings of the Royal Society of London. Series A, Mathematical and Physical Sciences*, *314*(1519), 529–548. doi:10.1098/rspa.1970.0021

Hidayanto, B. C., Muhammad, R. F., Kusumawardani, R. P., & Syafaat, A. (2017). Network intrusion detection systems analysis using frequent item set mining algorithm FP-max and apriori. *Procedia Computer Science*, *124*, 751–758. doi:10.1016/j.procs.2017.12.214

Hijazi, S. (2017). *Big Data and Knowledge Management: A Possible Course to Combine Them Together*. Association Supporting Computer Users in Education.

Holden, H., & Karlsen, K. H. (Eds.). (2012). *Nonlinear Partial Differential Equations*. Springer.

Hounnonkpe, R. A., & Minguzzi, E. (2019). Globally hyperbolic spacetime can be defined without causal condition.

Hummers, W. S. Jr, & Offeman, R. E. (1958). Preparation of graphitic oxide. *Journal of the American Chemical Society*, *80*(6), 1339–1339.

Ibrahim, K. M., Irannejad, M., Ramadhan, A., Alayak, W., Sandeson, J., Cui, B., & Yavuz, M. (2014, August). Ultrafast light interaction with Graphene oxide aqueous solution. In *14th IEEE International Conference on Nanotechnology*, (pp. 830-831). IEEE.

Idhammad, M., Afdel, K., & Belouch, M. (2018). Distributed intrusion detection system for cloud environments based on data mining techniques. *Procedia Computer Science*, *127*, 35–41. doi:10.1016/j.procs.2018.01.095

Idrus, S., Pauzi, N. M., & Munir, Z. A. (2014). The effectiveness of training model for women entrepreneurship program. *Procedia: Social and Behavioral Sciences*, *129*, 82–89.

Irngartinger, H., Oeser, T., Jahn, R., & Kallfaß, D. (1992). Pyramidalization of the sp2-Hybridized Carbon Atoms in 1, 4-Bridged 2, 5-Cyclohexadienes. *Chemische Berichte*, *125*(9), 2067–2073.

Islam, A., & Shin, S. Y. (2019). BUAV: A blockchain based secure UAV-assisted data acquisition scheme in Internet of Things. *Journal of Communications and Networks (Seoul)*, *21*(5), 491–502.

Iyer, V. R., Dey, N., & Chakraborty, S. (2015). Advent of Information Technology in the world of Tourism. In *Emerging innovative marketing strategies in the tourism industry* (pp. 44–53). IGI Global.

Jain, U., & Hussain, M. (2020). Underwater Wireless Sensor Networks. In *Handbook of Computer Networks and Cyber Security* (pp. 227–245). Springer. doi:10.1007/978-3-030-22277-2_9

Jangirala, S., Das, A. K., & Vasilakos, A. V. (2019). Designing secure lightweight blockchain-enabled RFID-based authentication protocol for supply chains in 5G mobile edge computing environment. *IEEE Transactions on Industrial Informatics*, *16*(11), 7081–7093.

Jang, J., & Lee, S. S. (2000). Theoretical and experimental study of MHD (magnetohydrodynamic) micropump. *Sensors and Actuators. A, Physical*, *80*, 84–89.

Jan, S. U., Lee, Y. D., Shin, J., & Koo, I. (2017). Sensor fault classification based on support vector machine and statistical time-domain features. *IEEE Access: Practical Innovations, Open Solutions*, *5*, 8682–8690.

Jat, D. S., Bishnoi, L. C., & Nambahu, S. (2018). An Intelligent Wireless QoS Technology for Big Data Video Delivery in WLAN. *International Journal of Ambient Computing and Intelligence*, *9*(4), 1–14. doi:10.4018/IJACI.2018100101

Jayakumar Loganathan, J., & Subbiah, J. (2020, February). Energy aware Dynamic Mode Decision for Cellular D2D Communications by using Integrated Multi-Criteria Decision Making model. *International Journal of Ambient Computing and Intelligence*, *11*(3), 7.

Jayakumar, L., & Dumka, A., & Janakiraman, S. (2020). Energy Aware Dynamic Mode Decision for Cellular D2D Communications by Using Integrated Multi-Criteria Decision Making Model. *International Journal of Ambient Computing and Intelligence*, *11*(3), 131–151.

Jayakumar, L., Dumka, A., & Janakiraman, S. (2020). Energy aware dynamic mode decision for cellular D2D communications by using integrated multi-criteria decision making model. [IJACI]. *International Journal of Ambient Computing and Intelligence*, *11*(3), 131–151.

Jiao, L., Wu, H., Bie, R., Umek, A., & Kos, A. (2018). Multi-sensor golf swing classification using deep CNN. *Procedia Computer Science*, *129*, 59–65.

Kahate, A. (2014). *Cryptography Techniques. In Cryptography and Network Security* (3rd ed.). McGraw.

Karaveli, S., Gaathon, O., Wolcott, A., Sakakibara, R., Shemesh, O. A., Peterka, D. S., & Englund, D. (2016). Modulation of nitrogen vacancy charge state and fluorescence in nanodiamonds using electrochemical potential. *Proceedings of the National Academy of Sciences of the United States of America*, *113*(15), 3938–3943.

Kesavan, S. (2003). *Functional Analysis and Applications*. New Age Inter. Pvt. Ltd.

Khalili, A., & Sami, A. (2015). SysDetect: A systematic approach to critical state determination for Industrial Intrusion Detection Systems using Apriori algorithm. *Journal of Process Control*, *32*, 154–160. doi:10.1016/j.jprocont.2015.04.005

Khan, K., Mehmood, A., Khan, S., Khan, M. A., Iqbal, Z., & Mashwani, W. K. (2020). A survey on intrusion detection and prevention in wireless ad-hoc networks. *Journal of Systems Architecture*, *105*, 101701. doi:10.1016/j.sysarc.2019.101701

Kodali, R. K., & Chougule, S. (2013). Hybrid key management technique for WSNs. In *International Conference on Heterogeneous Networkingfor Quality, Reliability, Security and Robustness*. Springer. doi: 10.1007/978-3-642-37949-9_74

Koucham, O., Mocanu, S., Hiet, G., Thiriet, J. M., & Majorczyk, F. (2018). Efficient Mining of Temporal Safety Properties for Intrusion Detection in Industrial Control Systems. *IFAC-PapersOnLine*, *51*(24), 1043–1050. doi:10.1016/j.ifacol.2018.09.719

Kowalewski, T. A., & Blonski, S. (2009). Turbulence in micro-channels. *Fluid Mechanics and Thermodynamics, 7th World Conference on Experimental Heat Transfer*, Krakow, Poland.

Kowalewski, T. A., Blonski, S., & Korczyk, P. M. (2006). Turbulent flow in a micro-channel. *Microchannels and Minichannels, Fourth International Conference on Nanochannels*, Limerick, Ireland.

Kratou, M. (n.d.). *Three solutions for a semilinear elliptic boundary value problem*. https://www.ias.ac.in/public/Volumes/pmsc/forthcoming/PMSC-D-17-00359.pdf

Kronheimer, E. H., & Penrose, R. (2008). *On the structure of causal spaces*. Cambridge University Press. doi:10.1017/S030500410004144X

Krzysztoń, M., & Marks, M. (2020). Simulation of watchdog placement for cooperative anomaly detection in bluetooth mesh intrusion detection system. *Simulation Modelling Practice and Theory*, *101*, 102041. doi:10.1016/j.simpat.2019.102041

Kumari V, Raman TCK, (2013) On characterization of b-open sets in a topological space. *International Journal of application or innovation in Engineering & Management*, *2*(12), 229-235.

Kumar, J., Singh, A., & Bhadauria, H. S. (2021). Congestion control load balancing adaptive routing protocols for random waypoint model in mobile ad-hoc networks. *Journal of Ambient Intelligence and Humanized Computing*, *12*(5), 5479–5487.

Kumar, K. (2018). An Overview on the Importance of Chemical Vapour Deposition Technique for Graphene Synthesis. Advanced Science. *Engineering in Medicine*, *10*. Doi:10.1166/asem.2018.2241

Kumar, S. A. (2008). *Entrepreneurship development*. New Age International.

Kuusi, T., Mingione, G., & Sire, Y. (2015). Nonlocal equations with measure data. *Communications in Mathematical Physics*, *337*(3), 1317–1368.

Laster, D. J., & Santiago, J. G. (2004). A review of micropumps. *Journal of Micromechanics and Microengineering*, *14*, R35–R64.

Lee, J. (2011). *Introduction to topological manifolds. Graduate Texts in Mathematics*. Springer.

Lee, K. S., Lee, S. R., Kim, Y., & Lee, C. G. (2017). Deep learning–based real-time query processing for wireless sensor network. *International Journal of Distributed Sensor Networks*, *13*(5).

Leray, J. (1953). *Hyperbolic differential equations*. The Institute for Advanced Study.

Levine, N. (1963). Semi-open sets and semi-continuity in topological space. *The American Mathematical Monthly*, *70*, 36–41.

Li, D., Chen, R., Liu, D., Song, Y., Ren, Y., Guan, Z., & Liu, J. (2022). Blockchain-based authentication for IIoT devices with PUF. *Journal of Systems Architecture*, *130*, 102638.

Li, G., Dong, M., Yang, L. T., Ota, K., Wu, J., & Li, J. (2020). Preserving edge knowledge sharing among IoT services: A blockchain-based approach. *IEEE Transactions on Emerging Topics in Computational Intelligence*, *4*(5), 653–665.

Li, H., & Olsen, M. G. (2006). Examination of large-scale structures in turbulent microchannel flow. *Experiments in Fluids*, *40*, 733–743.

Lima, E., Moraes, J., Oliveira, H., Cerqueira, E., Zeadally, S., & Rosário, D. (2021). Adaptive priority-aware LoRaWAN resource allocation for Internet of Things applications. *Ad Hoc Networks*, *122*, 102598.

Ling, E. (2020). Aspects of C^0 causal theory. *General Relativity and Gravitation*, *52*(6). Doi:10.100710714-020-02708-9

Lions, P. L. (1985). The concentration-compactness principle in the calculus of variations. The limit case, part 1. *Revista Matemática Iberoamericana*, *1*(1), 145–201.

Liu, Y. Jian, Y., & Tan, W. (2018). Entropy generation of electromagnetohydrodynamic (EMHD) flow in a curved rectangular microchannel. *International Journal of Heat and Mass Transfer*, *127*(A), 901-913.

Liu, M., & Yang, J. (2009). Electrokinetic effect of the endothelial glycocalyx layer on two-phase blood flow in small blood vessels. *Microvascular Research*, *78*(1), 14–19. doi:10.1016/j.mvr.2009.04.002

Liu, Y., & Jian, Y. (2019). Electroviscous effect on electromagnetohydrodynamic flows of maxwell fluids in parallel plate microchannels. *Applied Mathematics and Mechanics*, *40*, 1457–1470.

Loganathan, J., & Subbiah, J. (2020). Energy aware Dynamic Mode Decision for Cellular D2D Communications by using Integrated Multi-Criteria Decision Making model. *International Journal of Ambient Computing and Intelligence*, *11*(3). doi:10.4018/978-1-5225-5643-5.ch028

Lou, C., & Zhuang, W. (2016). Energy-efficient routing over coordinated sleep scheduling in wireless ad hoc networks. *Peer-to-Peer Networking and Applications*, *9*(2), 384–396.

Luque, A. M. A. (2016). *Brief Introduction to Knowledge Management in Organizations*. Academic Press.

Maden, C. (2015). A gendered lens on entrepreneurship: Women entrepreneurship in Turkey. *Gender in Management*.

Mahdavi, E., Fanian, A., & Amini, F. (2020). A real-time alert correlation method based on code-books for intrusion detection systems. *Computers & Security*, *89*, 101661. doi:10.1016/j.cose.2019.101661

Majumder, S., & Dey, N. (2022). Artificial Intelligence and Knowledge Management. In *AI-empowered Knowledge Management,* (pp. 85–100). Springer.

Majumder, S., & Mondal, A. (2021). Are chatbots really useful for human resource management? *International Journal of Speech Technology*, *24*(4), 969–977.

Manan, J., Ahmed, A., Ullah, I., Merghem-Boulahia, L., & Gaïti, D. (2019). Distributed intrusion detection scheme for next generation networks. *Journal of Network and Computer Applications*, *147*, 102422. doi:10.1016/j.jnca.2019.102422

Mandhare, V. V., Thool, V. R., & Manthalkar, R. R. (2016). QoS Routing enhancement using metaheuristic approach in mobile ad-hoc network. *Computer Networks*, *110*, 180–191.

Manzanera-Román, S., & Brändle, G. (2016). Abilities and skills as factors explaining the differences in women entrepreneurship. *Suma de negocios, 7*(15), 38-46.

Martinez Dy, A., & Jayawarna, D. (2020). Bios, mythoi and women entrepreneurs: A Wynterian analysis of the intersectional impacts of the COVID-19 pandemic on self-employed women and women-owned businesses. *International Small Business Journal*, *38*(5), 391–403.

Martínez, J. F., Castillejo, P., Zuazua, M., García, A. B., López, L., DaSilva, A., & Hernández, V. (2010). Wireless sensor networks in knowledge management. *Procedia Computer Science*, *1*(1), 2291–2300.

Mashhour, A. S., Abd El-Monsef, M. E., & El-Deeb, S. N. (1982). On pre-continuous and weak pre-continuous mapping. *Proceedings of the Mathematical and Physical Society of Egypt*, *53*, 47–53.

Masliyah, J. H., & Bhattacharjee, S. (2006). Electrokinetic and Colloid Transport Phenomena. Wiley–Interscience. https://doi.org/10.1002/cjce.5450840612.

Mathew, R. V., & Panchanatham, N. (2011). An exploratory study on the work-life balance of women entrepreneurs in South India. *Asian academy of management journal, 16*(2).

Mathew, V. (2010). Women entrepreneurship in Middle East: Understanding barriers and use of ICT for entrepreneurship development. *The International Entrepreneurship and Management Journal, 6*(2), 163–181.

Matin, M. A., & Islam, M. M. (2012). Overview of wireless sensor network. *Wireless Sensor Networks-Technology and Protocols, 1*(3).

Mazini, M., Shirazi, B., & Mahdavi, I. (2019). Anomaly network-based intrusion detection system using a reliable hybrid artificial bee colony and AdaBoost algorithms. *Journal of King Saud University-Computer and Information Sciences, 31*(4), 541–553. doi:10.1016/j.jksuci.2018.03.011

Minguzzi, E. (2019). Lorentzian causality theory. *Living Rev. Relativ., 22*(3). DOI:10. 1007/s41114-019-0019-x.

Minguzzi, E., & Sanchez, M. (2008). In H. Baum & D. Alekseevsky (Eds.), The causal hierarchy of spacetimes. Recent developments in pseudo-Riemannian geometry, ESI Lect. EMS Pub. House, (pp. 299–358). doi:10.4171/051-1/9

Minguzzi, E. (2009). K-Causality coincides with stable causality. *Communications in Mathematical Physics, 290*, 239–248.

Minguzzi, E. (2019). Lorentziaan causality theory. *Living Reviews in Relativity, 22*, 3. https://doi.org/10.1007/s41114-019-0019-x

Mishra, T. K., Dass, A. K., & Panda, S. K. (2018, December). Enhanced Path Planning Model for Anchor-Free Distributed Localization in Wireless Sensor Networks. In *2018 Fifth International Conference on Parallel, Distributed and Grid Computing (PDGC)* (pp. 430-435). IEEE. 10.1109/PDGC.2018.8745802

Mishra, T. K., Kumar, S., & Dass, A. K. (2018, December). An Enhanced Path Planning Model for Anchor-Free Localization in Wireless Sensor Networks. In *2018 International Conference on Information Technology (ICIT)* (pp. 204-209). IEEE.

Mohajan, H. (2017). *The roles of knowledge management for the development of organizations*. Academic Press.

Mohajan, H. K. (2016). Global hyperbolicity in spacetime manifold. *Int. J. Prof. Studies, 1*(1).

Mohammad, M. N., Sulaiman, N., & Muhsin, O. A. (2011). A novel intrusion detection system by using intelligent data mining in weka environment. *Procedia Computer Science, 3*, 1237–1242. doi:10.1016/j.procs.2010.12.198

Mosconi, S., & Squassina, M. (2016). Nonlocal problems at nearly critical growth. *Nonlinear Analysis: Theory. Methods & Applications*, *136*, 84–101.

Moses, C., & Mordi, C. (2010). Women entrepreneurship development in Nigeria: The effect of environmental factors. *BULETINUL Universitatii Petrol-Gaze din Ploieti, 62*(4), 43-52.

Mukit, M. M. H., Nabila, N. J., Islam, M. S., & Abdel-Razzaq, A. I. (2020). Valuing women entrepreneurship in the informal economy in Bangladesh. *ECONOMICS-Innovative and Economic Research*, *8*(2), 37–54.

Mukkamala, S., Sung, A. H., & Abraham, A. (2005). Intrusion detection using an ensemble of intelligent paradigms. *Journal of Network and Computer Applications*, *28*(2), 167–182. doi:10.1016/j.jnca.2004.01.003

Muller, O., Sanchez, M. (2014). An invitation to Lorentzian geometry. *Jahresbericht der Deutschen Mathematiker-Vereinigung, 115*(3).

Munkres, J. R. (2000). Topology. Pearson Education.

Nadiammai, G. V., & Hemalatha, M. J. E. I. J. (2014). Effective approach toward Intrusion Detection System using data mining techniques. *Egyptian Informatics Journal*, *15*(1), 37–50. doi:10.1016/j.eij.2013.10.003

Nakao, K., & Nishide, Y. (2020). The development of social entrepreneurship education in Japan. *Entrepreneurship Education*, *3*(1), 95–117.

Natrajan, V. K., & Christensen, K. T. (2010). The impact of surface roughness on flow through a rectangular microchannel from the laminar to turbulent regimes. *Microfluidics and Nanofluidics*, *9*, 95–121.

Navarro, J. J. B., & Minguzzi, E. (2011) Global hyperbolicity is stable in the interval topology.

Navarro, J. J. B., & Minguzzi, E. (2011). Global hyperbolicity is stable in the interval topology.

Nawaz, F. (2009). Critical factors of women entrepreneurship development in rural Bangladesh.

Nazarov, V. U. (2012). *Band-structure of graphene above the vacuum level.*

Ngan, T. T., Tuan, T. M., Son, L. H., Minh, N. H., & Dey, N. (2016). Decision making based on fuzzy aggregation operators for medical diagnosis from dental X-ray images. *Journal of Medical Systems*, *40*(12), 1–7.

Nguyen, B. H., & Nguyen, V. H. (2016). Promising applications of graphene and graphene-based nanostructures. *Advances in Natural Sciences: Nanoscience and Nanotechnology*, *7*(2), 023002.

Niethammer, C. (2013). Women, entrepreneurship and the opportunity to promote development and business. *Brookings blum roundtable policy brief, 37*, 1-10.

Niu, R., Moore, M., & Klamer, D. (2004). *Decision fusion in a wireless sensor network with a large number of sensors*. Academic Press.

O'Neill, B. (1983). *Semi-Riemannian Geometry*. Academic Press.

Obaji, N. O., & Olugu, M. U. (2014). The role of government policy in entrepreneurship development. *Science Journal of Business and Management*, *2*(4), 109–115.

Obitade, P. O. (2019). Big data analytics: A link between knowledge management capabilities and superior cyber protection. *Journal of Big Data*, *6*(1), 1–28.

Omairi, A., & Ismail, Z. H. (2016). Modeling battery state of charge in wireless sensor networks based on structured multi-layer perceptron. *Journal of Advanced Research in Applied Sciences and Engineering Technology*, *5*(2), 36–45.

Omotayo, F. O. (2015). Knowledge Management as an important tool in Organisational Management: A Review of Literature. *Library Philosophy and Practice (e-journal).*

Osei, C. D., & Zhuang, J. (2020). Rural poverty alleviation strategies and social capital link: The mediation role of women entrepreneurship and social innovation. *SAGE Open*, *10*(2), 2158244020925504.

Özyer, T., Alhajj, R., & Barker, K. (2007). Intrusion detection by integrating boosting genetic fuzzy classifier and data mining criteria for rule pre-screening. *Journal of Network and Computer Applications*, *30*(1), 99–113. doi:10.1016/j.jnca.2005.06.002

Pan, J., Wang, J., Hester, A., Alqerm, I., Liu, Y., & Zhao, Y. (2018). EdgeChain: An edge-IoT framework and prototype based on blockchain and smart contracts. *IEEE Internet of Things Journal*, *6*(3), 4719–4732.

Pan, L., & Li, J. (2010). K-nearest neighbor based missing data estimation algorithm in wireless sensor networks. *Wireless Sensor Network*, *2*(02), 115.

Patel, M. M., & Aggarwal, A. (2013). Security attacks in wireless sensor net- works: A survey. *Intelligent Systems and Signal Processing (ISSP), International Conference on 2013*, 329-333.

Patsariya, M., & Rajavat, A. (2021). Node capability-based route selection on mobile ad hoc network. *Materials Today: Proceedings*.

Peng, C., Luo, M., Wang, H., Khan, M. K., & He, D. (2021). An Efficient Privacy-Preserving Aggregation Scheme for Multidimensional Data in IoT. *IEEE Internet of Things Journal*, *9*(1), 589–600.

Penrose R (1969) Gravtational collapse: the role of general relativity. *Revista del Nuovo Cimento, 1*, 252-276.

Penrose, R. (1972). Techniques of Differential Topology in Relativity. CBMS-NSF Regional Conference Series in Applied Mathematics. SIAM.

Penrose, R. (1965). Gravitational collapse and spacetime singularities. *Physical Review Letters*, *14*(3), 57–59.

Penrose, R. (1972). *Techniques of differential topology in relativity*. Society for Industrial and Applied Mathematics.

Penrose, R. (2002). Gravitational collapse – the role of general relativity. Gen. Rel. Grav, *34*, 1142–1164.

Perlick, V., & Tausk, D. V. (2003). Gravitational lensing from a spacetime perspective. *Living Reviews in Relativity*, *7*. www.livingreviews.org/Irr-2004-9

Phoemphon, S., So-In, C., & Niyato, D. T. (2018). A hybrid model using fuzzy logic and an extreme learning machine with vector particle swarm optimization for wireless sensor network localization. *Applied Soft Computing*, *65*, 101–120.

Pirzado, A. A., Le Normand, F., Romero, T., Paszkiewicz, S., Papaefthimiou, V., Ihiawakrim, D., & Janowska, I. (2019). Few-layer graphene from mechanical exfoliation of graphite-based materials: Structure-dependent characteristics. *ChemEngineering*, *3*(2), 37.

Prasanna, S., & Rao, S. (2012). An overview of wireless sensor networks applications and security. *International Journal of Soft Computing and Engineering*, *2*(2), 2231–2307.

Qu, H., Qiu, Z., Tang, X., Xiang, M., & Wang, P. (2018). Incorporating unsupervised learning into intrusion detection for wireless sensor networks with structural co-evolvability. *Applied Soft Computing*, *71*, 939–951. doi:10.1016/j.asoc.2018.07.044

Rahman, M. A., Hossain, M. S., Loukas, G., Hassanain, E., Rahman, S. S., Alhamid, M. F., & Guizani, M. (2018). Blockchain-based mobile edge computing framework for secure therapy applications. *IEEE Access: Practical Innovations, Open Solutions*, *6*, 72469–72478.

Rajani, N., & Sarada, D. (2008). Women entrepreneurship and support systems. *Studies on Home and Community Science*, *2*(2), 107–112.

Rana, A., Reza, M., & Shit, G. C. (2022). Electromagnetohydrodynamic thermo-fluidic transport in a porous microchannel with wall roughness. *Colloids and Surfaces. A, Physicochemical and Engineering Aspects*, *650*, 129336.

Raschellà, A., Bouhafs, F., Mackay, M., Shi, Q., Ortín, J., Gállego, J. R., & Canales, M. (2020). A dynamic access point allocation algorithm for dense wireless LANs using potential game. *Computer Networks*, *167*, 106991. doi:10.1016/j.comnet.2019.106991

Rasheed, I., & Banka, H. (2018, March). Query Expansion in Information Retrieval for Urdu Language. In *2018 Fourth International Conference on Information Retrieval and Knowledge Management (CAMP)* (pp. 1-6). IEEE.

Rashid, M., Shahzadi, I., & Nadeem, S. (2018). Corrugated walls analysis in microchannels through porous medium under Electromagnetohydrodynamic (EMHD) effects. *Results in Physics*, *9*, 171–182.

Rathore, H. (2016). Case study: A review of security challenges, attacks and trust and reputation models in wireless sensor networks. In *Mapping Biological Systems to Network Systems* (pp. 117–175). Springer.

Rathore, S., Kwon, B. W., & Park, J. H. (2019). BlockSecIoTNet: Blockchain-based decentralized security architecture for IoT network. *Journal of Network and Computer Applications, 143*, 167–177.

Rezaei, O. Akbari, O.A. Marzban, A. Toghraie, D. Pourfattah, F. and Ramin Mashayekhi, R. (2017). The numerical investigation of heat transfer and pressure drop of turbulent flow in a triangular microchannel, *Physica E: Low-dimensional Systems and Nanostructures*. doi:10.1016/j.physe.2017.06.013

Reza, M., & Rana, A. (2020). *Analysis of Exact Solutions of Electromagnetohydrodynamic Flow and Heat Transfer of Non-Newtonian Casson Fluid in Microchannel with Viscous Dissipation and Joule Heating*. Advances in Fluid Mechanics and Solid Mechanics. doi:10.1007/978-981-15-0772-4_11

Reza, M., Rana, A., & Patra, R. (2020a). Thermal Characteristics of a casson fluid in microchannel through porous medium with an interfacial slip with electromagnetohydrodynamic effect. *ISHMT Digital Library, 2019*, 959–964. doi:10.1615/IHMTC-2019.1610

Reza, M., Rana, A., & Patra, R. (2020b). Thermal performance for electromagnetohydrodynamic flow of non-Newtonian Casson fluid through porous microtube. *IACSIT International Journal of Engineering and Technology, 9*(3), 804–811.

Reza, M., Rana, A., & Shit, G. C. (2021). Thermo-fluidic transport of electromagnetohydrodynamic flow in a corrugated porous medium microchannel. *The European Physical Journal Plus, 136*(5), 496.

Ribino, P., Augello, A., Lo Re, G., & Gaglio, S. (2011). A Knowledge Management and Decision Support Model for Enterprises. *Advances in Decision Sciences*.

Ringstrom, H. (2009). The Cauchy problem in general relativity. ESI Lectures in Mathematics and Physics. EMS.

Rizvi, D. R., Amjad, M., & Doja, M. N. (2017). A Study of Wireless Sensor Networks. *Int. Res. J. Eng. Technol, 4*(7), 2078–2081.

Robinson, Y. H., Julie, E. G., Saravanan, K., Kumar, R., & Son, L. H. (2019). FD-AOMDV: Fault-tolerant disjoint ad-hoc on-demand multipath distance vector routing algorithm in mobile ad-hoc networks. *Journal of Ambient Intelligence and Humanized Computing, 10*(11), 4455–4472.

Rossi, P. S., Ciuonzo, D., Kansanen, K., & Ekman, T. (2014). On energy detection for MIMO decision fusion in wireless sensor networks over NLOS fading. *IEEE Communications Letters, 19*(2), 303–306.

Roy, S., Karjee, J., Rawat, U. S., & Dey, N. (2016). Symmetric key encryption technique: A cellular automata based approach in wireless sensor networks. *Procedia Computer Science, 78*, 408–414. doi:10.1016/j.procs.2016.02.082

Rumelt, R. P. (2005). Theory, strategy, and entrepreneurship. In *Handbook of entrepreneurship research* (pp. 11–32). Springer.

Sadi, M. A., & Al-Ghazali, B. M. (2010). Doing business with impudence: A focus on women entrepreneurship in Saudi Arabia. *African Journal of Business Management, 4*(1), 001-011.

Sahingoz, O. K. (2013). Large scale wireless sensor networks with multi- level dynamic key management scheme. *Journal of Systems Architecture, 59*(9), 801–807. doi:10.1016/j.sysarc.2013.05.022

Saito, R., & Kataura, H. (2001). Optical properties and Raman spectroscopy of carbon nanotubes. In *Carbon nanotubes,* (pp. 213–247). Springer.

Samann, C. (2019). Global hyperbolicity for spacetimes with continuous metrics.

Samet, A., Lefevre, E., & Yahia, S. B. (2014, November). Belief function classification with conflict management: Application on forest image. In *2014 Tenth International Conference on Signal-Image Technology and Internet-Based Systems* (pp. 14-20). IEEE. 10.1109/SITIS.2014.34

Sanchez, M. (2006) A note n causally simple and globally hyperbolic spacetimes.

Sanchez, M. (2006). On causality and closed geodesics of compact Lorentzian manifolds and static spacetimes. *Differential Geometry and Its Applications, 24*(1), 21–32. doi:10.1016/j.difgeo.2005.06.008

Sanders, K. (2013). A note on spacelike and timelike compactness. *Clas. And Quan. Grav, 30*(11), 115014. Doi:10.1088/0264-9381/30/11/115014

Sarfaraz, L., Faghih, N., & Majd, A. A. (2014). The relationship between women entrepreneurship and gender equality. *Journal of Global Entrepreneurship Research, 4*(1), 1–11.

Sarkar, S. Ganguly, S., & Chakraborty, S. (2017). Influence of combined electromagnetohydrodynamics on microchannel flow with electrokinetic effect and interfacial slip. *Microfluid. Nanofluid, 21*(3), 1-16.

Sarkar, S., & Datta, R. (2016). A secure and energy-efficient stochastic multipath routing for self-organized mobile ad hoc networks. *Ad Hoc Networks, 37*, 209–227.

Sarkar, S., Ganguly, S., & Dutta, P. (2017). Electrokinetically induced thermofluidic transport of power-law fluids under the influence of superimposed magnetic field. *Chemical Engineering Science, 171*, 391–403.

Schiff, L. I., & Snyder, H. (1939). Theory of the quadratic zeeman effect. *Physical Review, 55*(1), 59–63. https://doi.org/10.1103/PhysRev.55.59

Schirhagl, R., Chang, K., Loretz, M., & Degen, C. L. (2014). Nitrogen-vacancy centers in diamond: Nanoscale sensors for physics and biology. *Annual Review of Physical Chemistry, 65*(1), 83–105.

Seifert, H. J. (1967). Global connectivity by timelike geodesics. *Z. Naturforsh, 22a*, 1356–1360.

Sharifabadi, A. M., & Zadeh, E. M. (2020). The evolution of university entrepreneurship over the past 20 years: A bibliometric analysis. *Entrepreneurship Education, 3*(4), 457–488.

Sharma, A. (2019, November). Entrepreneurship and Role of AI. In *Proceedings of the 2019 2nd International Conference on Signal Processing and Machine Learning,* (pp. 122-126).

Sharma, P. (2013). Women entrepreneurship development in India. *Global Journal of Management and Business Studies, 3*(4), 371–376.

Sharma, P. K., Chen, M. Y., & Park, J. H. (2017). A software defined fog node based distributed blockchain cloud architecture for IoT. *IEEE Access: Practical Innovations, Open Solutions, 6,* 115–124.

Sharp, K. V., & Adrian, R. J. (2004). Transition from laminar to turbulent flow in liquid filled microtubes. *Experiments in Fluids, 36,* 741–747.

Shiyal, S. M., Garg, A., & Rohini, R. (2019). Usage and Implementation of Artificial Intelligence in Entrepreneurship: An Empirical Study. *Seshadripuram Journal of Social Sciences (SJSS), 4.*

Silva, A. (2011). Multiple Solutions for the p (x)– Laplace Operator with Critical Growth. *Advanced Nonlinear Studies, 11*(1), 63–75.

Silwattananusarn, T., & Tuamsuk, K. (2012). *Data mining and its applications for knowledge management: a literature review from 2007 to 2012.* arXiv preprint arXiv:1210.2872.

Sima, W., Shi, J., Yang, Q., Huang, S., & Cao, X. (2015). Effects of conductivity and permittivity of nanoparticle on transformer oil insulation performance: Experiment and theory. *IEEE Transactions on Dielectrics and Electrical Insulation, 22*(1), 380–390.

Singh, J., Rathi, A., Rawat, M., & Gupta, M. (2018). Graphene: From synthesis to engineering to biosensor applications. *Frontiers of Materials Science, 12*(1), 1–20. Doi:10.100711706-018-0409-0

Singh, P., Masud, M., Hossain, M. S., & Kaur, A. (2021). Cross-domain secure data sharing using blockchain for industrial IoT. *Journal of Parallel and Distributed Computing, 156,* 176–184.

Sorkin, R. D. (1990) Spacetime and causal sets, Relativity and gravitation. Classical and quantum, 150-173.

Sorkin, R. D., & Woolgar, E. (1999) A causal order for spacetimes with C^0 Lorentzian metrics: Proof of compactness of the space of causal curves.

Spivak, M. (1979). *A comprehensive introduction to differential geometry.* Berkeley: Publish or perish, Inc.

Spyrou, K., & Rudolf, P. (2014). An introduction to graphene. *Functionalization of graphene,* 1-20.

Sriranga, N., Nagananda, K. G., Blum, R. S., Saucan, A., & Varshney, P. K. (2018, July). Energy-efficient decision fusion for distributed detection in wireless sensor networks. In *2018 21st International conference on information fusion (FUSION)* (pp. 1541-1547). IEEE. 10.23919/ICIF.2018.8454976

Steen, L. A. Seebach, J. A. Jr. (1995) Counter examples in Topology. Berlin, New York: Springer-Verlag.

Stone, H. A., Stroock, A. D., & Ajdari, A. (2004). Engineering flows in small devices microfluidics toward a lab-on-a-chip. *Annual Review of Fluid Mechanics*, *36*(1), 381–411. doi:10.1146/annurev.fluid.36.050802.122124

Strauman, A. S., Bianchi, F. M., Mikalsen, K. Ø., Kampffmeyer, M., Soguero-Ruiz, C., & Jenssen, R. (2018, March). Classification of postoperative surgical site infections from blood measurements with missing data using recurrent neural networks. In *2018 IEEE EMBS International Conference on Biomedical & Health Informatics (BHI)* (pp. 307-310). IEEE.

Struwe, M. (1991). Three non-trivial Solutions of anticoercive boundary value problems for the Pseudo-Laplace-operator. *Journal fr die reine und angewandte Mathematik, 325*, 68-74.

Sum, W. C. Y. (2012). *Towards magnetometry with nitrogen-vacancy center in diamond* [Undergraduate Thesis]. Department of Physics, National University of Singapore.

Sun, Z., Liu, Y., & Tao, L. (2018). Attack localization task allocation in wireless sensor networks based on multi-objective binary particle swarm optimization. *Journal of Network and Computer Applications*, *112*, 29–40.

Sutton, G. W., & Sherman, A. (1965). *Engineering Magnetohydrodynamics*. Mc Graw-Hill.

Swain, R. R., & Khilar, P. M. (2017). Composite fault diagnosis in wireless sensor networks using neural networks. *Wireless Personal Communications*, *95*(3), 2507–2548.

Swain, R. R., & Khilar, P. M. (2017, November). Soft fault diagnosis in wireless sensor networks using PSO based classification. In *TENCON 2017-2017 IEEE Region 10 Conference* (pp. 2456–2461). IEEE.

Syama, S., & Mohanan, P. V. (2019). Comprehensive application of graphene: Emphasis on biomedical concerns. *Nano-Micro Letters, 11*(1), 6.

Tambunan, T. (2009). Women entrepreneurship in Asian developing countries: Their development and main constraints. *Journal of Development and Agricultural Economics, 1*(2), 027-040.

Tambunan, T. (2007). Entrepreneurship development: SMES in Indonesia. *Journal of Developmental Entrepreneurship*, *12*(01), 95–118.

Tang, N., Tang, T., Pan, H., Sun, Y., Chen, J., & Du, Y. (2020). Magnetic properties of graphene. In Spintronic 2D Materials, (pp. 137-161). Elsevier.

Thaseen, I. S., & Kumar, C. A. (2017). Intrusion detection model using fusion of chi-square feature selection and multi class SVM. *Journal of King Saud University-Computer and Information Sciences*, *29*(4), 462–472. doi:10.1016/j.jksuci.2015.12.004

Thiagarajan, R., Babu, M. R., & Moorthi, M. (2021). Quality of Service based Ad hoc On-demand Multipath Distance Vector Routing protocol in mobile ad hoc network. *Journal of Ambient Intelligence and Humanized Computing*, *12*(5), 4957–4965.

Tiwari, A. (2007). *Entrepreneurship development in India*. Sarup & Sons.

Tiwari, S., & Tiwari, A. (2007). *Women entrepreneurship and economic Development. Sarup*. Sons.

Ullah, M. M., Mahmud, T. B., & Yousuf, F. (2013). Women entrepreneurship: Islamic perspective. *European Journal of Business and Management*, *5*(11), 44–52.

Vangala, A., Sutrala, A. K., Das, A. K., & Jo, M. (2021). Smart contract-based blockchain-envisioned authentication scheme for smart farming. *IEEE Internet of Things Journal*, *8*(13), 10792–10806.

Vargas, C., Arcos, J., Bautista, O., & Mendez, F. (2017). Hydrodynamic dispersion in a combined magnetohydrodynamic-electroosmotic driven flow through a microchannel with slowly varying wall zeta potentials. *Physics of Fluids*, *29*, 092002.

Venkatesan, C., Karthigaikumar, P., Paul, A., Satheeskumaran, S., & Kumar, R. (2018). ECG signal preprocessing and SVM classifier-based abnormality detection in remote healthcare applications. *IEEE Access: Practical Innovations, Open Solutions*, *6*, 9767–9773.

Vimal, S., Khari, M., Dey, N., Crespo, R. G., & Robinson, Y. H. (2020). Enhanced resource allocation in mobile edge computing using reinforcement learning based MOACO algorithm for IIOT. *Computer Communications*, *151*, 355–364.

Vinay, D., & Singh, D. (2015). Status and Scope of Women Entrepreneurship.

Vishwakarma, L., & Das, D. (2021). SCAB-IoTA: Secure communication and authentication for IoT applications using blockchain. *Journal of Parallel and Distributed Computing*, *154*, 94–105.

Wagh, M. B., & Gomathi, N. (2018). Route discovery for vehicular ad hoc networks using modified lion algorithm. *Alexandria Engineering Journal*, *57*(4), 3075–3087.

Wald, R. M. (1984). *General Relativity*. The University of Chicago Press. doi:10.7208/chicago/9780226870373.001.0001

Wallace, P. R. (1947). The band theory of graphite. *Physical Review*, *71*(9), 622–634. doi:10.1103/PhysRev.71.622

Wang, L., Jian, Y., Liu, Q., Li, F., & Chang, L. (2016). Electromagnetohydrodynamic flow and heat transfer of third grade fluids between two micro-parallel plates. *Colloids and Surfaces. A, Physicochemical and Engineering Aspects*, *494*, 87–97.

Wang, S., Li, H., Chen, J., Wang, J., & Deng, Y. (2022). DAG blockchain-based lightweight authentication and authorization scheme for IoT devices. *Journal of Information Security and Applications*, *66*, 103134.

Wang, Y., Meng, W., Li, W., Li, J., Liu, W. X., & Xiang, Y. (2018). A fog-based privacy-preserving approach for distributed signature-based intrusion detection. *Journal of Parallel and Distributed Computing*, *122*, 26–35.

West, J., Karamata, B., Lillis, B., Gleeson, J. P., & Alderman, J. K. (2002). Application of magnetohydrodynamic actuation to continuous flow chemistry. *Lab on a Chip*, *2*, 224–230.

Weston, M. C., Gerner, M. D., & Fritsch, I. (2010). Magnetic Fields for Fluid Motion. *Analytical Chemistry*, *82*, 3411–3418.

Wiklund, J., Davidsson, P., Audretsch, D. B., & Karlsson, C. (2011). The future of entrepreneurship research. *Entrepreneurship Theory and Practice*, *35*(1), 1–9.

Witten, E. (2020) Light rays, singularities, and all that.

Wojdała, M. (2021). The Internet of Things in carriage of goods by road-application, opportunities, problems. *Procedia Computer Science*, *192*, 3235–3242.

Wolters, J., Schell, A. W., Kewes, G., Nüsse, N., Schoengen, M., Döscher, H., & Benson, O. (2010). Enhancement of the zero phonon line emission from a single nitrogen vacancy center in a nanodiamond via coupling to a photonic crystal cavity. *Applied Physics Letters*, *97*(14). https://doi.org/10.1063/1.3499300

Xiao, Y., Chen, H., Yang, S., Lin, Y. B., & Du, D. Z. (2009). Wireless network security. *EURASIP Journal on Wireless Communications and Networking*, *2009*(1), 1–3.

Xuan, Y., & Li, Q. (2003). Investigation on convective heat transfer and flow features of nanofluids. *ASME Journal of Heat Transfer*, *125*, 151–155.

Yadav, V., & Unni, J. (2016). Women entrepreneurship: Research review and future directions. *Journal of Global Entrepreneurship Research*, *6*(1), 1–18.

Yang, K. (2014). *Wireless sensor networks*. Academic Press.

Yang, W., Wang, X., Song, X., Yang, Y., & Patnaik, S. (2018). Design of intelligent transportation system supported by new generation wireless communication technology. In Intelligent Systems: Concepts, Methodologies, Tools, and Applications (pp. 715-732). IGI Global.

Yang, W., Wang, X., Song, X., Yang, Y., & Patnaik, S. (2018). Design of Intelligent Transportation System Supported by New Generation Wireless Communication Technology. *Intelligent Systems*, 715-732.

Yang, G., Dai, L., Si, G., Wang, S., & Wang, S. (2019). Challenges and security issues in underwater wireless sensor networks. *Procedia Computer Science*, *147*, 210–216. doi:10.1016/j.procs.2019.01.225

Yang, G., Li, L., Lee, W. B., & Ng, M. C. (2018). Structure of graphene and its disorders: A review. *Science and Technology of Advanced Materials*, *19*(1), 613–648.

Yang, J., Lu, Z., & Wu, J. (2018). Smart-toy-edge-computing-oriented data exchange based on blockchain. *Journal of Systems Architecture*, *87*, 36–48.

Yang, X., Yi, X., Khalil, I., Luo, J., Bertino, E., Nepal, S., & Huang, X. (2021). Secure and Lightweight Authentication for Mobile Edge Computing Enabled WBANs. *IEEE Internet of Things Journal*.

Yick, J., Mukherjee, B., & Ghosal, D. (2008). Wireless sensor network survey. *Computer Networks*, *52*(12), 2292–2330. doi:10.1016/j.comnet.2008.04.002

Yuan, Y., Li, S., Zhang, X., & Sun, J. (2018, July). A comparative analysis of svm, naive bayes and gbdt for data faults detection in wsns. In *2018 IEEE International Conference on Software Quality, Reliability and Security Companion (QRS-C)* (pp. 394-399). IEEE.

Yu, K., Tan, L., Yang, C., Choo, K. K. R., Bashir, A. K., Rodrigues, J. J., & Sato, T. (2021). *A blockchain-based shamir's threshold cryptography scheme for data protection in industrial internet of things settings*. IEEE Internet of Things Journal.

Yu, W., & Chio, S. U. S. (2003). The role of interfacial layers in the enhanced thermal conductivity of nanofluids: A renovated Maxwell model. *Journal of Nanoparticle Research*, *5*, 167–171.

Zeidan, S., & Bahrami, S. (2011). Women entrepreneurship in GCC: A framework to address challenges and promote participation in a regional context. *International Journal of Business and Social Science*, *2*(14), 100–107.

Zhang, J. (2020). Pedagogical alignment for entrepreneurial development. *Entrepreneurship Education, 3*(3).

Zhang, S., Liu, X. D., & Song, W. W. (2014). *Using Knowledge-based Information Systems to Support Management of Wireless Sensor Networking Systems*. Academic Press.

Zhang, W., & Zhang, Z. (2015). Belief function based decision fusion for decentralized target classification in wireless sensor networks. *Sensors (Basel)*, *15*(8), 20524–20540.

Zhang, Y., Li, B., Liu, B., Hu, Y., & Zheng, H. (2021). A privacy-aware PUFs-based multiserver authentication protocol in cloud-edge IoT systems using blockchain. *IEEE Internet of Things Journal*, *8*(18), 13958–13974.

Zhang, Z., Hao, Z., Zeadally, S., Zhang, J., Han, B., & Chao, H. C. (2017). Multiple attributes decision fusion for wireless sensor networks based on intuitionistic fuzzy set. *IEEE Access: Practical Innovations, Open Solutions*, *5*, 12798–12809. doi:10.1109/ACCESS.2017.2722483

Zhaofeng, M., Jialin, M., Jihui, W., & Zhiguang, S. (2020). Blockchain-based decentralized authentication modeling scheme in edge and IoT environment. *IEEE Internet of Things Journal*, *8*(4), 2116–2123.

Zhao, G. Y., & Jian, F. L. (2016a). Heat transfer of nanofluids in microtubes under the effects of streaming potential. *Applied Thermal Engineering*, *100*, 1299–1307.

Zhao, G., Jian, Y., & Li, F. (2016b). Streaming potential and heat transfer of nanofluids in microchannels in the presence of magnetic field. *Journal of Magnetism and Magnetic Materials*, *407*, 75–82.

Zhu, L., Kara, O., & Zhu, X. (2019). A comparative study of women entrepreneurship in transitional economies: The case of China and Vietnam. *Journal of Entrepreneurship in Emerging Economies*.

Zidi, S., Moulahi, T., & Alaya, B. (2017). Fault detection in wireless sensor networks through SVM classifier. *IEEE Sensors Journal*, *18*(1), 340–347.

About the Contributors

Santosh Kumar Das received his Ph.D. degree in Computer Science and Engineering from Indian Institute of Technology (ISM), Dhanbad, India, in 2018 and completed his M. Tech. degree in Computer Science and Engineering from Maulana Abul Kalam Azad University of Technology (erstwhile WBUT), West Bengal, India, in 2013. He is currently working as Assistant Professor at School of Computer Science and Engineering, National Institute of Science and Technology (Autonomous), Institute Park, Pallur Hills, Berhampur, Odisha, India. He has more than eight years teaching experience. He has authored/edited of eight books with Springer in series as Lecture Notes in Networks and Systems, Tracts in Nature-Inspired Computing and Studies in Computational Intelligence, Taylor & Francis, Apple Academic Press, and IGI Global. He has contributed more than 50 research papers. His research interests mainly focus on Ad-hoc & Sensor Networks, Artificial Intelligence, Soft Computing, and Mathematical modelling. His h-index is 16 with more than 840 citations.

Nilanjan Dey is an Associate Professor in the Department of Computer Science and Engineering, Techno International New Town, Kolkata, India. He is a visiting fellow of the University of Reading, UK. He also holds a position of Adjunct Professor at Ton Duc Thang University, Ho Chi Minh City, Vietnam. Previously, he held an honorary position of Visiting Scientist at Global Biomedical Technologies Inc., CA, USA (2012–2015). He was awarded his PhD from Jadavpur University in 2015. He is the Editor-in-Chief of the International Journal of Ambient Computing and Intelligence, IGI Global, USA. He is the Series Co-Editor of Springer Tracts in Nature-Inspired Computing (Springer Nature), Data-Intensive Research (Springer Nature), Advances in Ubiquitous Sensing Applications for Healthcare (Elsevier). He is an associate editor of IET Image Processing and editorial board member of Complex & Intelligent Systems, Springer Nature, Applied Soft Computing, Elsevier etc. He is having 35 authored books and over 300 publications in the area of medical imaging, machine learning, computer aided diagnosis, data mining, etc. He is the Indian Ambassador of the International Federation for Information Processing—Young ICT Group and Senior member of IEEE.

* * *

M. A. Ansari received B.Tech.(1998) in Electrical Engineering from AMU Aligarh, M.Tech.(2001) in Instrumentation & Signal Processing from Indian Institute of Technology, Roorkee and Ph.D. (2009) in the area of Digital Signal & Image Processing from I.I.T. Roorkee. Dr. Ansari has more than 20 years of vast experience in teaching, research and industry including 3 years overseas teaching experience. He is currently serving as head of the department & research coordinator in the Dept. of Electrical Engineering, School of Engineering, Gautam Buddha University (GBU), Greater Noida. Previous joining to GBU, he has worked in College of Engineering Roorkee, Galgotias College of Engg. & Technology, Salalah College of Technology, Faculty of Engineering, Islamic University, Madinah, KSA. During this period, he has worked at several administrative positions such as Head of the Department, Dean (I/C), Coordinator of International Relations, IEEE Branch Counselor, Coordinator of Alumni Section, Research & Projects, University Admissions and NAAC/NBA accreditation. Along with teaching responsibilities, he has organized and attended several workshops, seminars, FDPs, STTPs, QIPs, national and International conferences, delivered key notes, invited talks and chaired sessions. His research interest includes Digital Signal & Image Processing, Medical Image & Video Coding, Biomedical Signal Analysis, Feature Extraction & Image Interpretation, Biomedical Instrumentation & Control, ANN/Fuzzy/GA/AI Applications to Signal & Image Processing, Wavelets/AI/Soft Computing Applications to Bio/Renewable Energy etc. He has supervised 10 PhDs, 100+ M.Tech. dissertations and 150+ B.Tech. projects. He has published more than 200 research papers in reputed National/International Journals/Conferences. He has been contributing as Editor/Reviewer to various Journals/Conferences.

Shom Das is an Associate Professor, Dept of CSE, Birla Global University, Bhubaneswar, Odisha, India.

Sudhakar Das obtained his PhD degree in 2017 & M. Tech. degree in 2010 from Biju Patnaik University of Technology (BPUT), India and B.E. degree in 2001 from Berhampur University (India) in the field of Electronics and Communication Engineering (ECE). He has more than 20 years of teaching experience and is currently working as a Professor in the department of ECE in NIST Institute of Science and Technology (NIST), India. He is a life member of IETE and is a member of IEEE. He has published more than 13 papers in the journals of repute and has presented his work in more than 15 national/international conferences. His research interest includes modeling of homo/hetero-structure semiconductor devices to operate at high frequency for use in high speed VLSI circuits & communication networks and mathematical modeling of fault diagnosis in Wireless Sensor Network (WSN) and robust distributed estimation in faulty WSN.

Ashish Kumar Dass completed MTech. in Computer Science and Engineering from National Institute of Science and Technology (NIST), Berhampur, Odisha .He is perusing his PhD in Computer Science and Engineering under BPUT, Odisha. He has more than fifteen years of teaching experience as an Assistant Professor at School of Computer Science and Engineering, National Institute of Science and Technology (Autonomous), Institute Park, Pallur Hills, Berhampur, Odisha, India. His research interests mainly focus on Under-water Wireless Sensor Network, Ad-hoc & Sensor Network, Computer Security, Artificial Intelligence and Mathematical modeling.

Arka Ghosh has completed his MS degree in Nanoscience and Nanotechnology from University of Siegen, Germany, in the year of 2022 and completed his B. Tech. degree in Electrical Engineering from Maulana Abul Kalam Azad University of Technology (erstwhile WBUT), West Bengal, India, in 2018. He has authored of a book chapter with Springer in series as Springer Tracts in Nature-Inspired Computing. His research interests mainly focus on nanoscience and nanotechnology, photonics, optical design, quantum ptics, clean room fabrication, chip designing.

Bhabani Sankar Gouda received his M.Tech degree in Computer Science and Engineering from Biju Patnaik University and Technology, Rourkela, Odisha, India, in 2010. Currently, he is pursuing his Ph.D. in the department of Computer Science and Engineering at BPUT Rourkela, India. He worked as Asst. Prof. from last 16 years at NIST Institute of Science and Technology (NIST) Berhampur, Odisha. His research interest includes fault diagnosis in wireless sensor networks and robust distributed estimation in faulty WSNs.

Azriel Henry is currently working as a full-time Research Scholar in the Department of Engineering and Physical Sciences, Institute of Advanced Research, Gandhinagar, India. He received his Master in Engineering degree in Software Engineering from Government Engineering College, Modasa, Gujarat. He received his Bachelor in Engineering degree in Computer Engineering from Hasmukh Goswami College of Engineering, Vahelal, Gujarat. His research interest includes Cyber-Security, Intrusion Detection System, Machine Learning & Deep Learning.

Sanjay Kumar has worked as software engineer, Wipro Technology, Banglore currently Have more than 12 yrs of vast experience of R&D in Image Processing And 5G . Author of various national and international research papers published in on various prestigious platforms. Guided so many research projects on various Data driven Technologies like IoT, Images processing, 5G.

Vikash Kumar is pursuing PhD from Sarala Birla University, Ranchi, India and completed M. Tech degree in department of Computer Science and Engineering, specialisation with information security from IIT(ISM) Dhanbad, India in 2016. He has total six years teaching experience in different university like GITAM University Hyderabad, India, Doon university Dehradun, India and UCET VBU Hazaribagh, India. He is currently working as an assistant professor in University College of Engineering and Technology, Vinoba Bhave University, Hazaribagh, India. His research interests mainly focus on information security, IoT, Edge Computing and Blockchain.

Soumi Majumder completed her PGDM, specialization in Human Resource Management from All India Management Association (Institute of Business Management, Jadavpur University) in 2012. She did Diploma in Labour laws with Administrative laws (DLLAL) and Masters in Human Resource Management (MHRM) from Annamalai University, under Tamilnadu Govt in 2013 and 2019 respectively. Currently, she is a Ph.D. student of the Department of Business administration Dept. at Vidyasagar University, Midnapore, West Bengal, India. She is associated with the School of Business- Sister Nivedita University, WB, India and Dept. of Business Administration, Netaji Subhash Engineering College, Kolkata. She is an associate researcher of Universidad Internacional de La Rioja, Logroño, La Rioja, Spain. Previously, she was associated with Techno India College of Technology, NSHM College of Management and Technology, Dinabandhu Andrews Institute of Technology and Management, J D Birla Institute of Science and Commerce, Future Institute of Engineering and Management, West Bengal State Labor Institute, Siliguri, Department of Management Science etc. She is having 6 years of experience in academia and 2 years of industrial experience. She is having 2 books (Springer Nature, CRC press) and 20 publications in the area of quality work-life, decent work-life, work-life balance, stress management, employee engagement, job satisfaction, leadership, training, and learning, etc. She has presented a research article at DSMLAI 2021 organized by India –Namibia Centre of Excellence in Information Technology, Namibia University of Science and Technology, Namibia. Furthermore, she is a member of the All India Management Association, ACM, IS, NIPM, etc.

Dwaipayan Mishra is pursuing his Ph.D degree in Mathematics from Midnapore College(Autonomous) affiliated to Vidyasagar University, West Bengal, India. He completed his M. Sc degree in Applied Mathematics from Vidyasagar University in 2015.

Javed Miya obtained his engineering degree (Computer Science & Engg.) in 2003 from C.C.S. University Meerut, Master degree (Computer Science & Engg.) in 2006 from Department of Computer Science and Applications (DCSA) at Kurukshetra University Campus Kurukshetra and Ph.D. degree (Computer Science & Engg.) in 2021 from UTU Dehradun. He has published over twenty-five research papers in national and international journals/ conferences of high impact factor and SCI indexed and IEEE proceeding publication and supervised more than 20 projects/dissertation of M.Tech & B.Tech. Students. He has published 4 patents on Medical Image Compression, Sensor based smart face shield and on Sensor based Crop recommendation system. He has participated in various National, International Conference, Seminars, Technical Session chair, Jury member and Workshop such as Digital Image Processing, Database Management System, Software Engineering, Networking Technology conducted by Hewlett-Packard (HP), High Impact Teaching Skills Organised by WIPRO Limited, Bangalore-India, Embedded System, Intellectual Property & Innovation Management in Knowledge Era Organized by National Research Development Corporation (NRDC) Delhi. He is currently working as a Associate Professor at Galgotias College of Engineering & Technology Greater Noida, India.

Himadri Shekhar Mondal is currently working as Associate Professor of Mathematics in Midnapore college(Autonomous). He received his B. Sc in Mathematics(Hons.) and M. Sc in Mathematics from Jadavpur University. Dr. Mondal received his Ph. D from IIEST, Shibpur. He has published more than 5 research articles in reputed international journals. He is an eminent painter. He has taken part in many national and international art exhibitions.

Debendra Muduli is currently working as an Assistant Professor at C. V. Raman Global University, Bhubaneswar, Odisha, India.

Ruchika Padhi received her M.Tech degree in Information Communication Technology from Indian Institute of Technology (IIT) Kharagpur, India, in 2013 and continuing PhD from C V Raman Global University under the guidance of Dr.Debendra Muduli. She is currently working as Assistant Professor from last 9 years at School of Computer Science and Engineering, National Institute of Science and Technology. Her research interest mainly on Wireless Sensor Network.

Sudhir Ranjan Pattanaik has completed his PhD in Chang Gung University of Science and Technology, Taiwan. He has more than 15 years of experience in different engineering colleges in India and abroad. Now he is continuing as professor and head of the department in School of Computer Science and Engineering, National Institute of Science and Technology (Autonomous), Institute Park, Pallur Hills, Berhampur, Odisha, India. His research interests mainly focus on Wireless Sensor Network, MAC protocols, Ad-hoc & Sensor Network, Computer Security.

Shamimul Qamar is a professor and an eminent scholar in the field of Computer Science & Engg.. He has done his B.Tech from MMMUT Gorakhpur (http://mmmut.ac.in), M.Tech from AMU, Aligarh and earned his Ph.D in Computer Science & Engg. degree from IIT Roorkee with highly honorable grade. Prof. Qamar has a wide teaching experience in various Engineering colleges. He had been selected as a full Professor by selection committee of U.P Technical University, Lucknow, Uttar Pradesh and appointed as a Full Professor and Head in MIET Meerut w.e.f July 2007. He worked as a Director & Professor of Kanpur Institute of Technology, Kanpur. Presently He is serving as a University Professor at King Khalid University, Abha, and Kingdom of Saudi Arabia. He has research interests in Computer Network & Multimedia, Networking & Security, and Artificial Intelligence. He has published several research papers in reputed Journal like Elsevier, supercomputing. He had guided and awarded six PhD Thesis under his supervision. He has published several research papers in reputed national/international Journals and conference. He served as Consultant in Jackson state university, USA. He is a reviewer of IEEE/Elsevier/ MDPI/EUROSIP/ IJCSIS, USA. He has written some text books and chapters in the field of Electronics & Computer Engineering. He is also a technical program committee member in various international conference. He is a life time member of international association of Engineers and a life member of Indian Society of technical educational. His technical depth and interest resulted in setting up a research lab according to latest technical innovations. Along with this, he has actively participated in various technical courses workshops, seminar etc. at the IITS.

Amalendu Rana is pursuing his Ph.D. degree in Mathematics from Jadavpur University, Kolkata, India, and completed his M.Sc. degree in Mathematics from Vidyasagar University, West Bengal, India, in 2016. He has about to two years research experience as junior research fellow at Department of Mathematics, National Institute of Science and Technology (Autonomous), Institute Park, Pallur Hills, Berhampur, Odisha, India. He has contributed more than 10 research papers. His research interests mainly focus on microchannel flow, Electroosmosis, Electromagnetohydrodynamics, Heat transfer and Mathematical modelling. His h-index is 3 with more than 20 citations.

Motahar Reza is currently working as Associate Professor of Mathematics and Data Science, School of Science, GITAM University Hyderabad. Earlier he was working as Professor in School of Computer Science and Engineering, National Institute of Science and Technology, Berhampur, Odisha . Dr. Motahar Reza, received his B. Sc. in Math(Hons) from Culcutta University, M. Sc. in Applied Mathematics from Jadavpur University, Kolkata and Ph. D. from Indian Institute of Technology, Kharagpur, 1995, 1998 and 2004, respectively. Dr. Reza was working as guest Scientist at LSTM, Friedrich-Alexander-Universität Erlangen-Nürnberg, Germany during 2005-2006. He has been awarded most prestigipus Fellowship as young scientst ,, BOYSCAST Fellowship, by Department of Science and Technology, Govt. of India in 2004. He was awarded Orissa Young Scientist Award in Physical Science by Govt. Orissa in 2006. He received the Research fund/ other fund from SERB/DST/AICTE/ DRDO/CSIR/TEQIP-II BPUT under various schemes. He visited Germany, France and UK to avail the Internal Travel grant fund by SERB, Govt of India. He visited Centre of Mathematical Science Cambridge University, UK in 2016. Currently he is teaching Data Science and Machine Learning to PG and UG students at GITAM University. His research interests include Machine Learning, Data Science, Parallel Algorithm and Computational Science, CFD and Turbulence Modeling. He published many paper in highly reputed Internal Journal and Proceedings in the area of Fluid Dynamics, Parallel Algorithm and Machine Learning.

Sagar Samal has pursued his Bachelor in Technology in Computer Science and Engineering from National Institute of Science and Technology, Palur Hills, Berhampur, Odisha, India, in 2020. His research interests mainly focused on Machine Learning and Intrusion Detection System.

Index

I

K

L

M

N

O

P

R

S

T

W

Z

www.ingramcontent.com/pod-product-compliance
Lightning Source LLC
LaVergne TN
LVHW081528300926
849161LV00006B/1347